Six Stories
from the End
of Representation

JAMES ELKINS

Six Stories from the End of Representation

Images in Painting, Photography, Astronomy, Microscopy, Particle Physics, and Quantum Mechanics, 1980–2000

STANFORD UNIVERSITY PRESS

STANFORD, CALIFORNIA

2008

Library of Congress Cataloging-in-Publication Data
Elkins, James
Six stories from the end of representation : images in
painting, photography, astronomy, microscopy, particle
physics, and quantum mechanics, 1980–2000 / James Elkins.
p.   cm.
Includes bibliographical references and index.
ISBN 978-0-8047-4147-7 (cloth : alk. paper)—
ISBN 978-0-8047-4148-4 (pbk. : alk. paper)
1. Representation (Philosophy)   2. Image (Philosophy)
3. Art and science.   I. Title.
B105.R4E45 2008
701—dc22   2007010175

Contents

Color Plates

Acknowledgments

I owe special thanks to the patient scientists who helped me with their work: Sha Xin Wei, Andreas Albrecht, and Andrei Linde spent time considering images used in theories of the inflationary universe; Nora Brambilla of the Institut für Theoretische Physik, Vienna, and Gunnar Bali provided me with material on quark confinement and superconductivity. Marc Redfield, Ed Casey, Martin Donougho, and Raja Halwani helped me find sources on the sublime. Jim Hugunin and Catherine Edelman helped me locate several photographers. Martha Fleming and Beth Kessler both shared their interesting work on scientific images, unfortunately just too late to be incorporated into the argument; I refer readers to their work in Chapter 3. Cynthia Pyle, Teri Reynolds, and Jorgelina Orfila helped me think about the debates surrounding Sokal and Snow. Leon Lederman, Hans Bethe, Bruce Winstein, Michael Murtagh, Jeannie Therrian-Gottschalk, Gron Tudor Jones, Victor Regener, and Robert Crease helped me locate archival materials in particle physics. The staffs of the archives at Fermilab, Brookhaven, the Art Institute of Chicago, the Lowell Observatory, and the Argonne National Laboratory were very helpful, especially Catherine Foster, Angela Johnson-Waren, Henry Giclas, Patricia Meehan, Kristen Merrill, and Michiko Tanaka. Chapter 4, on microscopy, owes much of its material to Jan-Olov Bovin of Lund University, Sweden; Bruce Phillips of Modulation Optics; Jie Liu (on scanning probe microscopes); Martin Mach and Dave Walker (on images of diatoms); Hugh Crenshaw (on confocal microscopy and video microscopy); Gleb Finkelstein (maps of quantum Hall liquid); Christoph Böttcher and Friedrich Zemlin in the

Forschungszentrum für Elektronenmikroskopie in the Freie Universität Berlin (TEM techniques); Jürgen Rabe and Stefan Kirstein in the Humboldt Universität Berlin (NSOMs, AFMs); and Eduard Chilla and Reinhold Koch in the Paul-Drude-Institute for Solid State Physics, Berlin (acoustical microscopy). Thanks also to Peter Galison, whose *Image and Logic* provided an impetus to write the final two chapters of this book.

Any writer who strays outside his discipline relies on the help of forgiving readers: in this case David Kaiser (who read chapters 5 and 6); Martin Donougho (Chapter 1, and parts of Chapter 2); Brian Rotman (the pages on infinity in Chapter 3); Piero Rosati (galaxy clusters, Chapter 2); Michael Vogeley, Changbom Park, and Juhan Kim (the Hubble Deep Field, Chapter 3); Jeeva Anandan (quantum mechanics, Chapter 6); Zsolt Paragi and Hsiao-Wen Chen (high redshift objects, Chapter 3); and Rennan Barkana (gravitational lenses, Chapter 3). Bruno Latour read an earlier incarnation of this book, effectively sank it, and made crucial suggestions. Thanks also to Paul Guyer for some perspicacious comments on Kant's intentions. Criticisms and observations by Douglas Crimp, Erling Sjovold, and Carol Dolan helped me clarify my reading of Agnes Martin. And thanks also go to the many audience comments and suggestions I received when I lectured on this material in Sofia, Blagoevgrad, Prague, Budapest, Gustavus Adolphus University, Duke University, the University of Illinois at Urbana-Champaign, Rhodes College, Bucknell University, the University of Pennsylvania, the University of Victoria, the University of Washington at Bellingham, the University of South Carolina at Columbia, the Akademie der Künste in Berlin, ZKM Karlsruhe, Yale University, the National Gallery in Ljubljana, the University of California at Santa Barbara, and the University of Maryland at College Park.

And finally, thanks to the faculty in the physics departments at Johns Hopkins, Berkeley, and the University of Chicago, who let me attend their classes—surrounded by young physics students in their teens and twenties, who seemed to want the whole world to be calculable even if nothing in it made a bit of sense.

Earlier versions of parts of Chapter 6, dealing more directly with Galison's *Image and Logic*, appeared as "Logic and Images in Art History," *Perspectives on Science* 7, no. 2 (1999): 151–80, and "On Some Useless Images," *Visual Resources* 17 (2001): 147–63. Material from the account of Marco Breuer in Chapter 1 is adapted from *Tremors*, exh. cat. (New York: Roth Horowitz, 2000).

Preface

This book is about a confluence of images from six fields: painting, photography, and four branches of physics. Things are happening in parts of those fields that belong together, and this book is an attempt to let them mingle just enough to show how they are already intertwined, without losing sight of their irreducible differences.

The images I am interested in show us things we can't possibly be seeing: things so far away, so faint, so large or soft or bright that they couldn't possibly be contained in the rectangular frame of a picture—and yet they are. These are pictures of objects that literally don't exist—that couldn't exist as they are pictured—but somehow do. They are abstractions of abstractions, feeble symbols for objects that have no reality of their own. They are formless things, blurred until they are unrecognizable. They are images of almost nothing, of single sparks of light, of inexplicable blazing color, of thin lines and dots that stand for objects that cannot be understood.

First and last, this book is about images that I love. It began as a pile of photographs on my desk that seemed to belong together, even though some were reproductions of artworks and others were illustrations cut from science journals. I had no articulate theory about them: they were my miscellaneous category. Eventually it became clear that the images were the results of investigations into the limits of representation. Few of them contained sharp or well-defined objects, and in most it was difficult to understand what was being taken as adequate representation. They seemed to have a common theme: they were images that did not simply depict objects, but demonstrated how some objects resist depiction.

As the project took shape, I started collecting texts that theorize such limits, and the pile became a mix of texts and images. For a while it seemed as if this would be a book of art history, because the majority of the images date from the 1980s to the present. But as it stands only two chapters out of six concern painting and photography. For the same reason this book is not a contribution to the history of science. Historical sequences and meanings are only incidental to my exposition: it matters that the images were made between the 1980s and the present, but it does not often matter exactly when. Most important, this is not a book of philosophy or aesthetics, because I have not let philosophic concepts—prominent in chapters 1 and 2—guide the exposition of the scientific subjects.

The book's primary purpose is to think through the images, demonstrating that they are all at work on the limits of representation in their respective media and technologies. Each picture was made by someone uninterested in whatever counted, for that time and place and purpose, as ordinary representation. When they belong to physics, these images are made at the limits of the instrumentation; when they are art, they are at the ill-defined borders shared by abstraction, competing naturalisms, and conceptual art. For me these are among the most compelling visual products of the past several decades, because they are more rigorous about the ends of representation than any number of pictures in science and art that give us things in full color, resplendent in all their detail, as if the whole world is visible and there is no obstacle to putting it in a picture. I would even hazard a general claim here: it strikes me that among the most important images of the last three decades are those whose makers are most fully preoccupied by whatever cannot be put in a picture. That is my theme in this book. I have also been guided by a methodological agenda, which is set out in the Introduction.

A word about the science and equations, which may put off some readers. Nonscientists may find passages that don't make sense at first reading. I would emphasize, however, that it is not only the physics that is technical. Painting has its own discourse, very rich and involved, and philosophy can be just as exacting. When disciplines converge there are apt to be difficulties, but glossing them would only introduce a conceptual blur where it is important to be precise. These choices, too, are defended in the Introduction.

Because many science images are originally digital, it is possible to see them at full scale and detail on a computer. Parts of this book, in particular Chapter 3 on astrophysics, should ideally be read alongside a good-quality monitor, preferably seventeen inches or larger. If you visit the URLs in the Notes, you'll see the images as they were meant to be seen. Painting is different: it is falsified by the glossy flatness, the haze, and the wide value range of computer screens, and should be encountered in the original.

The Crescent, Galway—rue Payenne, Paris—
Duke Street, Durham—Hyde Park, Chicago
1997–2008

Last rays. And the moon. And Venus. Nothing left but black sky. White earth. . . . Nothing but black and white. Everywhere no matter where. But black. Void. Nothing else. Contemplate that. Not another word. Home at last.

—Samuel Beckett, *Ill Seen Ill Said*

   Methodological Premises
of This Book

This book has two purposes. The first is to bring together images made
in the last quarter-century that work at the limits of representation. The
second is methodological: this book is my contribution to the problem
that is still occasionally called the two cultures—the gulf of misunder-
standing that has opened between the sciences and the humanities.[1] I
hope the interpretive method I practice here will contribute to the difficult goal of speaking at once to both the sciences and the humanities.

It seems to me that it would be hard to find a more resistant subject as
far as university and intellectual life is concerned. The mutual alienation
of natural scientists and humanists runs very deep, though it tends to be
papered over by the ecumenical atmosphere that universities foster—the
many committees, social relations, interdepartmental initiatives, and
more-or-less successful interdisciplinary conferences—which can easily
give the impression that there is a healthy ongoing dialogue between the
humanities and the sciences. Certainly there is no lack of casual or non-
technical conversation between humanists and scientists; but there is precious little overlap in courses or reading lists, and virtually no detailed,
substantive discussion.

Graduate students in the sciences generally take no humanities
courses, and it is nearly universal for humanities students to avoid sciences
after the basic first- and second-year college distribution requirements.
(In other university systems, such as those in England, the division is even
more stark, and humanities students may take no science courses at all.)
Those proclivities extend into professional life. In my experience professors

in the humanities often depend on chance social relations (relatives, friends) for their expertise in science, and scientists are apt to assume that help from specialists isn't necessary to appreciate the arts. The idea that there is a gap between the humanities and the sciences seems to be contradicted by the ease with which faculty members can communicate the generalities of their disciplines, and the very idea that there are two cultures seems to be moot, simply because it was raised so long ago (in 1962): a polemic that old, so the reasoning seems to go, must be inherently false, or poorly put. On the contrary, it seems to me that the abyss between the faculties, to use the administrative term, is immense: it has reached the point where I do not think it is unfair to say that contemporary universities are unified in name only.[2]

For all his quirks, C. P. Snow was largely right when he named the two cultures and pointed to the disparity between them.[3] His famous test was to ask writers to define the Second Law of Thermodynamics; when they couldn't, he compared the importance of the Second Law to Shakespeare's *Hamlet*. The question, "Can you describe the Second Law of Thermodynamics?" is, he said, equivalent to the question, "Have you read a work of Shakespeare's?"[4] Snow's test was annoying and belligerent, and also brilliant. It remains the case that virtually no one outside the sciences reads unpopularized science, and hardly anyone in the sciences reads the professional literature of the humanities. "I now believe," Snow wrote in 1963, "that if I had asked an even simpler question—such as, What do you mean by mass, or acceleration, which is the equivalent of saying, *Can you read?*—not more than one in ten of the highly educated would have felt that I was speaking the same language."[5] Snow's rude questions continue to be unanswerable. I am especially interested in the fact that they bother so few people: to a degree that continues to surprise me, the two cultures do not find their mutual lack of interest to be a problem. One reason for that insouciance appears to be the collegiality of university life. Another reason may be the assumption on the part of those in the humanities that the philosophy and sociology of science provide an adequate description of science, and the reciprocal assumption on the part of scientists that the philosophic self-descriptions of the humanities are epiphenomenal on art and culture, and therefore dispensable. Neither assumption, it seems to me, survives Snow's rude questions, unless it seems those questions are themselves without consequences.

In the mid-1990s some of Snow's concerns were revived by Alan Sokal's critique of humanists' misuses of science. In one essay Sokal proposes his own variant on Snow's questions. He imagines asking an "average undergraduate" student, "Is matter composed of atoms?"— and then when the student says, "Yes," he proposes asking, "Why do you think so?"[6] That second question is trickier than either of Snow's questions, because it leads into a number of further problems. Would it be reasonable to reply, "Because a consensus of scientists think so"?[7] Should everyone be armed with a demonstration of the atomic nature of matter? Sokal's test is not as sharp-edged as Snow's, but it is just as telling.

The literature on Snow's lecture and Sokal's writings is huge. By the late 1970s, when the interest in Snow's lecture was dying down, there were over 800 essays, reviews, and books on "The Two Cultures."[8] Sokal's book *Fashionable Nonsense: Postmodern Intellectuals' Abuse of Science* has been translated into Catalan, Dutch, German, Italian, Japanese, Korean, Polish, Portuguese, Spanish, and Turkish, with Russian, Hungarian, and Chinese translations forthcoming.[9] From the sheer size of the bibliographies on Sokal and Snow it might seem that there is an ongoing dialogue between the humanities and the sciences. Unfortunately the literature is often schematic, marred by polemical intent, and conceptually impoverished.[10] This is not the place to describe the two bibliographies or the other recent encounters between science and art.[11] Instead I just want to register my general impression of the kind and quality of the debates by citing a single example: a letter written by the physicist Jonathan Katz, which appeared in 1999 in *Physics Today*, an official publication of the American Institute of Physics, the major American association of physicists. Katz claims that "postmodernism is nothing more than a scheme to obtain cushy university jobs for its practitioners." He wonders why humanities departments don't consider themselves guilty for not recognizing the "scheme" for the "transparent scam it is." "Don't they have a nose for cowpatties at all?" he asks.[12] Katz's letter is curt, but his thoughts are in line with the discussion in *Physics Today*, which was mainly on Sokal and the sociologist of science Bruno Latour. I take it as a sign of the level of exchanges between the sciences and the humanities that the editors of *Physics Today* thought Katz's letter was an appropriate contribution—whether or not they agreed with it, they did not consider it irrelevant, inappropriate, or irresponsible.[13] (I assume the editors thought it was funny, but it would not have been published unless it was worth reading.)

Opinions like Katz's can be taken as evidence that science studies in the humanities have not diffused throughout the sciences. It is now possible to get a Ph.D. degree in a number of science-related fields within the humanities, including the sociology, history, and philosophy of science, and even the anthropology and literature of science. Each has contributed to the discussion about science, but it remains an open question whether the emerging disciplines have diminished the distance between the sciences and the humanities.

It is true that there are a number of admirable texts on the subject of science and the humanities by writers such as Michel Serres, Barbara Stafford, Jacques Bouveresse, Paul Feyerabend, Kurt Hübner, Bruno Latour, Ian Hacking, Andrew Pickering, E. O. Wilson, and Abraham Pais. Books like theirs can be well argued, well informed, and ecumenical—certainly good qualities for writing, whether the subject is the sciences or the humanities. But as reflective as such books can be, they raise a further problem. As far as I can see, these authors speak principally either to audiences in the humanities or in the sciences, rather than to mixtures of the two.[14] Pais, for instance, is read by scientists, even though he has written on subjects of general interest. Most of the other authors are read principally by people in the humanities. At the center there is a void: with the partial exception of E. O. Wilson, no writer I know is read, and few are trusted, by readers on both sides. (I am not talking here about popularized science, but books that address the links between science and the humanities. Otherwise writers such as Stephen Hawking, Stephen Kay Gould, and Bill Bryson would be counterexamples.)

Sokal's book *Fashionable Nonsense*, which he coauthored with Jean Bricmont, would seem to sit squarely between science and the humanities, because it is a review of scientific and mathematical errors committed by various post-structuralist French thinkers. Just in terms of content, it would be reasonable to expect that it would attract readers from both "cultures." When it first appeared in French, and then later in English, it attracted a wide range of readers. Yet I wonder how many of those readers work in the humanities. Reviewing the book, the philosopher Thomas Nagel asks how many admirers of postmodern thought might actually read *Fashionable Nonsense*: will "teachers of cultural studies and feminist theory," he asks, "go through these patient explanations of total confusion about topology, set theory, complex numbers, relativity, chaos theory, and Gödel's theorem?"[15]

Even scholars such as Serres, who have spent their lives writing back and forth between science and art, find few readers outside the humanities. Serres escapes censure in *Fashionable Nonsense*; Sokal and Bricmont say his work is "replete with more-or-less poetic allusions to science and its history; but his assertions, though extremely vague, are in general neither completely meaningless nor completely false."[16] It appears they do not consider Serres to be an interesting writer on the art-science border, as for example Bruno Latour does.[17] Like Jacques Lacan, who also left some texts on science, Serres is judged to be neither obviously wrong nor importantly right, and so he gets overlooked—even in a book explicitly dedicated to comparing sciences and arts.[18]

Another example of a writer with unequal readership in the arts and the sciences is the historian Peter Galison, who writes some of the best-informed historical accounts of modern physics. His book *Image and Logic: A Material Culture of Microphysics*, which I discuss in the last chapters of this book, was reviewed by two physicists in the journal *Isis*; they wonder what it means that "even Galison, who understands so much about high energy physics and has thought so seriously about its power structures, fails to persuade one of the great doers, [W. H. K.] Panofsky, of the relevance of his postmodern discourse."[19] It is necessary, the reviewers say, to stop and ask who reads "postmodern" history of science if the scientists don't.

It goes without saying that Bricmont and Sokal have readers in the humanities, and that Serres and Galison have scientists among their readers; but I am aiming at an important general question here. I want to introduce the methodology of this book by proposing that scholars who cross between science and the humanities should take as one of their goals that their work be read *equally* by people on both sides. I do not mean that a scholar should aim to have exactly the same numbers of readers from the two "cultures": rather, that potential readers could come with equal probability from either side. The polemics that mar the literature on Sokal and Snow are mostly one-sided: the contributors identify themselves as apologists for one side or the other. The more balanced literature, such as Serres's or Galison's, comes closer to being acceptable to readers from either side. When the writing reaches a certain level the problem named by the *Isis* reviewers comes to the fore: if an author's readers are mostly from one side or the other, in what sense is the work a bridge?

The central question, I think, is how to write in such a way as to create something that is meaningful both to working scientists and to humanists. Let me try to answer this question, and introduce the methodology of this book, by describing three fundamental problems that occur in writing that mixes the humanities and the sciences.

1. *The sciences are different from the humanities because scientific writing includes equations; hence humanist accounts of science that are set in prose risk omitting a central feature of scientific writing.*

Any interpretation opens a distance from the thing that is interpreted, but that difference is specially marked in writing about science because—at an elementary, evidentiary level—scientists work with equations and interpreters respond with words. When people write about the sciences they are obliged to substitute "literary" prose for what the classicist Wesley Trimpi calls "geometric" prose—mathematics.[20] As Trimpi stresses, literary and geometric discourses have been separate since the Hellenic period, and that makes the interpretation of science importantly different from the narrative interpretation exemplified by literary texts.[21]

I stress this simple fact to suggest that part of what inflames misunderstandings between the two "cultures," and prevents them from talking to one another in a fuller sense, may be a deep dissimilarity between modes of writing. The very word *writing* means something different when it includes symbolic configurations as well as the Roman alphabet. Reading, too, is very different in the humanities and the sciences, and for an equally basic reason: science textbooks usually have problem sets at the end of each section or chapter. Both students and professionals are expected to stop when they get to the problems and—in the telling phrase—*work* some of them. That means a scientific text cannot normally be read straight through. It involves stops and starts, and long pauses for calculation as well as cogitation. The same may be said of scientific and mathematical papers, which routinely truncate their lines of reasoning so that the reader has to work a little to reconstruct the full argument. It is appropriate at all levels of scientific writing to say something like, "I leave the full demonstration to the reader," or to proceed in such a way that the reader knows where that phrase might have been inserted.

That is not to say that I don't read Montaigne or Foucault with a pencil in hand: I do, and my copies of books are often filled with marginal arguments. But all that is *optional*, in the sense that Montaigne and Foucault both wrote in a manner that can loosely be identified with the speed of a speaking or silently reading voice. Like most everyone since St. Jerome, I read silently, which means I normally read faster than I would if I read out loud, and without the cadences that are typical of the spoken word. Yet there is an underlying similarity between reading Montaigne out loud and reading him to myself: the speed of his thought, his choice of words, his changes of pace and direction all appeal to an experience that is *continuous*, that moves through his ideas and words at a varying but ongoing rate. Things are different when the flow of words is periodically interrupted by numbered problems, propositions, and theorems. It is no more possible to read Newton's *Principia* without stopping on each page than it is to read any freshman biology textbook in one gulp as if it were a paperback novel. Accounts of reading that gloss over this point, and propose a generalized reading or writing, tend to be written by authors who have not read books like the *Principia*: for those who have, I think the difference is clear, deep, and fairly consistent across different scientific disciplines and literary forms.

It is not irrelevant to the subject of this book that the same can be said about the *images* in scientific texts. Images in humanities texts tend to have minimal impact on the rates of reading: I can glance at an illustration in an art history monograph in order to verify something the author says, but I am not often expected to interrupt my reading altogether to make a concerted study of the image. (The notion is that ideally, I will go and see the original.) A typical image in a science text, on the other hand, comes with its own caption, which instructs me how to read it: the caption tells me what the symbols mean, what is measured on the x and y axes, and so forth. Reading science, I often find I have to stop reading the text altogether and spend some time with the image in order to learn to interpret it; and the same process recurs repeatedly in the course of reading an illustrated science essay or textbook.

Writers in the humanities don't tend to stop their readers short with each new picture. Commonly the captions in humanities texts are minimal, because the continuous narrative in the body of the text does the work of interpretation. Paragraph-long captions are the exception, but they prove the rule because they make use of extended narratives (called, in the publishing trade, *discursive captions*), which are an attempt to repeat, in miniature, the continuously legible text that surrounds the image. In the sciences, on the other hand, it is often thought appropriate and even necessary to provide images with long, nondiscursive, densely technical captions. A problem arises when scientific images are placed in literary contexts, or vice versa, disturbing the normative relation between image and the flow of narrative.[22]

Assumptions about the appropriate explanation of images, and about numbered problems and equations, can be summed up in a phrase: they constitute what I will call the *problem of the pencil*. I can read Montaigne without a pencil if I choose, but I absolutely cannot read an introduction to quantum mechanics without a pencil and lots of extra paper. If I try to read it without stopping, as if it were a humanist narrative, most will go in one ear and out the other. I will not really understand the text: I will get its flavor, but unfortunately flavor does not count when it comes to quantum mechanics, at least not in the way I might claim such things as flavor, gist, or tone can matter when it comes to Montaigne. Hence an initial challenge for interdisciplinary work that tries to bridge the sciences and the humanities is what to do with the problem of the pencil: the assumption that the reader will have a pencil in hand, and that reading will be interrupted, without warning and repeatedly, by equations, technical illustrations, and unsolved problems. The fact that theories of reading and writing in the humanities continue to assume that equations, pictures with discursive captions, and problem sets are simple *additions* to "literary" narrative is a sign, perhaps the most fundamental, of the distance between the sciences and the humanities.

2. *Contemporary interdisciplinary scholarship, including work that links the arts and the sciences, does not always specify which of the disciplines under study is being used to illuminate which others.*

Writing about Sokal, the critic Teri Reynolds observes that "the point of any interdisciplinary endeavor is that work done in one field may be used to elucidate material in another."[23] The question is then a matter of deciding *which* field is to serve as the one that can "help us see new things about the other." For Sokal, the illuminating discipline is physics (or some other branch of science), and the unilluminated discipline is some post-structuralist theory. In work such as Serres's, a third discipline serves to illuminate two or more other disciplines that are themselves being compared to one another: again, it is a matter of choosing which discipline sheds light on which others.

In the history of art, interdisciplinary research that includes science typically focuses on the scientific sources for particular movements and methods in Western art. In that case science helps explain the art: so, for example, notions of the fourth dimension shed light on ideas in Duchamp's *Large Glass*, or Chevreul's color theories reveal how Seurat constructed *La Grande Jatte*. Let me label this operation, in which science helps explain art, as the first of several possible configurations in art history.

Other art historians prefer to go in the opposite direction and let the art explain the science—as it can in the history of anatomical illustration, which was intermittently influenced by contemporary painting, sculpting, printmaking, and even architecture. Call that the second configuration. It is also possible—the third configuration—to explain both the art and the science in terms of some other discipline. Among the many candidates, by far the most common is philosophy. There are several limit cases, which do not often appear as such but are helpful in thinking about the forms philosophy has taken when it is utilized in this way. Scholarship that depends on the reception of Hegel need not specify causal links between disciplines that are being studied, because the final purpose of such scholarship is to demonstrate the natural affinities between the parts of a culture. A Marxist scholar, on the other hand, might want to make explicit links between the elements of culture and the conditions of the society: the latter would always cause the former, whether the subject is art or science. The vicissitudes of the Hegelian and Marxist approaches could account for a large percentage of writing on cultural history; what matters in this context is that they both introduce philosophic or theoretical discourse as the third discipline, taking a relevant philosophic practice as the ultimate ground of explanation and adjudication.[24]

This third configuration helps elucidate the fourth, which is the preferred one in much contemporary historical writing: to let explanation itself become evanescent. In that practice science and art may be put side by side, but without drawing a direct link between a particular scientific idea and a particular artistic work. Jonathan Crary's work, for example, juxtaposes paintings, photographs, labor conditions, theories of government and capitalism, advances in physiological optics, Kantian philosophy, contemporaneous psychology, and leisure technologies such as panoramas, dioramas, and stereoscopes.[25] Crary makes sure that his examples fit narrow geographic and temporal bounds—mainly western Europe in the late nineteenth century—but he does not normally attempt to prove that a particular painter looked at a particular kind of optical device or read a particular text. For example, in one chapter of a book on later-nineteenth-century vision and visuality he discusses several paintings by Manet, the Kaiserpanorama (an optical diversion popular in Berlin in the 1880s), photographs by Muybridge, and theories of optical perception and attention. Michael Baxandall has called that kind of historical writing "inferential criticism": it is important for such work that influences existed, but it is not important to find direct links.[26] Manet might have been interested in popular optical diversions like the Kaiserpanorama, or he might not—but they pertain to the same decade and are symptoms of the same culture.

In one sense inferential criticism is sensible, because causal links can't always be found. In another sense it avoids larger questions: in general terms Crary belongs to the wider tradition of Marxist criticism, although his Marxism is tempered and enriched by readings of late-twentieth-century writers such as Michel Foucault and Guy Debord. In that matrix, the ultimate cause of cultural phenomena—if the word *cause* still has purchase—is the configuration of class, labor, and society that constituted bourgeois culture. That culture was a complex and evolving whole, making it unnecessary or unrewarding to spell out causal connections. In the most fundamental sense, Hegel's model of history gives this kind of historical writing its meaning: the very concept of a zeitgeist implies that late-nineteenth-century European bourgeois culture was a web of associated phenomena, making the search for cause-and-effect links inadvisable or unrewarding, and the reliance on any single direction of explanation (from social class to taste, for example) misguided.[27] Thus explanation can be evanescent. In some passages Crary juxtaposes technology and art; in others, he practices inferential criticism; and in others, he implies that his larger project is an inquiry into the workings of nineteenth-century bourgeois culture. That flexibility can be an asset, because it avoids the narrow search for direct causes that can be so vexing to historians in search of crucial pieces of evidence, and also because it avoids an unproductive adherence to theoretical models that can become incompatible with the material at hand. Some of the richest and most evocative texts on science and the humanities remain uncommitted to any specific explanatory model. In other cases the same strategy can seem evasive; for example, when it seems that the reason Crary declines to explain an image directly in terms of some technology or vice versa—in the way that much older scholarship does—is because avoiding a causal narrative might make the overall account denser and more rewarding.

These are four configurations for the relation between disciplines: where the scientific discipline explains the nonscientific one; where the roles are reversed; where a third discipline, ultimately philosophy, guides the interpretation of both the scientific and nonscientific material; and where various disciplines are put in ambiguous conjunction. The four possibilities quickly lead into formidably complex problems, and interdisciplinarity itself is the subject of a growing literature.[28] It is seductive to create "unexpected associations that reorganize a familiar conceptual field and allow us to behave differently within it,"[29] as Reynolds says, but it matters how that is done: by using science to explain art or vice versa; by shifting the onus of explanation back and forth; or by proposing an underlying conceptual problem that is illuminated by both. Reading an interdisciplinary text, it is usually not difficult to tell which discipline is doing the explaining, or if the author is moving back and forth between different explanatory systems. Sometimes—and this is especially true in some of the most sophisticated and influential texts—it is difficult to be sure what configuration, or set of configurations, is at work, and in that case the entire concept of interdisciplinarity becomes conceptually unclear. The result can be rich and evocative, but it can also be impressionistic and ill-structured, so that readers in the sciences especially may not feel that their disciplines are adequately represented.

3. *Interpretation operates by metaphors and models that are themselves not necessarily part of the originating discourse.*

Interpretation can be put as an abstract question of representation, because an interpretation is a discourse that re-presents something—a concept, text, image, process, or act—that has already been presented. In linguistic terms, the engine of that re-presentation is metaphor. For a sociologist of science, the illuminating metaphor might be power relations among scientific institutions, so that an account of the activity of a particular laboratory might be told in terms of grant proposals, rank, gender, or relations with other departments and laboratories. For a philosopher of science, the leading interpretive metaphor might be taken from epistemology, resulting in narratives about the ways that experience is gathered and segmented into data, and how that data is assumed to be related to phenomena. For a popular-science writer, the interpretive metaphor might be biography, so that Einstein's life might become the story of his friendships, family, politics, and professional appointments. I take the word *metaphor* very capaciously here: it might be a heuristic concept, an extended model, a thought experiment, an analogy, an image, or even a fully articulated metaphysical program. What I mean to suggest is that an interpretation cannot work—it cannot be perceived *as* an interpretation—unless it can be understood as a *particular* way of projecting a discourse that is already present onto a new discourse. The particularity of the re-presentation is then perceived as a metaphor or model.

Why put things so abstractly? Because seeing interpretation as an engine constructed of metaphors makes it clearer why scientists can find it unrewarding to read humanists' accounts of their work. For a reader who knows the actual science, sooner or later the metaphors, models, or analogies proposed by sociologists or philosophers of science will begin to sound like something other than science. A scientist may read patiently on, mentally translating the various metaphors back into their original contexts, until he or she reaches the point where the interpretation seems to lose touch with its subject—where the metaphor engine appears to break down. Even if nothing in the account is overtly wrong, the interpretation may seem increasingly and finally irreparably irrelevant. The metaphors or models, which seemed so fruitful and illuminating to the scholar who wrote the account, will appear to lose their grip on the subject at hand—on the actual practice or whatever is construed as part of the original presentation of the science, before its interpretation. At that point, the scientist may give up reading and go back to doing science, or—if the interpretation seems pernicious or egregiously misinformed— write a text like Sokal's or Snow's.

This problem of the failure of metaphor pertains as much to detailed studies like Pickering's or Galison's as it does to popularizations like Stephen Hawking's *A Brief History of Time*.[30] It has no simple solution: any writer who sets out to interpret another discipline proposes and remains within a program of interpretation, whether it is spelled out or just implied. The issue is not the power of the metaphor or its flexibility or nuance: it is more a matter of the inherent structure of interpretation itself.

I mention these three issues in order to set up the approach I am taking in this book. It strikes me that to move forward with the project of talking about the humanities and the sciences together, it is necessary to take a radical position in regard to narrative, explanation, and interpretation. The methodology I have in mind can be put as three answers to these three problems. In order, they are:

1. *Regarding narrative: let geometric discourse remain geometric.*

The historian Kevin Mulligan, reviewing Sokal and Bricmont's book *Fashionable Nonsense*, pointed to two texts by Robert Musil, who was following writers such as Wilhelm Dilthey in worrying about the distinction between the *Naturwissenschaften* and the *Kulturwissenschaften*. Mulligan comments, "Where Sokal and Bricmont betray their exasperation with writers who simply could not be bothered even to consult scientific popularizations, Musil [in 1921] suggests that the important thing is to 'go to the end of the trampoline of science before jumping off.'"[31] To me, Musil's answer is optimal, even though he never wrote the kind of text that his argument implies.[32] The problem of the pencil certainly cannot be solved by adding problem sets to books in the humanities, but it does not follow that it is a good idea to minimize the number of equations in books on the history, sociology, or philosophy of science. (In particular it does not appear that omitting equations widens the audience for books in the history and philosophy of science. Hawking's publisher supposedly told him that every equation in *A Brief History of Time* would halve his readership: but that problem only affects genuinely popular books.)

In this book I have tried to resist the inevitable slide toward what Trimpi calls literary discourse by keeping as close as possible to the specific textures of writing in the different sciences. There is no way to include *all* the equations, graphs, and technical arguments of the source texts: that would not be interpretation, but repetition. Nor is there any such thing as pure geometric discourse, excepting perhaps *Principia Mathematica*. Yet it seems to me that as far as possible, equations and even problems should be included in accounts of scientific texts simply to address the damage caused by changing geometric writing into literary writing.

Some scholarship on science tries for a middle ground; for example, by inventing very simple problems that are still quantitative or by adopting new kinds of qualitative schematics. Humanities-style illustrations—that is, without technical captions or quantitative content—can also help re-present quantitative "geometric" discourse. Popular books on string theory and topology, for example, tend to have more illustrations than the primary texts. But given that Western writing has been divided between two discourses for so long—effectively, in Trimpi's argument, since the inception of Greek philosophy, mathematics, and rhetoric—it does not necessarily make sense to invent a third kind of writing between the arts and the sciences. In particular I have avoided writing a semi-technical pidgin that might mediate between the scientists' writing and a literary discourse.[33]

Because there is no such thing as full literary fidelity to geometric discourse, it is essential to work continuously to break down the illusion that geometry can be melted into literary discourse. (That illusion drives popular science, making periodicals like *Scientific American* seem adequate to their subjects. But in what sense does an article in *Scientific American*, with no equations, represent a discipline that can only represent truth *through* equations?) The best choice, I think, is an *uneven* form of writing that occasionally breaks the flow of narrative to make a place for equations, technical illustrations, and problems. It does no harm if the reader occasionally needs to grab a pencil, linger over an incomprehensible illustration, or ponder the sudden intrusion of an equation bristling with new symbols.

2. Regarding explanation: avoid making one discipline dependent on others by abjuring cause-and-effect relations between disciplines.

I suggested that humanistic writing about science comes in four configurations: texts in which art explains science; those where science explains art; those in which a third discipline, normally philosophy, is called upon to provide the explanation of both science and art; and those in which science and art explain one another in an evanescent or inconsistent fashion, with ambiguous implications. There is a fifth and final possibility: it is possible to imagine a kind of scholarship in which explanation itself is avoided. I call that kind of narrative *noncausal*, and it is the one I follow in this book.

The painter Giorgio De Chirico provides a ready example. He talked about some of his own early metaphysical canvases as if they were inspired by a concept he called "the enigma." As far as it is possible to tell, "enigma" for De Chirico was a personal mixture of his memories of the city of Turin, mingled with his thoughts about Nietzsche (who had lived there) and De Chirico's own feelings of solitude and romantic oblivion, which he got partly from Giacomo Leopardi. Art historians have noted De Chirico's idea, but they have tried to interpret his canvases in other terms. Some scholars have concentrated on his biography; others have written on the sociology and politics of art in his time and place, on the breakdown of perspectival painting, and on the Freudian elements of his pictorial fantasies. There is a sense in which *all* such interpretations miss the mark, because De Chirico's main concern was a specific kind of conceptual opacity. "Enigma" was an apotropaic magic word, meant to ward off interpretations that would dissect his allusions. In a case like De Chirico's, it seems to me that it might be interesting to re-conjure the word and leave it intact so it can repel psychoanalytic, social, and formal interpretations. That does not mean De Chirico's early paintings should be proof against art-historical scholarship: it means that the scholarship could focus on understanding how the concept of "enigma" was built and exactly what kinds of clarity it was meant to repel. Writing of that sort would be noncausal in the sense I mean it here: it would decline to explain De Chirico's paintings in terms of *any other discipline.*[34]

Noncausal narrative is not always called for. Most studies of art-science links have clear causal direction: Galileo influenced the depiction of the moon in paintings by Velázquez, Rubens, and others; and at least two art historians have claimed that Galileo's Counter-Reformation tastes in art influenced the direction of his science.[35] It is also possible to study the early-seventeenth-century culture of art and science by switching back and forth between cases where art influenced science and vice versa.[36] In such cases there is no particular reason to avoid causal connections. Noncausal writing starts to make sense when it becomes apparent that cause-and-effect links are not doing full justice to the material. Moons and other celestial phenomena are minor elements of late-Renaissance paintings, even those specifically influenced by Galileo's discoveries, so an account such as Eileen Reeves's *Painting the Heavens,* which studies the moons in several paintings of Galileo's generation, works more as a footnote to the history of astronomy than an adequate account of the paintings.[37] Instead of saying that Galileo's discovery of mountains on the moon prompted Velázquez to paint ambiguously translucent and solid mountains on a moon in his *Assumption of the Virgin,* it might be better to look at how the painting presents translucency and opacity, in order to discover meanings that are independent of scientific causes.

Most of the images I study in this book were made either by artists who were largely uninterested in science or by scientists unaware of the relevant art. That in itself would not preclude a causal analysis—historical scholarship does not need to be restricted to *conscious* emulation—but it means that if I tried to present the scientific images as products of late modernism, my argument would quickly become counterintuitive. I would have to emphasize, for example, the few examples of scientists who are actually influenced by art. Several of the scientists I discuss in this book are interested in modern art, and a few also paint or make artworks by manipulating scientific or mathematical images. But even in those cases, it would not help much to list the artistic influences, because they account for so little of what goes into the scientific images. The converse is also true: if I were to describe the paintings and fine-art photographs in chapters 1 and 2 as part of a larger culture informed by scientific illustration, I might succeed in finding a few interesting links, but I would end up explaining only a little of the paintings and photographs in question, and I would go against the grain of artists', critics', and historians' understanding of the images.

Presenting each of the six fields in this book as an independent discipline has the advantage of letting each speak its own language, using its own concepts and interests. If I do not say that a kind of scientific image was made possible by the scientists' awareness of abstract painting, then I avoid explaining away the science as an *effect* of some nonscientific practice.

That is why each chapter of this book concerns a separate subject: the idea is that the form of the book supports its noncausal agenda. The images are contemporaneous (they were all made in the last quarter of the twentieth century and the beginning of the twenty-first), and I think they are part of the same phenomenon, which I will describe, differently in each context, as a growing fascination with the last moments of ordinary representation. But I do not offer a unifying account of the phenomenon: no philosophic frame, no common ground in visual culture, no shared aesthetic or artistic qualities.

At one point while writing this book, it seemed that the images *could* be united if I made use of the aesthetic concept of the sublime. The same period that produced the images in this book also saw a renascence in studies of the sublime in psychoanalysis, Continental philosophy, and postmodern literary theory. The coincidence seemed irresistible, especially because sublimity is the quality many historians and critics would assign to images of deep space or elementary particles. The first draft of this book was an extended study of the postmodern sublime, taking the images as examples. Essentially that version was a philosophic discourse on visual forms of the postmodern sublime. The book was tightly organized, because it is not difficult to find images that illustrate many of the critiques of the Kantian sublime or that enact the epistemological disappointment that characterizes recent attempts to fashion a postmodern sublime. Yet I paid a high price for philosophic unity: the sublime is virtually unknown outside the humanities, so I lost a lot of the detail of what was happening in the scientific images. Eventually I decided the book had to be entirely rewritten, excising the sublime from the chapters on scientific images and sequestering it in the discussions of painting and photography, where it is already part of the critical conversation. The first draft was a philosophic and art critical text that would have worked to bring several sciences into the fold of humanist discourse on postmodern images. As it stands, this book is a contribution to no particular discipline, and I hope it entirely fails to bring any image-making practice into the domain of any other.

It might be said that this book is still basically philosophy because it is about representation, which is itself a philosophic concept. Philosophy is sophophagic—it eats other disciplines—and inevitably what I have to say has taken on a philosophic or analytic cast. But I have not introduced any of the philosophic vocabulary of representation (*Vorstellung, Darstellung,* Idea, simulacrum, *eidolon,* imago). I hope the book is only philosophic in a weaker sense: the title announces a common theme that could be philosophic, except that I have not pursued it *as* philosophy. The first draft of this book depended on a philosophic theme at the expense of most of the nonart images. In a noncausal narrative, it is the reader's task to decide what threads might tie the images together and just how tightly they should be pulled.

3. *Regarding interpretation: defer the problem of inappropriate metaphors by avoiding metaphors that are not found in the original discourse.*

To the extent that interpretation is metaphor, it will always drag a discourse off in the direction of some other discourse. Again, there is no definitive solution—only strategies to avoid. Part of the point of not introducing the philosophy of representation is to avoid placing the various practices I describe under the sign of some foreign metaphor before they have had a chance to speak in their own voices. I do not want to claim that scientists work on "the limits of representation" when they are really working on the calculation of blur circles, Fourier transforms, or pixel thresholds. On the other hand, I have not tried for some pristine interpretation that would somehow speak in the exact language it is meant to interpret, without metaphors—and therefore without interpretive power. The two methodological principles that have guided me here are: not to rely on metaphors that are not found in the primary texts, and to mark the places where metaphors begin to guide the inquiry, so that they can at least become objects of discussion and self-criticism.

I believe that the clearest, most fruitful response to the abyss between the humanities and the sciences is to set out the disciplines, in detail, side by side, and let them tell their stories in their own languages. As far as I can see, that is the only way to produce a book that can be read by scientists and humanists without the creeping feeling that their disciplines are being explained—or explained away—by someone who does not really understand them.

Needless to say, this methodology has its limitations. The theme of the book has to be discovered and rediscovered by the reader in each new context, and there is no conclusion in which the whole is neatly packaged as a thesis about representation at the beginning of the twenty-first century. (I indulged in some talk of that kind in the Preface, because it is true to the genesis of the book. In that respect the book is an extended criticism of the synthetic ideas in the Preface.) I hope the open-ended quality of this book can be experienced as a gift, because it allows each reader to discover what counts as commonality and difference. As a reader in any of these disciplines, or in none of them, you can decide which are the best ways of describing the limits of representation. It is possible that you may conclude that I have written six separate stories, not linked by any particular theme, problem, trope, analogy, or common history. I am willing to risk fragmentation or incoherence in the hope of writing something potentially acceptable to people working in different disciplines. What could be more important, in this age of specialization—in this age overrun by messy popularizations of all sorts—than a text that could serve as a meeting place of disciplines?

And finally, it is worth saying that this is not a programmatic book. The images came first, then the naive overarching theory of the sublime, and finally the thought that it is better to risk fragmentation than to write yet another theory of the humanist meaning of science or yet another popular-science exposition of a partly incomprehensible scientific discipline. This Introduction was written last, and these lines last of all. That is as it should be: after all, starting from first principles and writing a book to fit would have been yet another capitulation to philosophy.

 Painting

Much of the history of painting can be said to be involved with the limits of the representable, but there is an essential distinction that helps me
limit this chapter to just a few images. In the broader history of Western
painting, artists concerned with objects and ideas that could not be put
into paint were also interested in objects and ideas that could be adequately shown. The history of Christian representations of the divine
is adequate testimony to that.[1] The vestments God wears in heaven, the
scrolls of the heavenly vault, the white dove that represents the Holy
Spirit, the iridescent wings of angels—those are the kinds of objects that
have long counted as discrete accomplishments within the history of attempts to picture what cannot, in the end, be pictured. The same can be
said, closer to the period I am concerned with, about romantic paintings
such as Caspar David Friedrich's *Cloister Graveyard in the Snow*, a study of
ancient oak trees in the dim milky light of a winter snowstorm, with
powdery snow drifting around gravestones. Friedrich's work was driven
by interests that he did not represent directly, such as the limitations of
conventional faith, the effects of the passage of time, and the immanent
spirituality of old trees and lonely places. Those concerns were tropes in
the poetry of Friedrich's generation, and one might ask whether their
appearance as verbal images relies on a similar sense of the inadequacy of
language. What matters in this context is that paintings like *Cloister
Graveyard in the Snow* are replete with objects that are meant to be taken
as adequately represented. To say it the other way around, representation
is understood to be adequate to objects like snow, gravestones, and trees,

whose very visibility can be understood as a sign of the unrepresentability of the concepts or feelings that make them worth representing.

The paintings I am interested in here are fundamentally different because they concentrate on the inadequacy of representation. Little or nothing in the images in this chapter is depicted in a satisfying, clear, comprehensible fashion, like Friedrich's snowy pine trees and gravestones. It is possible to put this in terms of ease and difficulty: in the nineteenth century Friedrich was new, but now much of his work is easy in the sense that his motives and strategies have been taken into common post-romantic practice from painting to Hollywood. His romanticism is familiar, and so is his pining for a kind of elusive spirituality (no matter how hard it might still be to find the right words for it). Nineteenth-century romantic landscapes that descend from painters including Friedrich are a dime a dozen in auctions, and even thoughtful transcendentalism like George Inness's has become easy to understand.[2] Yet what I miss in painters from Friedrich to Inness is an awareness that—to contradict a phrase of Barnett Newman's—the sublime isn't "now." Landscape paintings that claim to actually give us spirituality or sublimity are no longer convincing, and that is largely, I think, because they reason too simply about what can and cannot be put in a picture.[3]

At the end of the twentieth century painters were still working with the echoes of that same problematic, resisting a sense of painting in which realism can be at once adequate to any conventional task and also the proper or natural conduit for concepts that are taken to be unrepresentable. Some late-twentieth-century painters avoided being complacent about representation by trying to be weak, letting lapse whatever counted as academic or traditional representational skill. Luc Tuymans's apparently insouciant pictures are an example: they demonstrate a mistrust of the adequacy of what was taken as conventional realist representation. Tuymans displays an interest in demonstrating that mistrust by painting with an apparent lack of care and effort.

In the 1980s and 1990s painters developed a number of strategies for not presenting naturalistic skill as an adequate mode of displaying the world. Some said very little or put what they said without exactitude. (Jo Baer's nearly white paintings are an example.) Others made counterfeits of illusion (Vik Muñiz's work is an instance of this), lavished attention on nearly nothing (Simon Hantaï's impatient abstractions can be understood this way), or opted for nearly liquid figuration (Marlene

Dumas's lovely figures are an example). These painters do not form a group; they work in different manners and have different objectives, but I would like to suggest that they share an aversion to whatever seems clear or sufficient in naturalistic representation. In the last decade of the twentieth century, the painter whose work came to stand for many of these possibilities was Gerhard Richter. His paintings from photographs are painted as if smeared, ruining the viewer's access to detail and convincing realism. In the literature, Richter's smearing is understood as a sign of his political and biographical concerns. (His facsimile of photographic *blurring* is a different strategy, which I will take up in the next chapter.) The series *October 17, 1977*, on the Baader-Meinhof gang, has been especially well studied in this regard. Here Richter's paintings from photographs, along with work by the other artists I have named, are understood more as paintings that are in dialogue with previous painting, on the subject of painting's inadequacy.[4]

In art-historical terms, the painters in this chapter come in the wake of minimalism and work in various ways against minimalism's purity. It is as if minimalism created a dilemma for painting by presenting it with two unpalatable possibilities: either keep to the object itself (the canvas or the paper) as minimalists did, or else return to a sense of realism that minimalism itself had closed off. The painters in this chapter feel the unhelpful pull of minimalism and the useless pressure of more traditional realist painting.

Edging toward transcendental themes, and also away from them. Mulling over the role that traditional realism might still have in painting. Wondering about minimalism, but refusing it at every turn. In this chapter I am going to approach these gigantic themes very narrowly, concentrating on just two painters, Agnes Martin and Günter Umberg, and on two artistic strategies: one I will call the bracket, and the other the ladder. But first it is necessary to spend a few pages introducing elements of the sublime, which will serve as the chapter's central critical concept.

1   *Adequate Representation in Ruins*

The word *sublime* means "up to the threshold"—*sub* is "up to," and *limen* is "lintel" or "threshold." Up to it, and no further: the sublime is the encounter between what can be thought and what cannot. In images, it is the point where the picture gives way to what is taken to be unrepresentable.

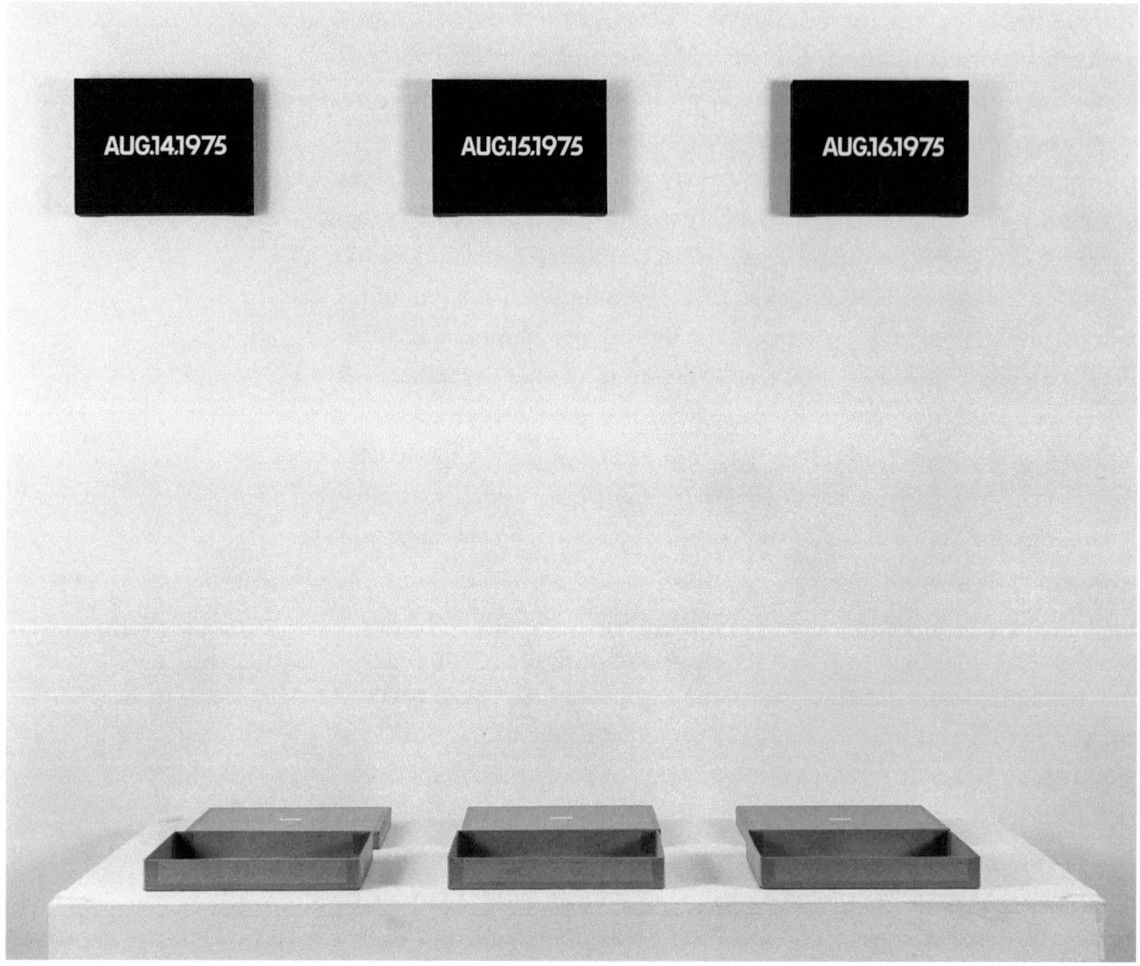

FIGURE I

On Kawara, *Today Series, 1966 . . . (Saturday),* from the series *Date-Painting,* 1975. Liquitex on canvas, cardboard box, newspaper. CNAC/MNAM/ Dist. Réunion des Musées Nationaux /Art Resource, New York.

The sublime is distinct, so it is said, from the beautiful, from aesthetics, and from taste. In philosophic terms, experiences that do not contend with the sublime are self-content, self-enclosed, adequate to their own means: they propose questions and dispose answers with no remainder. So, in painting, a hackneyed panorama of a fishing village probably has nothing to say about the sublime, but a blank painting with a date painted on it might (Figure 1). The artist On Kawara's "date paintings"—medium-sized black canvases with dates carefully painted in white—are gestures in the direction of things that he couldn't, or wouldn't, represent. They were each made along with a box that contains some of the day's events, typically a newspaper from the city where (so we are invited to imagine) On Kawara was at the time, and a map of the itinerary of his travels on that day (or the travels we are asked to imagine that he made). Nothing on the maps is labeled. Nothing about the choice of headlines is explained. There is no diary or artist's statement to explain the choice of dates. An all-too-ordinary painting of a fishing village is also all-too-beautiful, too easily understood, too confident about representation. A fishing boat, grounded by low tide and listing to one side, looks just like *this*. A pile of wooden lobster traps, with green nylon netting thrown on top, looks just like *that*. Such paintings can be relentlessly complacent about what painting can do, and can be.[5] On Kawara's project has remained both everyday and mysterious since he began it forty years ago, and each new painting again fails to explain itself *as* painting. The sublime, in these terms, is the encounter between what can be readily put into paint and what cannot.

A simple picture of balls—let me make here the first of many intrusions of science, without apology or preamble—on a billiard table probably won't make contact with the sublime, but the same picture, labeled as subatomic particles, may (see Figure 69). Particles are like billiard balls, but only to a point; so the picture is a knowing dissimulation, a gesture in the direction of objects that can never be pictured.

What fascinates me about On Kawara's paintings is how they imply so much and say so little: they are like unused diary pages, each one printed with a date and nothing more. He paints the numerals and letters fastidiously, in a font he invented, and he goes over them with a fine brush until they are perfectly shaped.[6] If he doesn't finish before midnight, he discards the painting, leaving no trace. Even a finished painting says almost nothing and conjures uncertain thoughts about the painter's hidden life. What was he doing on January 2, 1966? Why was it worth making a painting on that day? What had happened in the days between that painting and the one before it? What solace, what pleasure, could the painting possibly have given him?

A physics textbook with a billiard-ball picture makes me think of real billiard balls and how they make a sharp sound when they click against one another; but that thought only impresses on me even more strongly that the particles these balls represent are like nothing I will ever hear or see. I know that particles aren't spherical, that they aren't solid, and that their collisions create events fantastically more complicated than the collision of two billiard balls as it's described in introductory mechanics courses. Thinking of the balls' glossy enamel paint and the soft green of the billiard table only makes me feel even stranger about what lies beyond this picture.

On Kawara's painting and the simple image of billiard balls are sublime because they totter on the edge of solipsism or meaninglessness. Somehow, by balancing in that precarious place, they manage to be more suggestive, more hypnotic, than ordinary oil paintings or diagrams of ordinary objects. Neither picture is beautiful—in fact, the computer graphic is fairly ugly—but for me they are compelling because they concentrate on what cannot be shown. Nothing is more interesting, I find, than an image where the possibility of adequate representation itself is in ruins. When representation itself gets lost or becomes weak on its way to what the pictures want to show us, the effect is sublime in the proper sense.

2   A Brief History of the Sublime

The sublime is a concept with a long, sad, and complicated history. It begins, more or less, with the ancient writer known as Longinus, who used a word translated as "sublime" in a book of literary criticism. Everything is in doubt with Longinus: he probably lived in the third century A.D., though it has been claimed he lived in the first century; Longinus may or may not be his name; and his notion of sublimity can only sometimes be identified with what gets called "sublime" in books written after the Renaissance. (That last area of ambiguity causes difficulty in recent texts on painting that cite Longinus, such as T. J. Clark's *Farewell to an Idea*.[7]) Longinus's text is somewhat pedantic and scholarly, and it is "permeated with unsureness," as the philosopher Jean-François Lyotard says, about its own subject.[8] Even so, Longinus is the usual jumping-off place for histories of the sublime, as if he were the first to write on it (which he wasn't), and as if he shares concerns with Enlightenment and modern writers (which he often doesn't).

The next principal sources are the erudite, sometimes obscure late-Renaissance historian Giambattista Vico, and the French classicist Nicolas Boileau-Despéaux, who translated Longinus as an unlikely support for his own theory of poetry.[9] Boileau, as he is known, largely initiated the Enlightenment interest in the sublime, and a half century later there is the Irish essayist and historian Edmund Burke, the author of the wild and often poetic book *A Philosophical Enquiry into the Origin of Our Ideas of the Sublime and Beautiful* (1757). From there the history of the sublime goes straight to the center of Western thought, in Immanuel Kant's *Critique of*

Judgment (1790), the final and crucial third volume of his trilogy on reason. (It completes the *Critique of Pure Reason* and the *Critique of Practical Reason*.) Afterward, the sublime was invoked by a widening circle of romantic philosophers, poets, and artists, and it became part of the general cultural conversation on art and aesthetics.

The reason I call the history of the sublime "sad" is that it was abused, or at least overused, by painters and critics, most recently in regard to abstract expressionism. It didn't help, for example, that Newman waxed philosophic about the sublime, along with "the penetration into the world mystery," religious art, and even the "basic truth of life."[10] As a result of his overheated writing and the criticism that followed it, the sublime has largely fallen out of favor in the art world. Sublimity sounds old-fashioned, a throwback to early-nineteenth-century romanticism and Victorian transcendentalism.[11] Yet at the same time the concept is alive and well in literary theory: there it gathered one of the most intricate and difficult critical traditions in all twentieth-century Continental philosophy. Lyotard helped make the "postmodern sublime" a central term in the discourse of postmodernity, and the subject is discussed in some exceptional texts by Jean-Luc Nancy, Philippe Lacoue-Labarthe, Thomas Weiskel, and Neil Hertz. People who study literary criticism know the sublime as a touchstone—flawed, often misread, but indispensable—to any understanding of modernity and postmodernity. Versions and excerpts of that discourse have made their way into art criticism and art history, and that is why I begin with the sublime—as it turns out, a rejection of elements of the Kantian sublime is a crucial part of the interpretation of the paintings that interest me here.

3   Kant's Sublime

Far and away the most influential description of the sublime experience is Kant's.[12] No account of the sublime can bypass Kant's, so let me pause for a moment to consider his original description.[13]

Kant divides sublime experience into two types: the mathematical and the dynamic. The latter comes about when an experience is overwhelmingly, irresistibly powerful. One of Kant's examples, and still the standard one, is a stormy ocean. By sheer force, the ocean denies us the freedom to act or to move as we will. It is Kant's idea that a person confronting a tumultuous sea will first be paralyzed and humbled, and then begin thinking of how people avoid drowning: how they navigate the ocean, and how they are, in the end, independent of it. That train of thought is essentially a defense, and it brings with it a comforting sense of detachment. The person then experiences the freedom that Kant claims is an inalienable part of human experience: all of us are free to think, and so to extract ourselves from the objects we encounter. We realize, at the minimum, that the danger posed by the ocean is restricted to the domain of the body and its senses, and that the mind belongs to another domain—the "stratum" of the "supersensible."[14] For Kant, the feeling of the dynamic sublime arises from the contrast between two mental states: first the abject dependence on unmasterable forces, and then the freedom that comes with the awareness that thinking is a different kind of experience.

The image of a person standing on a beach, looking out at a pounding surf, feeling a sense of mingled freedom, paralysis, horror, loneliness, humility, confusion, and comfort, is an age-old trope of peril and salvation. The philosopher Hans Blumenberg has written a book on the variations on the theme, some of them framed long before Kant.[15] Kant's theory is an unusually exacting variation. Later it was abused in various ways; one of several points missed in romantic interpretations of Kant is that the sense of sublimity comes not from the transcendence of the object, but from the person's awareness that thought is itself transcendent. The sense of freedom is outside the purposeless thrashing of the ocean—the waves cannot touch it.

I will not have anything further to say about the dynamic sublime. It is the other sublime, the mathematical sublime, that is bound up with the history of painting. An argument could be made that the dynamic sublime is at work in nineteenth-century romantic paintings of people looking at the ocean, or in eighteenth-century depictions of natural wonders such as Mt. Etna. (Volcanoes are another of Kant's examples.) Today I might look for the dynamic sublime in IMAX films and thrill rides like the ones at Universal Studios theme parks. But those are the garish exceptions to the rule. Most pictures do not set out to make a viewer feel helpless: most images aren't that aggressive—but it's an arguable point.

The mathematical sublime, on the other hand, is a perfect description, *avant la lettre*, of central developments in twentieth-century image making. Kant defines it as an experience of something unencompassable, so large that it exceeds our capacity for comprehension.[16] His example, the starry sky, is also one of my subjects in this book.[17] At first, confronted with the starry sky, a viewer is dazzled, confused, and—again—humbled. Then she tries to employ some concept to help her comprehend the incomprehensible heavens: say for example the concept of infinity. The viewer says to herself, "This is infinite," and for a moment she is comforted. But there follows a third moment (Kant's analysis is typically intricate) when it becomes clear that the "manifold" object is too expansive to be understood. It cannot be gathered under a single concept, a single intuition. And further: the concept itself, in this case infinity, is not directly experienced, but somehow known. There follows a fourth and final moment when the viewer realizes that her innate capacity to reason is what drives the desire to encompass an unencompassable object with an inadequate concept. Even though the attempt to understand the object fails, the viewer becomes aware of a mysterious, inbuilt capacity to *try* to match her imagination to objects: a capacity that includes the very idea of a fully adequate concept even though no such concept can be imagined. The game is lost, but she knows that she has a "supersensible" faculty, which allows her to think *about* such things as infinity and the correspondence between inadequate concepts and unknowable objects. "*The bare capacity of thinking this infinite* without contradiction," Kant says, "requires in the human mind a faculty that is itself supersensible [*das selbst übersinnlich ist*]."[18] That is the sublime: a pleasure that comes from displeasure—the displeasure of realizing that the imagination has been, and will always be, defeated.

Kant intended the sublime to apply to natural phenomena, but it has often been remarked that it applies equally well to art.[19] I'll give a concrete example that fits one of the subjects of this book. I have a small reflecting telescope, big enough to see star clusters and bright nebulae. A telescope is not a source of continuous amazement, because there are so many things to distract the attention: confusing star charts, guidance problems, weather reports, night dew, light pollution from nearby towns, and the inevitable creeping cold of the late night. Most of the time familiarity breeds blindness, so I don't really see what I am looking at. But it can still happen that I feel a sudden vertigo, wondering at the fact that what I have been calmly studying through the lens of my telescope is actually entirely and hopelessly beyond my comprehension. It is nothing I know or ever will know: it has no name, it is immeasurably distant and irreparably meaningless. Those moments are distinctly unpleasant.

But immediately my mind scrambles to understand what has happened, and I attach some label to what I have seen. It is really the Big Dipper, I say, or just Andromeda. I remind myself that the heavens are mapped, that the sky has named parts and a library of books to describe it. While I say that, I know that it is an empty thought, but still I feel it as a comfort. Usually a label like "Andromeda" acts like a tranquilizer, inducing a kind of placid amnesia, and I go on with what I had been doing. Other times it makes me nostalgic for the weird dizziness I had just felt, and I try to recapture the feeling by looking again without thinking of the comforting words. Usually that doesn't work; looking again, I see "Andromeda," and not the nameless object that had frightened me a minute before.

Kant says that the concepts I bring with me—the ideas and terms of astronomy—are a way of comprehending something "manifold" as a "unity." They domesticate the "manifold," taming the unsettling thing I had seen—a thing that has no name, and really no place in my life—by calling it "Andromeda." An existentialist might stop there and stress the hopelessness of trying to understand. Jean-Paul Sartre in particular was fascinated with the queasy feeling that comes from thinking too directly about what any object really *is*. He says a round stone held in the hand becomes inexpressibly slimy and horrible: it metamorphoses into an object so impossibly alien to any human intuition that it can only appear as a monstrous growth, oily and heavy. (That is what happens in the novel *Nausea* when the protagonist suddenly realizes how everyday words have sheltered him from the naked awareness of existence.) From Kant's perspective, Sartre's melodrama is only part of the story. For Kant, the moment I realize that a phrase like "Big Dipper" is only an inadequate convention—the moment I see that the whole of astronomy is the pitiful best that people can do—I sense that I have also been trying to comprehend an unimaginably large and distant object *as* a single, unified concept. Inadequate as the concept has to be, it is evidence that human reason can come to terms with the inhuman, and with unreason. That is the moment of the mathematical sublime, because I then understand how a finite concept, my own, can be allied to a transfinite object: I am uncomfortably conscious of the capacity and limits of my own thought.

4　The Visual Itself as Sublime

Kant's description makes it sound as if the sublime might be a rarefied or elusive experience. It can be both, but in a wider and partly non-Kantian sense it is common, and since romanticism, it has become one of the common conditions of a viewer's interest in an image. Looking at a painting, I may be struck by the fact that the picture continuously and effortlessly escapes my attempts to describe it. In such an encounter I acknowledge I am entangled in an ongoing project, called art history or criticism, whose purpose could be said to be an attempt to capture parts of images in words. That, I take it, is one of the principal concerns of Louis Marin's endlessly suggestive book *To Destroy Painting*, and it can be read in Jacques Derrida's "The Truth in Pointing" and other texts.[20]

One of the optimal models for these negotiations of the limits of the verbal representation of pictures (and the limits of representation *in* pictures) is the sequence, leading from the known to the unknowable, which is given in the analytic of the sublime. In that account, the only time the experience of a painting is *not* sublime is when I become complacent about my words and decide that they are somehow adequate. Kant wouldn't recognize much of his analytic in the day-to-day operation of art history or criticism, except the recurring focus on epistemological limits. Yet I agree with Marin that the sublime remains a centrally important concept for the intermittent epistemological disappointment that is art writing. Certainly it is essential in considering what has happened when the philosophic discourse of the sublime has been enlisted in art history and criticism. I mention this here at the beginning of the book, because in later chapters not only will the sublime have no place, but there will be little opportunity to register the idea that the content of pictures might not be exhausted by description.

5   A Glance at the Beautiful

In Kant, the beautiful and the sublime are complements, equals and opposites: the beautiful produces pleasure, while the sublime incites confusion, displeasure, and even "horror"; the beautiful belongs to taste and aesthetics, while the sublime exceeds both. In late-twentieth-century art, things were not that simple, but it helps to preserve Kant's and Lyotard's strong distinction between the sublime and the beautiful. It's important, in this context, that the beautiful has no threshold: it is self-contained, so its prettinesses, its harmonies and inner coherences, are all of a piece.[21] No matter how the two experiences mingle in recent painting, the sublime is always broken: it entails aporia, uncertainty, inadequacy, and even an unpleasant or painful disorientation.

Among painters active in the last two decades of the twentieth century, Ross Bleckner was often cited as a painter of the sublime. Yet I agree with the critic Joseph Masheck that Bleckner's paintings are mostly beautiful, and not sublime: Bleckner paints chandeliers, hummingbirds, orchids, urns, and a whole paraphernalia of glittering rococo interiors. Those subjects, Masheck observes, are too ornate, too precious, to be sublime.[22] On the other hand, Bleckner also paints ominous empty interiors, glistening abstract stripes, overwhelming monumental domes, and deep cavernous spaces pierced by shards of light. A few of his paintings forego decorations altogether and become more single-mindedly sublime. His *God Won't Come* (1983) is empty except for a steel-colored double reflection, like the moon shining off a metal plate. *Mister, Mister* from the same year is a searchlight, hidden somewhere over a dark ridge, pointing straight up into foggy skies: the same mysterious composition as Goya's cryptic *Dog* (an icon of sublimity, made into a friendly mascot in the Prado gift shops).[23] In 2000–2001, Bleckner made paintings with soft bursts of color, streaks, stains, and little more.

Flora and the Future is especially unrelenting about the inadvisability of the beautiful (Color plate 1). It has a cold, glistening surface—it could be a body of water at night—with two arrows in deep relief, one pointing up and inward, the other down and away. Other arrows, immured in the painting's surface, point upward. The surface of the painting is heavy, and badly scored like scraped aluminum. (As Thomas Crow says, Bleckner treats "the flat plane of the picture as possessing a substance of its own, like a thickened film or skin."[24]) The whole is burdened by dense paint. "By *putting* darkness *into* his image with black," Masheck says, "Bleckner only confirms Burke," who recommended "a judicious obscurity."[25] With its inexplicable arrows and its hidden source of light, the picture literally points beyond itself. There is a strong break, in other words, between what is shown and what is absent, and there are no lively colors or glittering ornaments to make the scene prettier. In the terms I am developing here, this is a painting of the sublime threshold.

More typically, Bleckner mixes sublime and beautiful effects, smorgasbord-fashion, and at times his work draws close to first-generation romanticism. The painting *Chamber*, Masheck says, "might really date from circa 1800"; Crow and the art critic Peter Schjeldahl have made similar remarks.[26] It is not easy to disentangle Bleckner's preciousness—his beauty—from his encounters with the sublime, especially when those encounters depend on such a close approach to the past of beautiful paintings. At least there is more of the sublime in paintings like *Flora and the Future* than in Bleckner's rococo fantasies filled with swirling apparitions, stars, ghosts, and silverware. Despite his admixtures of prettiness and stark sublimity, Bleckner's paintings preserve the essential choice: either there is a threshold, and the painting is *about* that threshold, or else the picture is a simpler source of pleasure stocked with easier delights for the eye.

I will stop there with Bleckner, because I am not interested in the remnants of the sublime that mixed, especially in the 1980s and 1990s, with signs of beauty. The philosopher Philippe Lacoue-Labarthe, one of the most acute observers of the sublime, sees part of the sublime as "a minor tradition," attesting "to an exhaustion of the sense of the beautiful."[27] It does seem, in artists such a Bleckner, that there is such a thing as a depleted sublime, one that readily sinks back into the beautiful. In the 1980s that pallid, pliant sublime mingled with what were taken as beautiful, lovely, or exquisite moments in painting. Bleckner's work demonstrates that it is possible to be reasonably exact about how those mixtures worked and what remained of the sublime—principally, I have suggested, the threshold itself. What concerns me, however, is another strain of the sublime operative in the same years—one involving an increasingly disorienting passage up to a limit of understanding, producing, in its best moments, the queasy feeling of looking forward into blankness.

6 *Agnes Martin*

I now come to the first of my two examples in this chapter: the abstract painter Agnes Martin (Figure 2). Her canvases beginning in the 1970s are severely disciplined images, done with a rigorous self-imposed restraint. In the simplest cases they are monochromes, overlaid with closely ruled grids drawn in pencil.

But then there is also Agnes Martin the landscape painter, whose paintings are really about the light of the New Mexico desert or the flat plains of Saskatchewan where she was born. If Martin were a landscape painter in the traditional, expected senses of that expression, then she could have been in the tradition of Georgia O'Keeffe—who preceded her in New Mexico—or any number of fairly abstract painters who were

directly inspired by the qualities of specific landscapes. It seems Martin is hinting that she paints landscapes when she gives her works desert colors or titles such as *Starlight, Hill, Blue Flower, Islands, Night Harbor, Mountain,* or *Lemon Tree.* When the pictures are untitled, as they often are, it seems that is only because a landscape metaphor didn't come immediately to mind. A handful of paintings hover between illusionistic scenes and untitled abstractions, for example one called (in a perfect equivocation) *Untitled/Grey Bird.*

Is it best to think of Martin as a pure abstract painter, working with grids, color fields, and ruled lines? Or do her paintings tell us something about the desert of the American Southwest, or the northern Canadian plains? This is not an uncommon quandary in abstraction, and it has been asked in relation to work by Pollock, Kandinsky, Diebenkorn, De Kooning, Rothko, and Mondrian, among many others. The preponderance of the literature on Martin's work has it that she makes abstract paintings somehow derived from her experiences in the landscape. Only a few texts— Martin's own talks from the 1970s and 1980s, an essay by the historian Rosalind Krauss, another by a critic named Kasha Linville—say otherwise.

Martin was a friend of Ad Reinhardt, the painter known for his nearly black paintings and for some indulgently paradoxical and dogmatic statements such as this one:

> The one work for a fine artist now, the one thing in painting to do, is repeat the one-size-canvas—the single-scheme, one color-monochrome, one linear-division in each direction, one symmetry, one texture, one formal device, one free-hand-brushing, one rhythm, one working everything into one dissolution and one indivisibility, painting everything into one overall uniformity and non-irregularity. No lines or imaginings, no shapes or composings or representings, no visions or sensations or impulses, no symbols or signs or impastos, no decoratings or colorings or picturings, no pleasures or pains, no accidents or ready-mades, no things, no ideas, no relations, no attributes, no qualities. . . . Everything into irreducibility, unreproducibility, imperceptibility. . . . The one thing to say about the best art is the breathlessness, lifelessness, deathlessness, contentlessness, formlessness, spacelessness and timelessness. This is always the end of art.[28]

Martin heard this energetic and partly disingenuous manifesto in a different way, I think, than Reinhardt intended. "My paintings," she said, "have neither objects, nor space, nor time, not anything, no forms. They are light, lightness, they are about merging, about formlessness, breaking

FIGURE 2

Agnes Martin, *Untitled #12,* 1977. Graphite and gesso. Chicago, Art Institute. Photo: author.

down form."[29] Here the self-contradictions serve a different purpose than they do in Reinhardt's texts, which—as I read them—are driven by an unslakable desire not to be discovered to be saying anything. Martin's purpose is more consistently to describe the *process* of her paintings. She says her paintings have "no forms," as Reinhardt also said—but then she claims they are "about formlessness," which sounds like a story about the thing rather than the thing itself, and that they are about "breaking down form," which makes them sound like way stations on the path to Reinhardt's implausible, perfectly paradoxical, contentless and conceptless purity. Reinhardt himself would never have written those last two clauses. His work, so he claims, *is* formless. Martin says her paintings have "no forms," but then she seems to have thought better of it, and she ends by imagining her painting is work in progress, "breaking down form," talking "about formlessness."

On the other hand, Martin is very much like Reinhardt in her adamant denials that her work comes from the New Mexico landscape. (Reinhardt was a hard-line art-for-art's-sake polemicist, and it is easy to imagine him denying similar charges.) "The landscape doesn't have any relation to my art," Martin said in 1989, "the paintings are not abstractions from nature." In her mind, the work is connected to experience in a more abstract fashion. She has spoken a great deal about beauty and has said that she wants to be remembered as an artist "in the classic tradition (Coptic, Egyptian, Greek, Chinese), as representing the Ideal in the mind."[30] It is not easy to enlarge on the meaning of her reference to "Coptic, Egyptian, Greek, Chinese"; in the received notion, Egyptian and Chinese artists were idealists, keeping their art apart from the things they saw.

In this reading, the paintings aren't about landscape after all. And yet there it can be hard to suppress the temptation to cease speaking of beauty or the ideal and look instead at the links between Martin's canvases and landscapes. After citing the same passages I have just cited, the critic Marja Bloem says Martin's work "often evokes the sense of landscape, because the paintings are essentially her expression of psychic scenery reflecting interior mental worlds of peace and beauty, an endless inner space."[31] That would be an account of how the purity of the "classic tradition," of perfect abstraction and formlessness, drifts back toward the landscape painting tradition, of impure abstraction and of talking "about formlessness."

The manner in which a critic or historian controls the temptation to reintroduce landscape is a measure of that writer's place in modernism and postmodernism. Dore Ashton, most prominent as a critic in the 1960s and 1970s, starts her essay on Martin by bringing some of her high-flown abstractions back into line with the landscapes that must have inspired them. "Years ago," Ashton writes, when Martin's paintings

> were close to their sources and her thinking about them less arcane, Agnes Martin often recalled the place of her birth, Saskatchewan, and her childhood in Vancouver. Her father's wheat farm and the sealike vastness of wheat fields were not forgotten. The great prairies had endowed her with an undying hunger for spaces. With time, Martin's relationship to nature, once so direct, has become more oblique and the metaphors have changed, but the sites in her imagination are ineffable.[32]

Ashton goes on in other directions, developing Martin's affection for Chuang Tzu, speculating on her affinities with a tenth-century text by Ching Hao, and then placing her in the "classical" tradition, with its "emphasis on qualities of mind," going back through Le Corbusier to William Blake and Leon Battista Alberti. Ashton concludes that it was Martin's own experiences with actual landscapes in Saskatchewan, Vancouver, Manhattan, and New Mexico that best explain her work. Ashton has no difficulty seeing how Martin shied away from overt references to landscapes and turned her work into a meditation on beauty, "qualities of mind," and the painter's "Idea." But landscape, Ashton says, is the root cause of the paintings, and their ultimate referent. She shows her cards right at the beginning, saying that Martin's landscape references were once "less arcane" and that her work has since "become more oblique"; in Ashton's account the paintings *belong* to landscape, and Martin's own later writing is all, ultimately, a figure for the "ineffable" landscape "in her imagination."

The fundamentally opposed viewpoint, according to which Martin's paintings are not landscape paintings, has been set forth most forcefully by the art historian Rosalind Krauss. For her, the work "turns its back on nature," and becomes "classical" in a different sense than Ashton imagines.[33] Krauss's argument steals the paintings away from landscape—and transcendence, and presence—in order to seal them into a project of self-referentiality.

Krauss cites an essay by the critic Kasha Linville, who describes what it is like to stand close to Martin's paintings and observe how the thin pencil lines skim over the canvas surface, becoming "dotted or broken" (Figure 3). At a distance of perhaps a foot or two, the paintings are just materials: irregular canvas weave, skipping and stuttering lines. The picture surface becomes nothing other than what it is: gesso (an acrylic material used to protect the canvas from the paint) and a little pigment. Over the surface, Martin has drawn a grid. Sometimes her vertical pencil lines follow the grooves of the canvas's warp, and other times they skip from one furrow to the next.

After considering the picture's fine structure, Linville says, a viewer will step back, and that is when something like illusion takes place: Linville doesn't see atmosphere "in the spatially illusionistic sense I associate with color field painting"; instead she senses "a non-radiating, impermeable mist." That is not to say Martin has *depicted* mist: "It feels like, rather than looks like, atmosphere." Martin's repeated lines "somehow . . . dematerialize the canvas, making it hazy, velvety." After taking in the "impermeable mist," Linville says a viewer would naturally want to step back even farther, to try to see the picture all at once, but then "the painting closes down entirely, becoming completely opaque." In the end, Martin's paintings are "impermeable, immovable as stone."[34]

Three viewing distances, three different experiences. Krauss sets Linville's reading against two more texts that at first seem very different. One is the art historian Hubert Damisch's book *Théorie du /nuage/* (*Theory of the /Cloud/*), where /cloud/ flanked by slashes means that "cloud" is to be taken as a word among other words, rather than as a symbol for an actual cloud. (The slashes are a symbol used in linguistics and semiotics.) What matters about /clouds/ is how they interact with other words in sentences and with other objects in pictures—how writers and artists use them, how they work in context. In the theory of semiotics, such relations are called syntactic (as opposed to semantic relations, which link signs to the things they denote); in the terms Krauss prefers, /cloud/ is a *signifier*, and it is "bracketed off from its 'content,' the *signified*" actual cloud. Damisch says clouds cannot be accommodated in the system of linear perspective because they move and cannot be measured, and he points out that the first perspective picture was topped by a silvered sky so it would reflect actual clouds passing overhead. In that way clouds were brought into the web of geometric constructions that constitute perspective. They were trapped between painted geometry and real life, confined within the syntactic system of the picture. In the same way, Martin's "clouds"—her

FIGURE 3

Agnes Martin, *Untitled #12*, detail. Photo: author.

elusive atmosphere, her "non-radiating, impermeable mist"—are contained, sealed in, between the close-up view where the paintings are all canvas and pencil lines, and the far-off view where the paintings are "impermeable, immovable" walls of paint. "The /cloud/," Krauss concludes, "remains bracketed within its peculiar system." [35]

Ashton's and Krauss's readings are nearly diametrically opposed. One of them accepts the paintings' references to landscape, and the other rejects those references and brackets /clouds/ into the system of /painting/. It seems to me that as a matter of perception the truth is in between. The experience of viewing Martin's canvases is less stable than either Krauss's or Ashton's readings imply. It has more to do, I think, with the gerunds and actions I have been noting: "breaking down," making pictures "about formlessness," becoming "oblique." Martin's work is peaceful and still, as many people have said, but it is also *about* the strain of moving away from landscape, toward abstraction, and back again, in a bracketed system of possibilities.

7   *The Abstract Sublime*

This is where the sublime enters in, because Krauss's principal purpose is to quarantine Martin's paintings, keeping them at a distance from a popular but sloppy tradition of criticism that sees the paintings as examples of the "abstract sublime." Ever since Newman insisted that his paintings were about such things as sublimity and the "world mystery"—not to mention tragedy, ecstasy, religion, and doom—critics have intermittently linked abstract painting to the sublime. (Masheck's piece on Bleckner is a late, and especially good, example.) A fundamental tenet of such criticism (Masheck's excepted) is that paintings like Newman's and Rothko's are really distilled landscapes, with the inessential details evaporated off according to the logic of the sublime.

The art critic Carter Ratcliff was not the first to propose this reading of Martin's work, but he was one of the most resourceful. This is Krauss's summary of Ratcliff's argument:

> Characteristically, Carter Ratcliff referred Martin's work to Edmund Burke's
> *Inquiry on the Sublime* which, in the mid-eighteenth century, laid down a
> recipe for satisfying the growing taste for "sublime effects," turning on ways
> art could produce a sense of limitlessness by abandoning the measure
> parceled out by traditional modes of composition and working instead with

forms "melted as it were into each other." Burke's description of "a perfect simplicity, an absolute uniformity in disposition, shape, and coloring," his call for a succession "of uniform parts" that can permit "a comparatively small quantity of matter to produce a grander effect than a much larger quantity disposed in another manner" seemed made for Martin's work, just as that work—as paired down and simplified as it might appear—could be thought nonetheless to smuggle within it diffused references to the repertory of natural "subjects" that followed from Burke's analysis: "the sea (Turner), the sky (Constable), foliage (Church), and, simply, light."[36]

In this critical tradition, Krauss notes, twentieth-century abstract paintings can always be "read out" through nineteenth-century paintings. ("Read out" is a lovely phrase: it sounds like the paintings' meanings are exhausted by the misreading, and that the reading itself is tiresome, as if the paintings are tired out after their interpretations.) The abstract sublime was codified by the work of the art historian Robert Rosenblum, beginning with an essay in 1961 and culminating in *Modern Painting and the Northern Romantic Tradition, Friedrich to Rothko* (1975).[37] Rosenblum saw Rothko's paintings as echoes of Friedrich's, and he saw Pollock's all-over paintings as transcendently enlarged versions of landscapes by Bierstadt and Turner. In that vein, Martin's paintings could be taken to be skies, deserts, or some other small part of a naturalistic nineteenth-century landscape painting, enlarged and purified beyond recognition. They are, fundamentally, engines of the abstract sublime.

This school of interpretation, sometimes including Rosenblum and Ratcliff, but more often presenting itself as independent, is still the predominant one when it comes to Martin, as well as Rothko, Pollock, and other abstract expressionists, color-field painters, and minimalists. It's tempting, it's satisfying, and it speeds past what really happens in front of the paintings. Krauss's brief polemic is like a splash of cold water in the face of this sometimes overheated tradition, because it shows, at the least, that Martin's paintings are not *only* nostalgic echoes of the desert Southwest or the cold, level plains of Saskatchewan. They are that, unavoidably, but they also offer experiences of compression, of possibilities blocked off, of imagination sequestered and turned back to the canvas itself.

The abstract sublime interpretation is wrong because it is not true to experience, as Linville's essay demonstrates. It entails an extremely self-interested interpretation of Burke, who had nothing like Martin in mind when he wrote about simplicity and uniformity (he was thinking of poetry, eighteenth-century landscapes, and garden design). It cuts through the interesting problems of looking at Martin's paintings; it clears a path from her work right back to early-nineteenth-century landscape painting, "reading it out," as if to say that the intervening developments are just so much irrelevant undergrowth.

In structural terms, the abstract sublime is too open-ended. Reinhardt's paintings try to give everything away: they want to *be* formless, to confidently *prove* that there is no meaning outside of art. By a similar logic Newman hoped his paintings would simply *be* sublime. Krauss's and Linville's readings impose brackets on the unlimited experience of the sublime: an inner bracket where the viewer is close to the painting's surface, and an outer bracket where the painting is seen all at once, as a whole. The "impermeable mist" Linville reports—the region where the viewing oscillates between the sense that the painting is a landscape and the sense that it is a surface—is only experienced between the near viewing distance and the far distance. The proposition might be notated this way:

$$| \, [\leftrightarrow]$$

(In this little formalism, the picture is on the left. Close, middle, and far viewing distances are to the right.) If the outer bracket is missing, the painting might be reminiscent of landscape from all distances greater than a few feet. Rothko can be experienced as an example of this possibility, which can be symbolized this way:

$$| \, [\rightarrow$$

because his paintings are cloudy and hypnotic from any distance, and they are only flat from up close.

It is also conceivable that the inner bracket might be missing, but the outer one may remain in place. John Andrews is an example: he paints tiny dots on aluminum, and from a long distance the paintings look as flat as the walls they are painted on. From close up and from middle distances the dots work like a screen, revealing a dreamy landscape beyond. The outer bracket is in place, but the inner bracket is missing:

$$| \, \leftarrow]$$

I don't want to propose that all paintings can be corralled into a formalism of brackets and slashes: I just want to say that the phenomenolog-

ical truth of Martin's work is somewhere between the open-ended accounts of the abstract sublime and the closed-off theory Krauss compiles
from Linville and Damisch. Clearly, Martin's paintings are not images of
starlight, hills, or blue flowers. Just as clearly, they are not semiological
systems that imprison the landscape within the walls of the canvas and
paint. (Krauss says, at the end of her essay, that it is "determined" and
"formalist" to speak the way she has been, but that it's what Martin has
been doing.) I think Martin was right when she said her paintings are
"about merging, about formlessness, breaking down form." What matters
in looking at them is the process of losing meaning and trying in vain to
recover it. The result is a mazy uncertainty, a kind of placid but severe co-
nundrum. By contrast the abstract sublime is too quick and wild, and
Krauss's semiological brackets are too tightly screwed onto the paintings.

Looking at one of Martin's paintings, I first see nothing, and then,
coming closer, I see the fine lines. They look like a graph of something or
a map of someplace. Yet if I were as close to the lines as I would be to a
map or a graph, I see that they are really "dotted or broken," and that
they skim across an irregular gessoed surface. On occasion I have stepped
back again and rehearsed Linville's three-step account of the close view,
the middle view, and the distant view; but even before I have done that, I
already knew the painting failed to make unambiguous sense. Martin's
paintings can imply many different brackets. At one point I may find myself inspecting the margins of the painting where the grid gives way, and
at another point I might spend a few moments looking up and down, following the vertical lines. Linville's reading is only part of the story, because the landscape meanings are not perfectly contained between the
close and far views. They are already present in the very closest view, as I
look onto, and also into, the slightly irregular surface. And as the art critic
Thomas McEvilley points out, landscape forms and colors are present
even in the very farthest views: the grids disappear into the "otherwise
formless" ground, he says, "where they reside always in a kind of latency,
giving the ground an appearance of floating vibrancy, of light-filled po-
tentiality, of invisible but active force."[38] I can never be entirely certain of
what I am seeing: an inaccurate grid (from up close), a "non-radiating,
impermeable mist" (from a middle distance), or a "latency" (from far
off). My wavering indecision about what I am seeing is a failed sublime,
or a truncated sublime. Looking at the paintings, I feel a kind of radical uncertainty, oscillating between thoughts of depicted transcendence—such

as Friedrich's apocalyptic sunsets, Newman's endless walls of blue—and thoughts of the absence of transcendence—as in Krauss's reading of Martin, where even the memory of landscape is clamped between the canvas and its "wall" of paint. Krauss's reading is sober and unforgiving (she needs it to be, to counteract Ratcliff and Rosenblum); I prefer a more porous account where the brackets and slashes may not work quite as efficiently. Either way, the brackets and slashes are markers of the breakdown of the Kantian sublime—an ontological disappointment that has long been nominated the postmodern sublime.[39]

8   The Ladder

Consider the idea that for a painter uninterested in making pictures where representation is adequate to its task, creating a painting can be like climbing down a ladder into a dark space. Rung by rung you go down, feeling your way, and with each step the techniques and skills used to represent the visible world become more distant. In the metaphor of the ladder, which I am pairing with the metaphor of the bracket, painting that is skeptical of the machinery of realism works by subtracting strategies until nearly nothing remains. I can simplify a picture by omitting some shapes and colors and gestural marks, making room for emptiness. I can pare away unnecessary ornaments, focus on essentials, even try to remove the subject matter altogether (as Reinhardt said). If the picture makes use of perspective, I can compress its space, hollow it out, or flatten it. Each of those steps, and no doubt many others, are ways of subtracting forms and at the same time gradually letting go of the idea of adequately realistic representation.[40]

The history of twentieth-century abstract painting is scattered with moments when painters tried to empty their paintings as if they were buckets of water, pouring out anything that could count as content. Reinhardt, Newman, and Martin all practiced variations on that theme. The traditional emblem of the desire for pure emptiness in painting is the monochrome canvas.[41] Monochromes were made repeatedly in the course of the twentieth century: Kasimir Malevich experimented with them, and so did Ad Reinhardt, Barnett Newman, Frank Stella, Robert Irwin, John McLaughlin, Allan McCollum, Yves Klein, Sam Francis, Boris Turetsky, Wladyslaw Strzeminski, Jean Degottex, and Ellsworth Kelly. Robert Rauschenberg made white paintings (and famously, he erased a drawing of De Kooning's), and Jean Dubuffet made near-monochromes by painting with dirt. A full list might run into the hundreds; it would include a number of post-painterly abstractionists such as Brice Marden, Jo Baer, Byron Kim, Frank Gerritz, Callum Innes, and Judy Ledgerwood. Minimalism, from Frank Stella to Robert Ryman, provides many further examples. Perhaps the most concerted efforts are those made in the last two decades of the century by the "radical painting" movement that includes Joseph Marioni, Marcia Hafif, and Günter Umberg.[42]

An art historian would not normally juxtapose all the artists I have mentioned, because their purposes and rhetoric were so disparate. From an art-historical standpoint it seems best to say that abstraction, at least as it was understood in the twentieth century, often turned toward a kind of purity that both mutes painting and appears to transcend or somehow "solve" it.[43] A monochrome canvas seems to end painting, at least for the moment, by eliminating its conventional sources of interest; yet that same canvas can be said to transcend painting by anticipating future possibilities. The very different historical contexts that have accompanied pure black, white, red, or blue monochrome paintings belie the fact that they all participate in a larger twentieth-century notion of the abstract.

As Mark Cheetham and others have noted, the ideal of purity—the hope that a painting is like a vessel, so that it can be completely emptied, leaving only its perfect shape, pregnant with nothing—is a recurrent trope in modernist painting.[44] From that vantage monochromes can be uninteresting as objects, because for them the game of painting is already over. (The same can be said of conceptual artists who make use of canvases, paint, and stretchers, such as On Kawara.) It is fascinating, on the other hand, to stand in front of one of Stella's "black paintings," which are not completely black, and feel the space being squeezed out by the lines of raw canvas that run back and forth, exactly four inches apart, across the uniform black field. Near-monochromes can articulate the unrepresentable in ways that monochromes cannot. In the 1990s, Callum Innes made paintings with three or four rectangular regions: one might be white, another pitch-black, another washed gray or blue. The colored rectangles are the most interesting because they are not only emptied of forms, as any monochrome is, but also blurred: he paints over his colors with a brush loaded in turpentine and lets the dissolved paint run off the canvas. That is a more intriguing strategy than simply painting (yet another) monochrome. Joseph Marioni effectively does the same with many wet, thin layers of paint that are allowed to pour off his paintings. With Innes, it's curious to see a rectangular area in the process of disappearing next to an area that is densely done in black, and another that is pure, thin white.[45] When one of Reinhardt's nearly monochrome paintings is in good repair and not too brightly lit, it can be just as curious to look at: the dark, faint colors come forward just a little and then recede back into uniform shadow.[46] The paint, and the painting, aren't simply *there*. They fade into the weave of the canvas: they aren't solid enough to be properly seen.

Near-monochromes, then, where forms and colors are still faintly present, can be compelling answers to the problem of subtracting from painting until almost nothing remains. The question is *how* empty, how nearly perfect the canvas can be. Günter Umberg is an example of this tendency at its furthest limit (Color plate 2). He paints small pictures, brushing dry pigment on aluminum and then spraying it with damar resin. By repeating the process forty or fifty times, he arrives at a "disconcerting" surface made of opaque, "porous, grainy, atomized material." The results are, in Henry Staten's words, "the most stunningly opaque and dry-looking paintings that have ever been made."[47] To my eye the pictures are more ravishing, more intensely seductive, than Reinhardt's somewhat careless surfaces (which show only dust and canvas threads from close up), or even Frank Gerritz's obsessively darkened graphite surfaces. An Umberg painting has the virtue of being interesting right up to the limits of vision: the paintings are still compelling when they are nearly perfectly black and when I look at them from as close as my eyes can focus.

Marioni, another "radical painter," has been praised by the art historian Michael Fried as "one of the foremost painters at work anywhere at the present." Marioni's painting, Fried says, has a new, "deeply founded integration of color, amateriality, and support" that "has something of the character of a new beginning" after minimalism. In the context of this book it is Umberg's painting that is more radical and promising, because it eschews that "integration" and focuses strictly on the remains of painterly marking.[48] Marioni's surfaces are dripped, turning viewers' thoughts to his gestures—the actual swoops and splashes. Umberg's are much closer to the bone: they do not permit that latitude of thought. They are done from farther down the ladder.

Umberg's idea is to make painting that is entirely about the paint, and not at all about illusion or anything outside the painting's materials.[49] His is not minimal art, which tries to be literal about the support (the stretchers and so forth), but "radical art" in the sense that it is literal, in Staten's phrase, "relative to pictorial illusionism."[50] Umberg speaks as if his works had achieved the break with illusionism that had been adumbrated by Mondrian and Stella. Some minimalist and *support/surface* works (that is, French minimalist paintings) propose such breaks, and they would be Umberg's closest parallels. Yet despite what Umberg says, his paintings do not quite break with the tradition of illusionism, and for me that saves them as painting.

What happens while looking at an Umberg canvas is a tightened, airless version of what takes place in front of a Martin painting. From a distance nothing can be seen except the total darkness of the rectangle of paint and the thinness of the aluminum sheet that serves as its support. A closer look reveals the surface, powdered in pitch-black colors (Color plate 3). It is a rich, compelling surface, "non-radiating" and "impermeable" to use the words Linville used to describe Martin, and yet deeply painterly. In order to perceive Umberg's paintings as obdurate flat surfaces, it is necessary to be so close that your nose is almost touching the surface. In that, they are unlike Martin's paintings, which are nothing but canvas and gesso when seen from less than a couple of feet away. At the closest feasible distance—say, six or eight inches—Umberg's paintings still possess an entrancing sense of "powdered" atmosphere. Umberg describes that fascinating effect in terms of the concept of painting: "I do not choose a green in order to bring it into the painting," he says (speaking of paintings that are nearly monochrome green, instead of his usual black), "but I must first paint the green so that it becomes a painting."[51] He lavishes attention on his ground, polishing it as in "Gothic and early Renaissance painting," covering it with dozens of layers of powdered pigment, mixing just the right combination of nearly black pigments, all in order to achieve the most concentrated effect of painting within the smallest possible compass of hues, gestures, and values. Looking at one of his surfaces from a half foot away (using my reading glasses, looking at the smallest details), I imagine that I can see all the gestures of Western painterliness, from Titian onward—the translucent glazes, the virtuoso turns of the brush, the gleaming colors—all dried, reduced, perfected, and compressed into a nearly impenetrable darkness. I do not find Umberg's paintings radical in the sense he imagines: they are not only about the object, its support, and its surface. Nor are they wholly radical in regard to illusionism, as Staten proposes. They are about airlessness: they breathe just the last few breaths of illusionistic painting. In the history of near-monochromes, Umberg's are the most intensely expressive and the nearest—just a hair's breadth—from the dull extinction of real monochromes.

Climbing down the ladder of representation is a strategy that works for a wide range of pictures, and it can also happen in representational paintings. In Rosenblum's account, it has been happening since Friedrich. Marie Krane's paintings are an example, and they are a bridge between the two themes I have been exploring (Figure 4). With the help of assistants, she paints large canvases covered with an extremely precise grid of small, gelatin-capsule-shaped ovals of paint. She bases her colors on natural phenomena, such as the gradual decay of a flower, so that a given painting might be almost perfectly monochrome. But within that near-perfection, the colors actually obey an intricate series of rules—they are not all the same color, but mathematically determined mixtures of several colors. Like Agnes Martin, Krane's paintings work within a bracket of viewing distances, and within that bracket, the paintings shimmer and shift as the viewer tries to decide if they are monochromatic. The little ovals of paint are so rigidly and skillfully laid down that they create only very slight moiré patterns; mostly, the paintings create a sense of motion so subtle it is nearly indefinable—the analogue, I think, of Martin's "impermeable mist." The ladder metaphor and the thematic of the monochrome are also at work in Krane's paintings, because they give up so much—virtually all color, nearly all apparent depth and motion, and all but the slightest increment of gestural freedom. They are also near-monochromes—asymptotically, infinitesimally close to perfect, machine-produced monochromes—and yet they are deliberately and elaborately constructed *not* to be monochromes.[52] At the beginning of the twenty-first century, the best paintings in this line of thought are the ones that manage to remain just a half step from the uninteresting oblivion of perfect blankness.

9 *The Bracket, the Ladder, and the Sublime*

The concept of the sublime is of some use when it is applied to Friedrich himself, along with painters such as Christian Købke, Théodore

FIGURE 4

Marie Krane, *(like April through October)*, 2005. Bottom: detail. Acrylic on linen. Museum of Contemporary Art, Chicago. Courtesy Marie Krane.

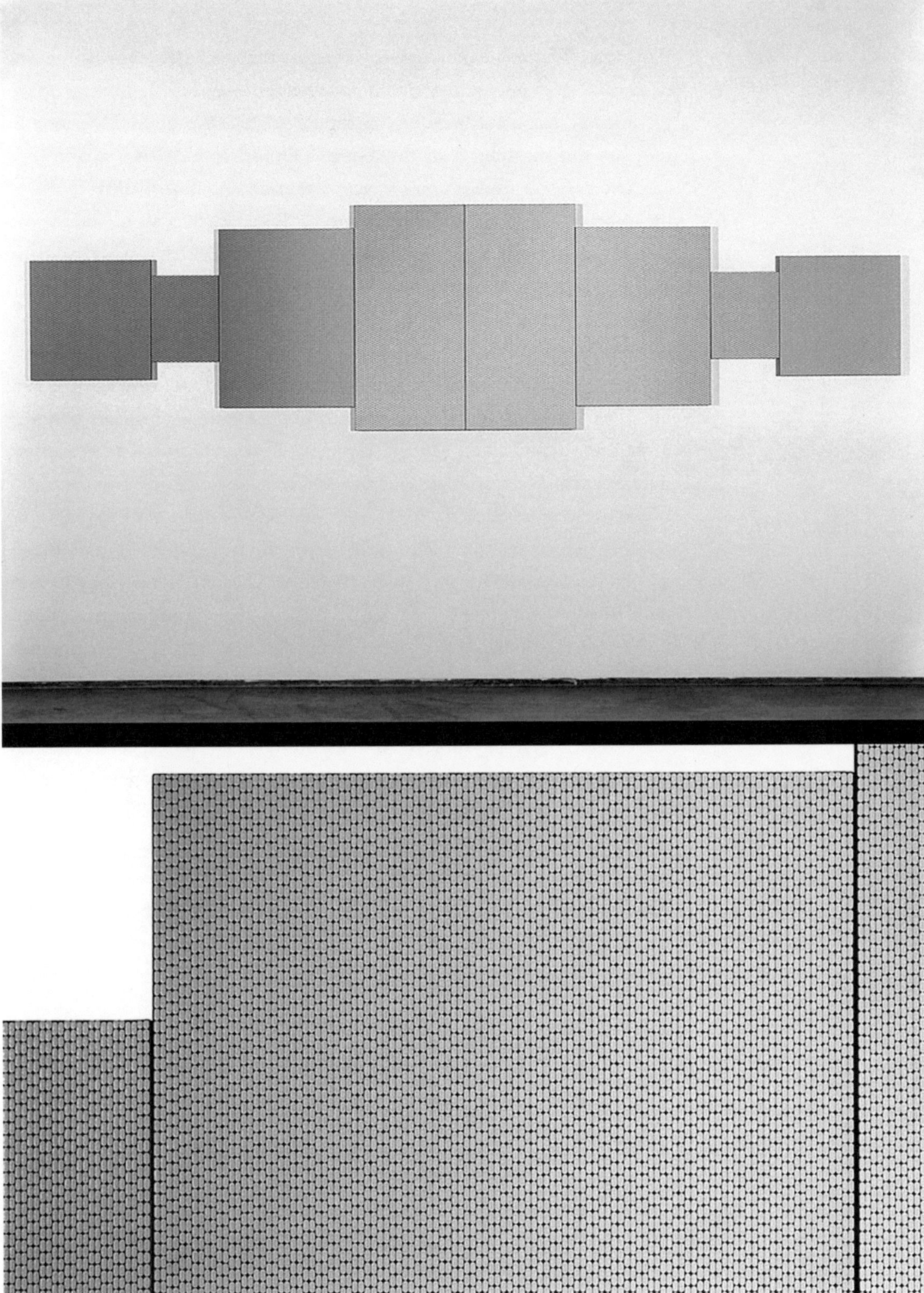

Rousseau, Thomas Cole, Alfred Bierstadt, Frederic Church, George Inness, and the whole line of their twentieth-century followers up to present-day landscape painters, including such artists as Alex Katz—but they are not the subject of this book.[53] Friedrich and other nineteenth-century painters were aiming at values that can be described in terms of the sublime, or of transcendence, but in doing so they were also setting themselves a variety of representational tasks, which involved many of the resources of naturalistic depiction. Bierstadt and Inness gestured toward transcendence—which they imagined as embodied in the limitlessness of the American landscape—but they were also taken by banks of cumulus clouds, the soft crowns of elm trees, the dark trunks of oaks, by deer, bear, squirrels and acorns, by boulders, scree slopes, marshes, hanging roots, and fallen leaves. Martin, Umberg, Innes, Marioni, and Krane are different from one another, but they share a lack of faith in what naturalistic representation in the nineteenth-century sense can do: they are more concerned with making pictures that enact the failures—the limitations—of representation. The purposeful thematizing of the inadequacy of naturalistic representation is the reason I have used the postmodern sublime, instead of the Kantian sublime, to help describe what happens in their pictures.

The sublime (Kantian or postmodern), transcendence, and naturalistic representation are big subjects, and it would not be an exaggeration to say that modernism in painting turns on what artists have decided about them. Here I have just tried to point out, using an exact phenomenology, what happens in a few of cases when painters have tried to negotiate the nearly barren ground between the uncompromising regime of minimalism, with its insistence on the object and disavowal of illusion, and the vitiated regime of illusionistic painting. This is part of what I mean by an interest in the limits of representation in late-twentieth-century painting.

t w o Photography

The ladder of disorder leaves conventional representation behind, in favor of images that explore inadequate representation. Each time I consider it, the ladder seems to get longer, like a dream descent into a bottomless well. In this chapter, as I move from painting to photography, I will add four more rungs: blur, darkness, the ruined grid, and the antioptical. These new terms do not constitute a full account of experiments with the limits of photographic representation in the period from 1980 to the present; rather, they are operative concepts, in use in the current literature, that will help me describe the work of just a few photographers. (As in the first chapter and all the chapters that follow, I am not trying to paint a portrait of a medium or a discipline, as much as I am selecting individual practices within media and disciplines.) I will also use this chapter to introduce the scientific photographs that will occupy the remainder of the book, and I will outline some similarities and differences in intention and interpretation.

10 *Formlessness*

Formless is one of Aristotle's terms; Kant used it to evoke the failure of intuition in the face of a sublime object. In the twentieth century, *formless* found its way into art criticism when the writer and philosopher Georges Bataille used it as one of the key concepts in a little dictionary of surrealism. The *informe*, as he called it, was revived in the last decade of the century by Rosalind Krauss and Yve-Alain Bois in an exhibition and a book called *Formless: A User's Guide*.[1] In between Bataille's dictionary and Krauss and Bois's book, *formless* was also used by Reinhardt, Rothko, Martin, and others, and by the philosopher Jean-François Lyotard, with reference to Kant and to painting, but not to surrealism.

The word *formless* is confusing from the outset because of its serpentine etymology, and it is also a difficult term because it can be a quality (Reinhardt's "formlessness"), a state (of having no form), or an operation (subjecting a painting to the *informe*). Bataille was ambiguous about this, or just undecided: he says the *informe* is a way "to bring things down in the world," implying that it is an operation; but then he also says that "affirming the universe resembles nothing and is only *formless* amounts to saying that the universe is something like a spider or spit."[2] In that metaphor the *informe* is a state, not an operation. Krauss and Bois note the vacillation, and they decide in favor of the formless as an operation. Otherwise, they say, it would be a "term" with a "genealogy," and one could write its history. It is best, they propose, to think of it as "neither a theme, nor a substance, nor a concept."[3] That is a radical move, because if they carried it through consistently, their book *Formless* could not be a history of art or of ideas.

To an extent, their book functions as a critical intervention—as a "user's guide"—rather than as an art-historical study of surrealism. In that respect, *Formless* is not primarily intended to excavate the history of art pertaining to Bataille's *informe*; instead it tells contemporary artists and viewers how certain interesting pictures work. At the same time, *Formless* is a book of art history—after all, it is structured around citations from Bataille's dictionary, and it is full of the apparatus of scholarly art history, including primary sources and notes referring to first-generation surrealist works and documents. At the heart of *Formless* there is a pendulum that swings between history and criticism. *Formless* is a study of ways the *informe* has been understood (principally by Bataille, and also within surrealism), and it is also a study of how the operation continues to produce interesting art even in times and places where artists and critics might never have thought of the word *informe*. (In this book, the pendulum swings differently, mainly between philosophy and history; but more on that later.) I want to be clear that *Formless*'s least important contribution, for this project, is its revival of a version of Bataille's concept. I am more interested in an aspect of the book that has not, as far as I am aware, been noticed in the reviews and the ensuing critical literature: the fact that the book can only partly be thought of as a contribution to art history. It is also a "user's guide," deliberately rethinking historical terms for critical purposes. The irresolvable mixture of art history and critical intervention is what is decisive: it makes *Formless* an exemplary text and one of the principal models for this book.

At the moment, however, I am concerned with the concept of the formless, which I want to put to work in describing some recent photography. For that reason I am also going to ignore *Formless*'s uneven reception and its dissemination in the later 1990s.[4] As an operation, then, the *informe* works to tear down good sense, clarity, reason, Cartesian order, system, and structure. Kant's use of *formlessness* follows Aristotle's, in that both were trying to give a name to a state of matter that is beyond apprehension, outside of concepts, or inaccessible to intuition. Krauss and Bois's *formless* is more radical in relation to representation, but less paradoxical than Bataille's *informe*. Krauss and Bois's *informe*, in turn, is more *formal* than Bataille's, because it depends more on the absence of clear forms. Bataille was not often interested in images that lack nameable forms—his writing is full of narrative and illustrational prints, paintings, and photographs. Krauss and Bois reproduce several works that share an absence of clear form: Robert Rauschenberg's dirt paintings, Cindy Sherman's photographs of mold, Lucio Fontana's punctured sculptures and slashed canvases. Almost all their examples are paintings, but as I will be arguing, a version of the *informe* is an apposite tool for understanding some recent photography.

11   *Blur*

Blurring is another key strategy for the production of art photography. An intentionally unsharp image, once an option specific to photography, became a common painter's device in the late 1980s and 1990s. Gerhard Richter is again probably the best-known example. His series *October 17, 1977* and other paintings enact photographic blur in paint, using fan brushes to ruin the crispness of the underlying painting, which is itself understood to be a facsimile of a photograph (Figure 5). In other paintings, Richter smears—usually horizontally—rather than blurs, which is interesting because it combines a fundamentally painterly technique (smearing) in imitation of a characteristically photographic possibility (blur). Another example is Jiri Georg Dokoupil, who has painted grainy black-and-white versions of news photos, a practice that derives from Warhol and Lichtenstein.[5] There, blur is not a result of defocus but of printing technology: the pictures are blown up so large—as if they were enlarged in a xerox machine—that both the halftone dots and the picture grain become visible.

(I want to distinguish here between five terms that could be unproductively conflated. The first four are *smearing*, *blur*, *defocus*, and *unsharpness*. The last two terms in this list will come up in later chapters, in connection with scientific instrumentation. *Smearing*, as I am using the term, is a characteristically painterly option; it involves effort, because the paint has to be pulled across the canvas to ruin the details of the picture. *Blur* is characteristic of photography, and it is important that it involves no effort—at least not more than turning the knurled focusing ring of the lens. In nonscientific photography, there is some overlap between *defocus* and *blur*, and both certainly produce unsharpness. But in scientific applications, especially in electron microscopy, *defocus* has another meaning, and in scientific image manipulation, *unsharpness* also has a special usage, as in "unsharp filters." I'll discuss these in greater detail later in the book. *Depth of field* is the fifth term in this set, and it has specific meanings both in fine-art photography and in science. In design and advertising, *depth of field* means *shallow depth of field*, either produced when the picture is taken or mimicked after the fact by selective blurring in Photoshop or other image manipulation programs. Elsewhere, *depth of field* is not necessarily shallow or deep; the expression just denotes the measurable, quantifiable range of distances that are considered to be in focus.)

A purer example of photographic blur mimicked in painting is Ed Ruscha's series of out-of-focus landscapes (Color plate 4). These paintings, done in the mid- and late 1980s, are meant to mimic out-of-focus photographs; they have the bleary look of something poorly remembered, seen through exhausted eyes, or peered at on a cloudy night. Some of the paintings also have conceptually blurry messages, as if the images were meant to communicate something urgent but ended up garbled. The painting *Strong, Healthy*, for example, shows two suburban houses, darkly silhouetted against a night sky. Under each is a bright white rectangle, and from other paintings in the series it is clear that "Strong" goes in one box, and "Healthy" in the other; but there is nothing to read in the boxes, so we are led to think that the people who live in the *Strong, Healthy* houses are anything but. Another painting in the series shows

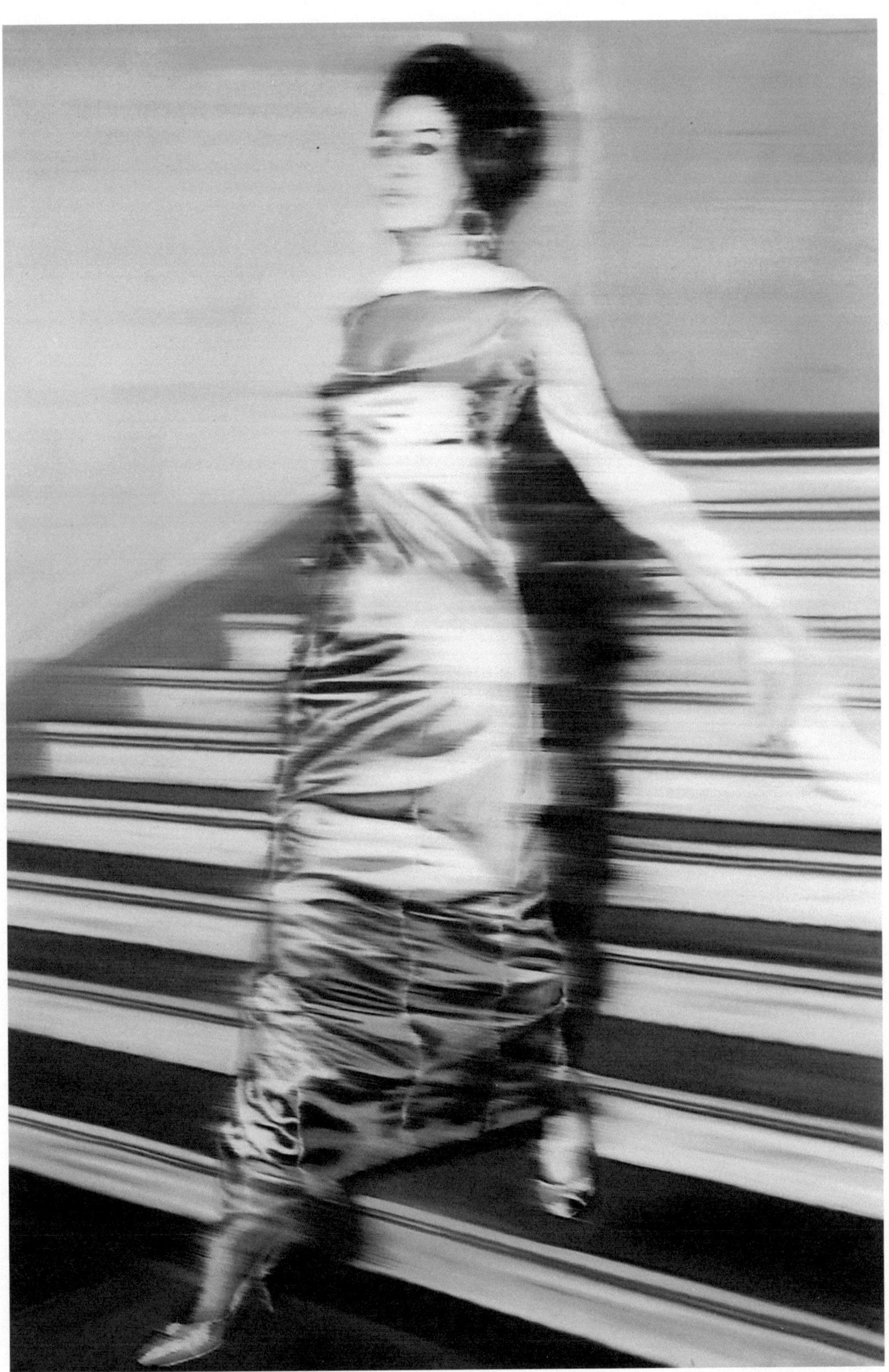

three or four suburban houses, simply seen, with a series of brilliant white bars superimposed, in this pattern:

The picture would be deeply enigmatic, if it weren't for the title: *Averages*. Looking again, it becomes clear that the houses are a sample of America, and the lines are a bar graph. It is not a proper graph—in effect it is an unmarked grid, like Martin's—but Ruscha's intention is clear enough: these are average houses, with typical faceless and nameless occupants.

These linguistic and graphic games exacerbate the visual blur. A third painting in the series shows three of the same nondescript, silhouetted houses, this time with "Name ____________" painted across the bottom in bright white. "Name ____________" can hardly be filled in when the houses are so utterly anonymous and when the picture is so nearly black. Ruscha says he has a "deep respect for things that are odd, for things that can't be explained," and paintings like *Name ____________* are certainly less explicable than *Averages* or *Strong, Healthy*.[6] The painting I am reproducing here is from another series where the houses are not quite so ill defined, but the labels and names are significantly more enigmatic. This *F* is a label, or maybe a dark joke (a scarlet letter), or even a sinister government classification. Whatever it is, the picture itself is nearly as dark as one of Reinhardt's, and except for the lights and the single letter, there is virtually nothing to see.[7]

The critic Peter Plagens has described Ruscha as a "blank" modernist—one who doesn't say whether he is putting you on, or maybe even patronizing you, and this picture, called *F House*, certainly raises those possibilities.[8] The dark and blurred paintings that Ruscha began making around 1982 have less cleverness than much of his earlier and more recent work: they are less suave or slick, more directly about loss.

They are "blank," unreadable, and "odd" (Ruscha's words), and blur is the instrument of their oddness.

These thoughts about blurring lead me toward one of the most diffuse questions of twentieth-century art: the flight from clear sense. "A clear idea is another name for a little idea," Burke wrote, praising obscurity and linking it with the passions.[9] Many of the paintings and art photographs I discuss in these opening chapters can be understood as attempts to avoid well-defined, unambiguous meaning, as well as attempts to avoid naturalism, detail, or what I called full and adequate representation. It can be argued that the measured retreat from obvious or unequivocal meaning is a central strategy of modernism and postmodernism. Scientific illustration does the exact opposite, trying to achieve the clearest and least ambiguous possible meaning in every instance—but it often ends, as I hope to show, in an equivalent state of uncertainty. One of the most incisive questions it is possible to ask of a late-twentieth-century image is how and why it evades whatever clear meaning it might have achieved. Likewise, it is often helpful to ask of scientific images how they manage and present their uncertainties.

It is also possible to put blurring in psychological terms. The painters and photographers could be said to "fear" clear meaning, as the physicists "fear" ambiguity or obscurity. Scientists flee anything they construe as unreason—that's a truism in scientific pedagogy and in accounts of academic writing, but its converse is less often noted: that some image making in the arts is intended to achieve just the right degree of unreason. Ruscha's paintings are precisely unfocused: they draw back from clear meaning just a certain amount, and then stop, like a professional photographer measuring the depth of field for each f-stop. A more radically blurred version of the painting *F House* might be too ambiguous to care about, and a fully "focused" painting of a house might be too much like a snapshot. These two wider meanings of the avoidance of fine focus— the flight from clear sense, or the "fear" of it—depend on what gets done, technically speaking, in the photograph (or in the painting mimicking a photograph).

In scientific imaging, blur—whether it is pictorial or conceptual—is something to be avoided. (Defocus, I will claim, is a different matter.) Some sample scientific images can show the disparity of intention in regard to blur. Some of the most interesting blurry images that astronomers produce are actually the results of efforts to remove blur. An instance is the discovery, in 1999, of a star that looks like a spiral (Color plate 5). (In describing these first scientific images, I will be using some technical detail that goes outside the norms of humanistic writing: that choice is theorized in the Introduction and developed in the next chapters.) Material ejected from the star actually spins like a pinwheel, and the astronomers made a film to prove it.[10] The object at the origin of the spiral, which is invisible in this picture, is a Wolf-Rayet star, a particularly intense kind of star that propels gases far enough away that they do not glow. The pinwheel was accounted for by positing a companion star that orbits the Wolf-Rayet star, shielding part of the ejected material. To verify that conjecture, the astronomers needed the highest resolution they could get. The solution was to use what was then the world's largest telescope, at the Keck observatory on Mauna Kea, Hawaii, in a mode called aperture masking interferometry.[11] The large primary mirror of the telescope was masked into thirty-six regions, and the interference between light arriving from those regions was reconstructed into an image. The angular diameter of the spiral is only 0.1 arcsecond, and the image resolves forms on the order of 40 astronomical units: this is an exceptionally tiny object, seen at a level of detail unsurpassed at the time. The image is detailed enough so that the astronomers could plot a best-fit curve—an Archimedean spiral—to the hazy pinwheel.[12] Without the interferometry, the image would have been a blur about ten times as large. Yet even with all that detail, the star itself and its companion remain invisible, and so does all the detail of the streaking and spiraling clouds of gas.

Here the unwanted blur of the astronomical image comes strangely close to the constructed blur of the *F House* painting. The photograph of Wolf-Rayet 104, as the star is called, is formally similar to some of Ruscha's paintings, especially a few in which pools of glue, syrup, or other liquids are depicted illusionistically against flat backgrounds. There is even a painting of Ruscha's that has this same spiral shape. This is the kind of coincidence that deserves, in virtually every context, to be called a coincidence; but it is not as clear to me as it once was what should count as fortuitous. The Wolf-Rayet star glows in the infrared, and the image I am reproducing here was published using an artificial shade of red (a "false color") that is very much like the red of the letter *F* in Ruscha's *F House*. Who is to adjudicate such "coincidences"? The scientists might enjoy their images and their choice of color, but they don't intentionally create the blur: in fact, they work to eradicate the same qualities that artists like Ruscha make on purpose. Yet despite the diametrically opposed intentions with respect to blur, the wholly different ways the two images are generated, and the entirely different institutional contexts, the two meet in on a middle ground between formlessness and crystal clarity. Without insisting on any influence from art to science or vice versa, and without wanting to reduce these two very specific image-making practices to their formal properties, I want to note that what interested both the team that made the film of Wolf-Rayet 104 and Ruscha, when he painted *F House*, was the construction and interpretation of an image that is just barely adequate to its subject—where most of the subject (its detail, its properties, its specificity, its meaning) evades the means they had at their disposal, and where little or nothing elsewhere in the image *is* adequate. That is the kind of "coincidence" that first got me started on this book, and it is the one I want to leave open throughout the book. As I argue in the Introduction, I have reasons for not pressing these analogies. Instead I want to set them out in separate chapters and let readers decide how they might best be understood: as six separate stories, as a single theme rediscovered in six fields, or—for example—as an illusory coherence brought on by certain habits of seeing.

But to return to my subject: astrophysicists are well practiced in "cleaning up" photographic plates by adjusting color and contrast, removing images of dust, correcting aberrations, restoring lost pixels, and balancing uneven background illumination.[13] When it comes to blur, the usual strategy is to specify what counts as "smooth" and what counts as "pointlike," and then refine the image until it exhibits the required pointlike properties. The problem is partly that astronomical images do not always have pointlike sources such as stars, so it isn't possible to focus the image until everything resolves into empty space and pinpricks of light. Some parts of the heavens are replete with blurred forms, "painterly" swatches of color, empty "envelopes," "tidal tails," bars, bubbles, streamers, jets, dark clouds, "worms" (wiggly gas filaments), "cold fronts," and any number of other unfocusable objects.[14] There are rust-colored emission nebulae with pink and orange smears, neon-blue absorption nebulae streaked in violet, and forms in every visible wavelength and false-colored invisible wavelength. The center of our own galaxy is a turbulent region mapped with features known as threads, lobes, and arcs; some have gotten traditional-sounding names like the Cane, the Pelican, the Snake, the Pistol, and the Sickle. All are intrinsically blurred.[15]

Figure 6 is a typical example: it is two views of the "anticenter shell," an enormous structure toward the center of our galaxy. The two images are of the same object: the left-hand image records emissions in the 21–centimeter wavelength range from gas moving at 71 kilometers per second; the right-hand image shows the same area, recording gas moving at 45 kilometers per second. Carl Heiles, the author of this study, says the anticenter shell "is visible, but only with difficulty."[16] These images were made in 1983, at the beginning of the period I am considering here, but no matter how sophisticated the imaging and computer processing become, these objects are still elusive, ill defined, and mobile.[17]

It is treacherous to apply image-enhancement routines to such pictures because there is no way to be sure how ill defined the objects are, or even if they should be counted as objects. In Figure 7, a small detail of a complex region in the constellation Norma, there is at least one extremely well-defined object: a pulsar, which if it were visible would appear as a perfect point source. In the detail it is marked PSR B1610-50. Yet even at this high resolution, the only thing to see is more supernova remnants, expanding into and past one another in an irresolvable soup.[18] A software routine to sharpen this image would run up against two potential sources of error: knowing what to aim for, and knowing when to stop the iterations leading from a known blur to an unknown definition.

In the late 1990s Frank Pijpers of Aarhus University in Denmark developed a method of reconstructing images that does not require users to guess what the final image should look like. He begins with the image's point spread function, a mathematical index to resolution inherent in each digitized image, and establishes a linear transformation between that

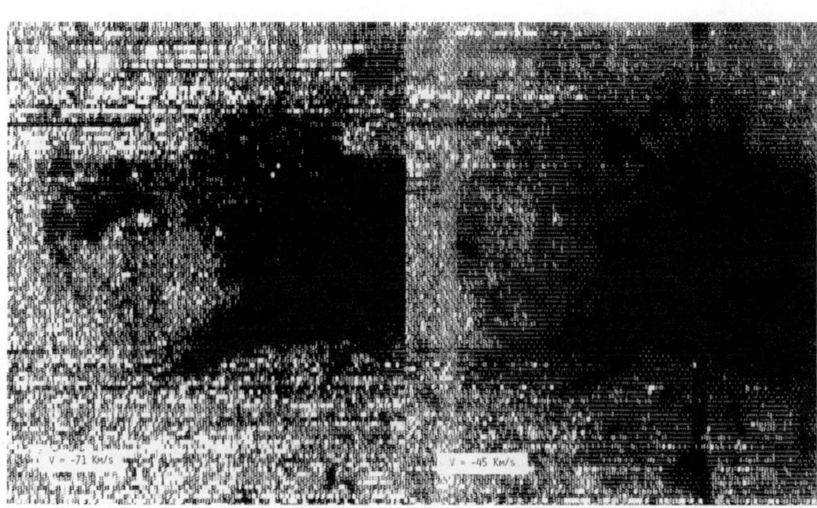

FIGURE 6

H I column densities in the "anticenter shell" at the Galactic Center, GS 174+02−64, at velocities of −70.7 km s^{-1} (left) and −45.4 km s^{-1} (right). From Carl Heiles, "H I Shells, Supershells, Shell-Like Objects, and 'Worms,'" *Astrophysical Journal Supplement Series* 55, no. 4 (August 1984): figs. 4a, 4b.

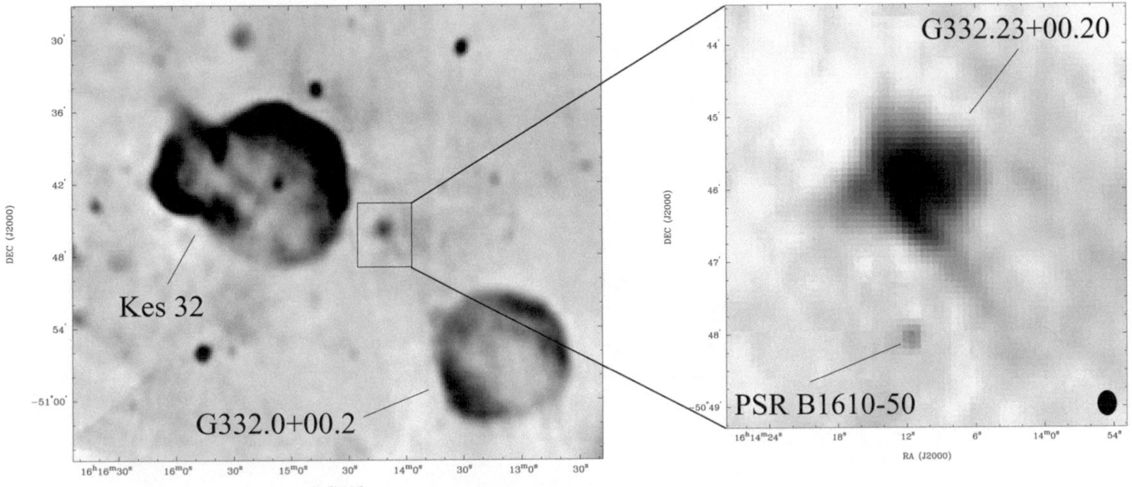

function and some desirable resolution.[19] In the terms I introduced at the beginning of this section (*smearing, blur, defocus, unsharpness*) Pijpers's routine reduces blur or increases sharpness without affecting focus. But notice how the meaning of words like *blur* and *sharpness* have suddenly changed, from properties that can be modified by turning a focusing ring on a lens (or applying a Gaussian blur or sharpening filter in Photoshop), to properties that are at once technical, specific, and intricate. In later chapters I will consider such "routines" in more detail.

Philosophically, Pijpers's method is interesting because it works without human interference, on a scale from abstract definition to abstract blur. The actual image that needs to be "deconvolved" and the image that is desired are only points on a continuous linear transformation of the point spread function. Thus "blur" is no longer a subjective matter of ill-defined forms on some actual picture, but a question of coefficients and parameters in a function that underlies any possible state of the image. In one demonstration, Pijpers begins with an image of a galaxy. He convolves the image with two Gaussian blurs—two functions that spread the image in different ways and work at different strengths. The operation simulates the kind of broad lack of focus that an actual image might suffer. The result is shown at the top of Figure 8. He then applies his method, deconvolving the image (center). The bottom image shows the difference between the deconvolved image and the original. The fit is nearly perfect, except for "side lobes," which are an artifact of the functions used in this example.[20] Pijpers's method shows that blur does not

FIGURE 7

Pulsar PSR B1610-50 and its surroundings, in the constellation Norma. From B. Stappers, B. Gaensler, and S. Johnston, "A Deep Search for Pulsar Wind Nebulae Using Pulsar Gating," arxiv.org/abs/astro-ph/9904056 (April 5, 1999), 8, fig. 5.

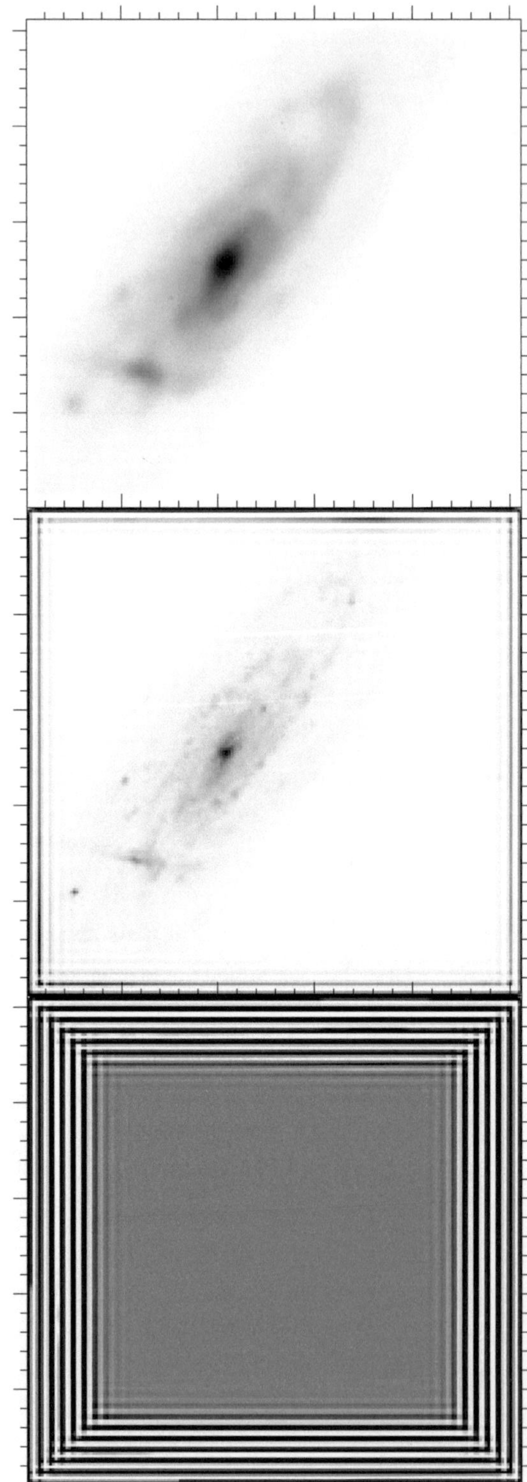

FIGURE 8

Top: Deliberately blurred image of galaxy; middle: improved image; bottom: match between the original image (not shown) and the improved image. From Frank Pijpers, "Unbiased Image Reconstruction as an Inverse Problem," arxiv.org/abs/astro-ph/9904075 (April 6, 1999), 7, fig. 7. Courtesy F. Pijpers.

need to be a matter of distance from some hypothetical optimal clarity: it can be a functional scale, independent of the viewer's notions of clarity and even of the image itself.

From these many possibilities I want to make just a simple conclusion. Blur is a strategy for some fine-art photographers and painters, and a nuisance for some astronomers; yet for their different reasons, the two groups of image-makers end up producing pictures that are intriguing because they show so little. For Richter or Ruscha, a blur is like a balm, soothing the image into a half-hidden state. For astronomers like Pijpers or Heiles, a blur is a bane that needs to be contained as carefully as possible. It is interesting, watching these two groups at work, to see how blur itself is an object of fascination, and how the images that result have an equivalent indistinctness.

12   *Darkness*

Naturally, whatever is down at the bottom of the ladder is folded in darkness. Just as naturally, darkness has been a trait of the sublime since Longinus. Burke is a maven of darkness. He names it as a principal source of the sublime, a species of "privation" along with vacuity, solitude, and silence.[21] When Burke was writing, the aesthetic of the dark and of night was already well under way—it had begun in the late Renaissance with the Caravaggisti, and arguably before that in Nativity scenes and depictions of the Rest on the Flight into Egypt. With the romantics, moonlight and even starlight became common tropes in poetry and painting. Kant's famous image of the starry sky is an early romantic trope, and so are the tenebrous landscapes in romantic poems by Goethe, Novalis, and Hölderlin. Goethe's "Erl-König," with its empty, misty, dark landscape—barely described but tremendously evocative—was a template for many that followed.

By the end of the twentieth century, the tide of nocturnes had long since ebbed, and darkness itself became a cliché. I think it was the poet Mark Strand who warned his fellow poets that "darkness" was "used up," no longer a viable word for serious writers.[22] But darkness is still an option and even a temptation for contemporary painters and photographers. Umberg's paintings are impossibly dark—beyond what can be reproduced, and nearly beyond what can be seen. Artists who follow that path need to be careful that their works do not lapse into dull invisibility. Bernard Borgeaud's wall-size assembly of photographs called *La Nuit, la pluie* (1990–93) frames the nearly featureless night sky in twelve panels of even darker, wetter rain (Figure 9).[23] His essay in darkness is saved by bright splatters, much as Rothko's somber monochromes in the Rothko Chapel are relieved by marks and stains, or Vladimír Kopecky's carefully painted bichromes (which superficially resemble Rothko's last paintings) are rescued by their blotted surfaces, or Robert Piesen's seething black monochromes are redeemed by the fact that the paint has been physically pressed in a die. Without such marks, Borgeaud's panels would be too soft on the eye, too easy to see and to forget. With their bright splashes and stains, the panels become intriguing, like a cloudy night sky seen through wet windowpanes.

With Bourgeaud (and with the scientific photographs) I am making a transition from paintings that mimic photographs to photographs themselves. This is an appropriate moment to introduce the problem of quality. There may be several hundred exhibiting photographers who use darkness, blur, and formlessness in their painting; these are very common strategies, and the first two in particular are very easy to achieve. While it is unarguable that some of the most ambitious paintings of the last quarter century are involved with themes I am developing here—Richter, as always, is the central instance—it is more problematic to associate these themes with interesting photography. On the one hand, the photographers who arrested public attention during the period in question—everyone from Catherine Opie to Thomas Struth, Rineke Dijkstra, Thomas Ruff, and Jeff Wall—are not involved in these issues. On the other hand, the number of exhibition photographers who make use of blur, darkness, formlessness, and other properties is enormous. The disparity is significant: the former are widely studied; the latter are—generally, with many exceptions—less well known but ubiquitous.

If there is a shortcoming in Bourgeaud's work, it may be his lack of interest in what I called the "bracket" in Chapter 1. Looking at one of Rothko's late deep-purple monochromes, I can never forget that I am in front of a slightly dilapidated surface.[24] Borgeaud forgets: he steps too quickly over the threshold, hoping that a photograph of night and darkness can be compelling by itself. I mention this to introduce the problem of quality: contemporary photographers who flirt with formlessness, blur, and darkness can end up making images a little too easily. Unlike Fox Talbot or Daguerre, they do not need to fight to keep a measure of

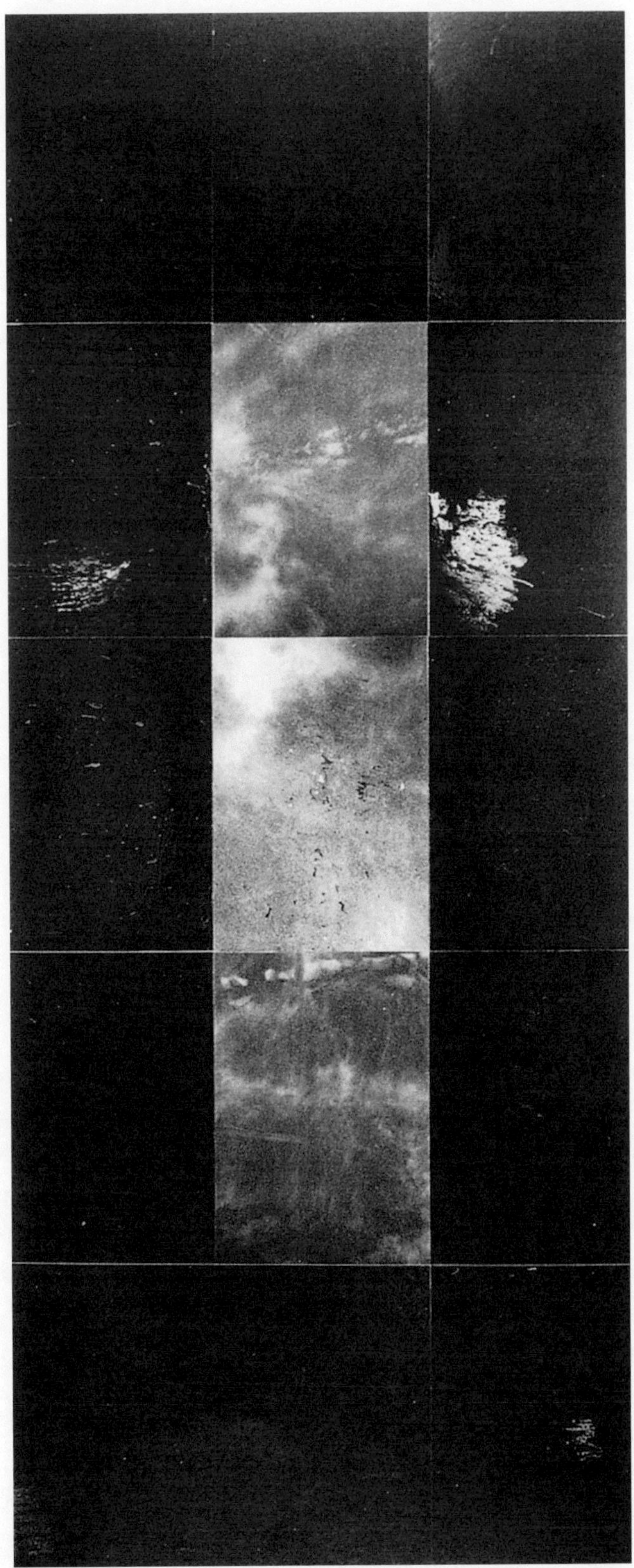

FIGURE 9

Bernard Borgeaud, *La Nuit, la pluie*, 1990–93. 300 × 120 cm. Photographs, Baryta coating paper mounted on PVC panel. Courtesy Toyko Metropolitan Museum of Photography.

clarity; and unlike Richter or Ruscha, they do not have to work to de-focus their images—and the result can be uninteresting darkness.

13   *The Ruined Grid*

Before I survey the field, I want to introduce two more critical concepts: the grid and the anti-optical. For Krauss it is the grid that saves Martin's paintings, making it impossible to perceive them as landscapes. Grids and sets of parallel lines take pictures out of lived experience and place them in a realm where visual phenomena are parceled out in discrete pieces. Thomas McEvilley says that Martin's gridded surface "functions as a kind of ontological ground . . . a threshold where energy passes from formless-ness to form." An empty gridded surface, like a sheet of graph paper, is a way of announcing that something is going to happen: an object is going to be measured, or an image is going to be transferred from one scale to another, or a form is going to be drawn and studied. But Martin's pictures have no forms aside from the grid itself, and the grid has no numbers or labels to help us read it. "So," McEvilley asks, "what is a grid standing empty, like graph paper on which nothing has yet been drawn?"[25]

For several decades, Krauss has been writing about grids and what they do to seal pictures away from their ostensive referents, encasing them in a self-referential world of picture making. A graph, after all, is a conven-tion that is distinct from realism and perspective. Ever since the medieval mathematician Nicole d'Oresme invented x-y graphs, they have been un-derstood as radical abstractions from reality.[26] They are not expected to correspond with normal habits of viewing or criteria of realism. The ob-jects in them are detached from real-world settings: they are distorted and transfixed by the gridlines, ready for sustained analytic attention.

I am not sure what Figure 10 would have reminded me of if I had not already known what it represents. The large round object is obviously rocklike: I would have preferred it not to be so clearly made of stone. Still, even knowing that it represents something astronomical, I find the photo-graph entrancing. It was taken on June 23, 1977, from the *Viking II* orbiter, looking down past the moon Phobos onto the Martian mountain called Ascraeus Mons.[27] The sense of unreality, of flatness, of pictorial possibili-ties restrained by some strict inner logic, is close to some painting and photography. The image is flat, monochrome, gridded with reference dots, and slightly asymmetric—all properties of fine-art images from the same

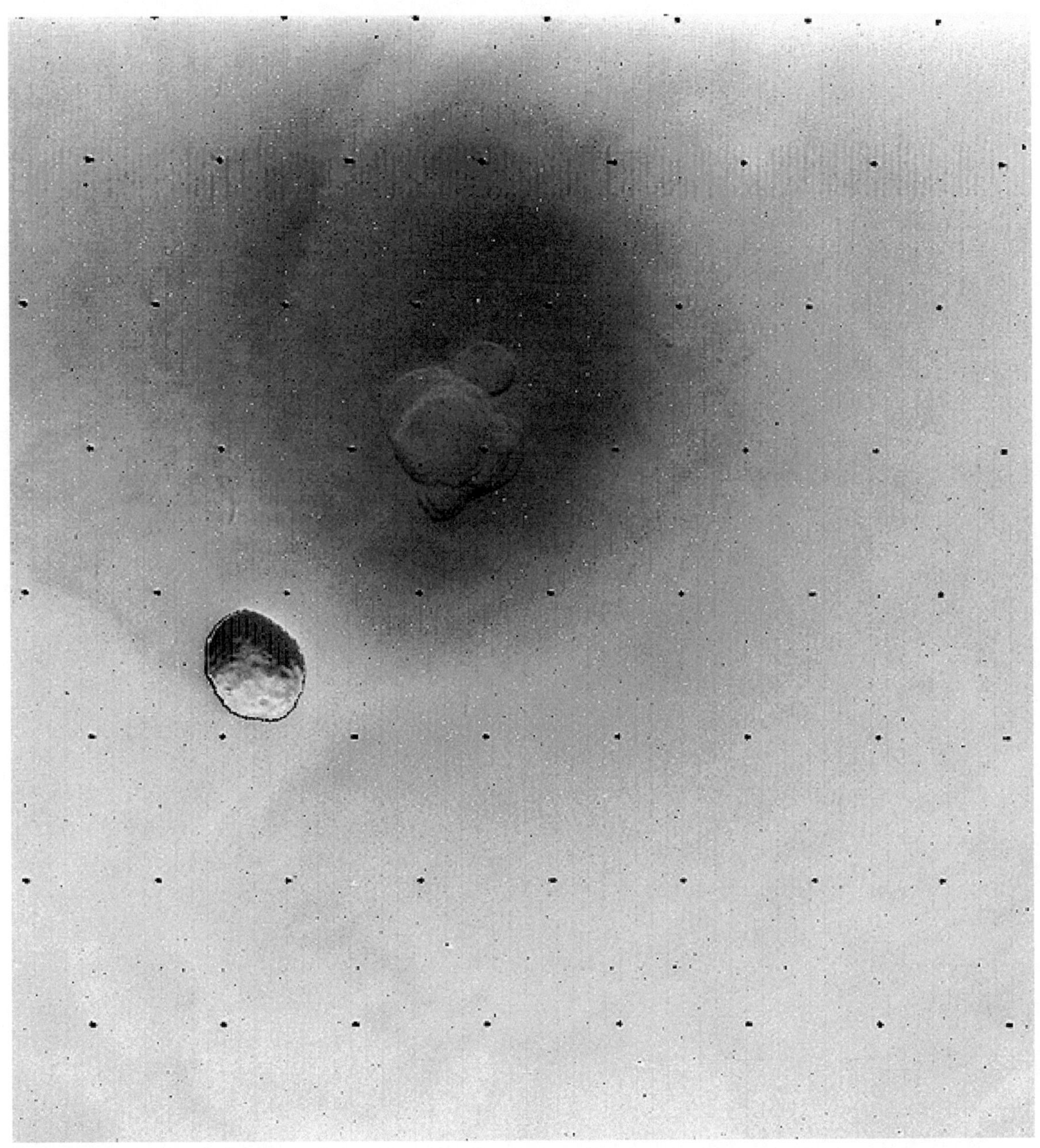

FIGURE 10

Phobos over the surface of
Mars, showing the Ascraeus
Mons. Photo taken June 23,
1977, from the *Viking II* orbiter.
Courtesy NASA.

decade. If it were a minimalist drawing, I could say it is about surface: the ground lighting is minimal, and Phobos is lit from below, undermining its three-dimensional effect. Even Ascraeus Mons (named after Ascra, Hesiod's birthplace in Beoetia) is dark rather than light, so it looks more like a stain than an eminence. I don't mean to claim that the people at NASA who programmed this exposure were thinking of minimalism, and even if they were, celestial mechanics would have prevented them from finding many opportunities to make pictures like this. On the other hand, it is always risky in art history to posit too strong a disconnection between contemporary imaging practices. This particular grid of reference dots, ranging without distortion across a virtually flat field, is especially close to the mapping impulse that attracted artists in the same decade.

Despite that, this photograph could not function in an art-world context, not least because—like every other scientific image in this book—it was intended to make objects visible with as much clarity and precision as possible. A gridded picture by Agnes Martin doesn't measure anything, and it only makes an indistinct reference to the concept of measurement. Marie Krane's offset grids (see Figure 4) are not made to measure anything, just to propose that *something* is measured, or—to put it in affective terms—to elicit the feeling of measurement. The minimalist artist Sol LeWitt is another example: he has made grids of different kinds, some without objects. Figure 11 is a close-up look at the edge of one of his wall drawings: it's the first five or ten inches of an eight-foot wall covered with a minute grid of pencil lines. The lines are so fine that from a distance of ten feet they become invisible and the wall looks hazy and off-white. The grids change configuration in different parts of the wall, so that some areas don't have horizontal lines and others don't have diagonals. The accuracy of the lines does not matter in this case, because this is one of a series of works that exists primarily as directions, written out by the artist. This drawing was conceived in 1971, and a team of volunteers in Chicago drew this example of it in April 1997.

It may seem that grids without quantitative precision and without objects would not be of interest in science, but scientists have done more with bare, nonreferential grids than artists have. Krauss distinguishes between grids and graphs: the latter have uses and are quantitative—a distinction that does not exist as such in science. In general relativity and cosmology, for example, there is ongoing discussion about the abstract nature of what is called "space-time." Philosophers and physicists debate the underlying

FIGURE 11

Sol LeWitt, *Wall Drawing #63*, detail, 1971, executed April 1997. Pencil. Chicago, Art Institute. Photo: author. Art copyright 2004 Sol LeWitt /Artists Rights Society (ARS), New York.

geometry—the graphs and gridlines—that determines the structure of space. In some accounts, the geometry of space-time determines all of physics, including the nature and development of the universe, making the bare grid utterly fundamental.[28] Even in other theories, where space-time geometry is not considered to be as fundamental as other physical properties, objects do not have to be placed in space-time models.

The same is true in everyday technology. Because computer code is organized in subroutines and packets, grids and other iterated structures are built into it at a basic level. Figure 12 is an interesting example: it is a FITS image (a graphics format used in astronomy) that I accidentally opened in a commercial graphics program without converting it.[29] The software on my computer tried to read the file and ended up creating an image of the packets of information, separated by fine white lines. At the top is the information on the file itself, which appears as a dark strip. The image is astonishingly complex: this is only about one-eighth of it (the original FITS file was 16 megabytes), and there are many patterns in the "noise." White borders move slowly up from left to right, creating a subtle pattern formally congruent with any number of minimalist and *support/surface* paintings, but far more intricate. No painting of Martin's has this surface complexity, and no drawing of LeWitt's has the same compositional rigor. In comparison, Martin's striped canvases or Frank Stella's black paintings might look like sketches, marred by "errors"—scratches, slips, small misconfigurations. My accidental FITS image captures a latent property of any digital graphic: it is organized in accord with a grid—using that term loosely, as Krauss does, to indicate a surface that is patterned and without referent.

The grid, fourth of the five concepts I am introducing in this chapter, is a device for abstracting from the world. Whether it appears in a scientific illustration, a painting, or a random printout, it sequesters the object, pressing data into /data/, and lines into /lines/ (in the semiotic formalism introduced in Chapter 1). Psychologically, a grid is an inward-turning device, moving the viewer's attention from real-world objects to representation itself. That is why I don't think the similarities I'm proposing are purely formal, or formal in a simple sense: they answer comparable desires to make realistic depictions into abstract maps. The grid is a way of pushing the picture toward one of its initial conditions *as* a picture: its capacity—before and apart from objects—to contain representations.

The purest and most blank grids, such as Sol LeWitt's drawing or the space-time geometries of relativity, are relatively uninteresting as images. So are grids chock-full of concrete objects (like the moon Phobos interrupting the clean grid that seems to be below it). Nor are pictures of grids interesting when they serve as the quantified armature for represented objects. Grids are intriguing, I think, when they *almost* give up on the objects they are to represent, as in the illegible but patterned FITS printout, Martin's reticent *Untitled #12*, or—to take an example outside the period under consideration in this book—the "counterfeit" grids that help organize analytic cubist paintings.[30]

In recent art, a good strategy for avoiding pure emptiness or everyday utility has been to make sure the grid is unlabeled, so it does not measure anything, and then let it become useless, so it can never measure anything. One of Jasper Johns's signature forms, for example, is a decorative mosaic of hatch marks, which he says he first saw painted on the side of a truck (Figure 13). The paintings he has made from that nameless motif are probably sloppier than the original, but aside from their careless attitude to design standards, they are empty and gridlike. In this painting, *Corpse and Mirror II*, the broken hatchwork grid is painted across another grid formed by the canvases Johns has joined together. The left half of this detail is composed of three canvases attached together, and the right half is a single canvas, painted so it looks like it is three separate canvases. The painting is made of two systems of dysfunctional, partly disassembled grids: the careless hatchings, and the carelessly assembled canvases. Neither grid is consistent across the surface, and in that respect the painting is the opposite of Martin's perfectly consistent grids. (In art-historical terms, Johns's canvas can also be read as an ironized, disengaged version of analytic cubism's more intricately dysfunctional grids.)

Artists also make use of nonfunctional grids by juxtaposing them on usable grids such as graph paper, ledger sheets, and map "nets." Thus Hannah Darboven scribbles senselessly on bookkeeping forms, adding "explanations" in the form of empty gibberish; and Anselmo Boetti colors in his grids with a ballpoint pen, expending huge labor on each picture without clarifying the picture's meaning; Václav Bostík creates faint arrays of dots. For such artists, the more badly broken the grid, the farther it is from ever having to carry an object or make sense as a scientist's grid might. Johns's painterly hatch marks belong in one and the same culture

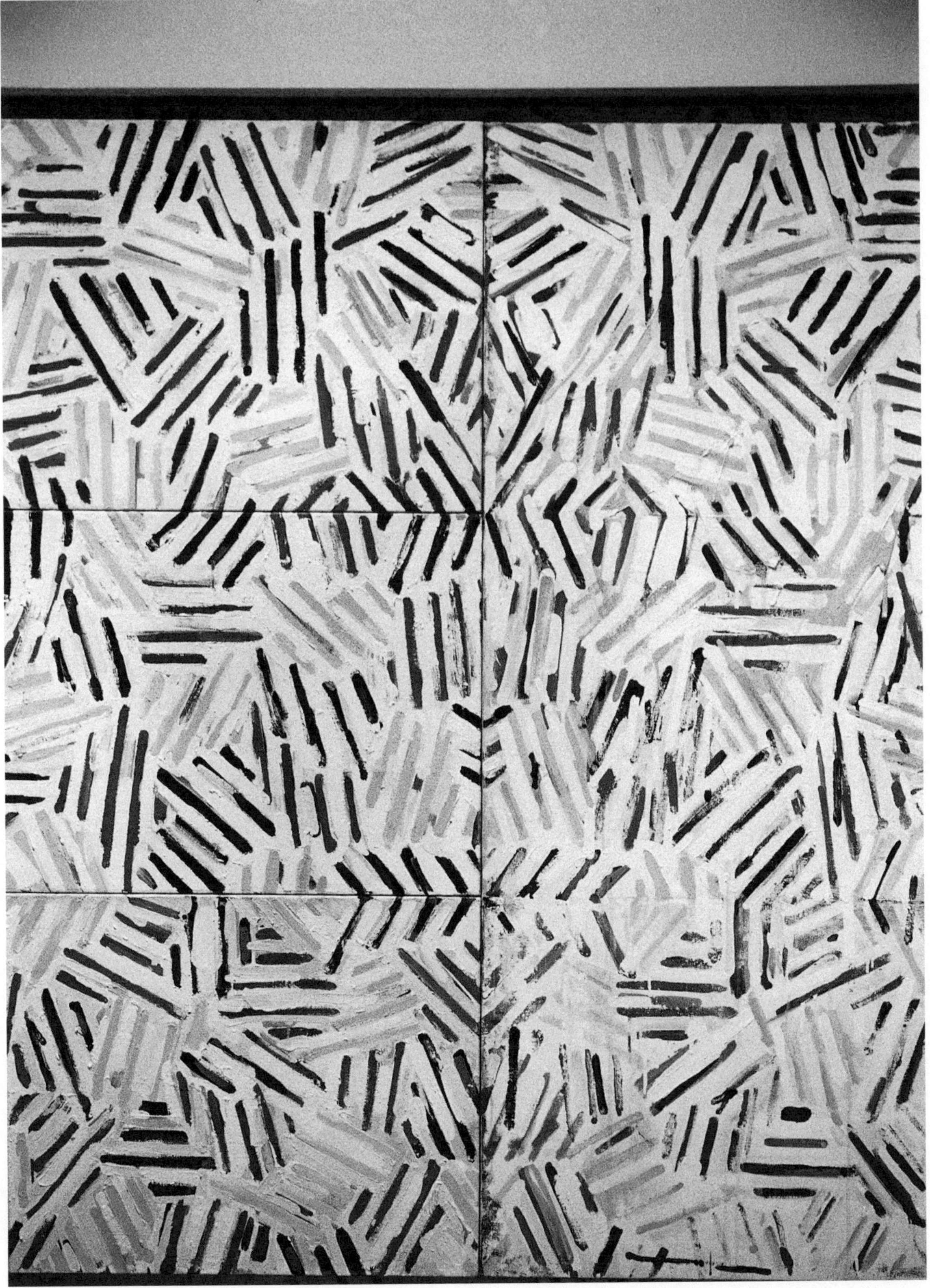

as any blank sheet of graph paper or any set of unused and overlooked reference dots.

14 *The Anti-Optical*

The anti-optical impulse is usually understood to be framed by modern art and its institutions. It does have a deeper history, known to medievalists and Renaissance specialists, involving recurrent crises in the status of religious images, but its immediate context is modernism and modernism's reverberations in late-twentieth-century practice.[31]

Marco Breuer's photographic practice is an exemplary case of this mixed—superficial and deeper—lineage. It is founded on a mistrust of the means of photographic representation—the camera, the negative, and the enlarger—and a complementary interest in discovering what photography can be when it gives up its habitual optical representation of the world.[32] From Breuer's point of view, cameras are parts of photography's weird aesthetic, which includes a dream of perfect optical correspondence between the print and the original object, a fetishization of equipment (lovely contraptions of metal and plastic), and the geometric simplicity of optical ray tracing. Breuer is tired of all that.

In the 1980s his work was self-reflexively photographic, experimenting with the range of photographers' tools. He also made photograms in which the camera and negative are not used (objects are put directly on the photographic paper and exposed to light, creating shadow pictures). Photograms are normally silhouettes, and the light is normally produced by the photographic enlarger in a darkroom. Breuer made photograms of matches struck on the surface of the paper, and burning fuses (Color plate 6). The fuses burned the paper and also exposed it by their own light. Technically, then, Breuer's photograms were more radical than ordinary photograms because they obviated the need for the enlarger. The only photographic tools left were the paper, its developing fluid, the stop bath, and the fixing solution.

In another series, Breuer did away with light altogether. He found that when photographic paper is scored with a knife in the darkroom and then developed, a thin gray line appears. The paper, the knife, and the chemicals comprise the photographic process, and light has nothing to do with it. Apparently the friction of the knife against the paper generates sufficient heat to react with the photographic chemicals in the paper. A handprint reacts with the paper's chemicals and develops to reveal a gray smudge. A spray of deodorant creates a halo of dull reds and greens. If photographic paper is scoured in the darkroom, the result after it is developed and fixed is an image crossed with webs of lines: some of the lines are scars from the steel wool, but there are also faintly colored lines produced by chemical reactions generated by the heat of the scrubbing.

At the end of the 1990s Breuer went a step farther, eschewing not only projected light but even the possibility of mechanical reproduction. He came across a book written in 1823 called *Observations on the Effects of Lightning on Floating Bodies* in which the individual plates were subjected to electric shocks in order to demonstrate, in each copy of the book and in each individual picture, what electricity can do. Breuer's *Pattern and Practice* (2000) is a book illustrated with photographs that have been individually treated, so each photograph is a record of its particular encounter with some object.

The series of photographs called *Tremors* is Breuer's most radical work to date: the photographic paper is burned, in the darkroom, by household appliances like hot irons and electric pans. The paper registers the burn just like any paper would, with carbon black streaks that sear the paper (they feel like old scars), and at the same time the paper registers the heat of the pan by changing colors when it is developed. In some of the *Tremors* series there is only an incremental difference between the photograph and an ordinary sheet of paper that has been burned. The rust color that follows some of the black burn marks might be chalked up to photosensitive chemicals, or it might not: *Tremors* makes a viewer wonder if the image is a photograph or just a seared piece of paper. The series *Untitled (Candy)* includes examples of several techniques, including pictures in which paperlike surfaces are represented on the surface of the paper (Figure 14). This is photography pressed close to several of its limiting cases. Nearly all the apparatus has been discarded: there are no cameras, no lenses, no filters, no negatives, no enlargers, and no lights. *Tremors* makes a viewer wonder if photography is as much like burning and scratching as it is about realism, resemblance, and optical representation.

A curious thing happens at this border between photographic representation and burning: the photographs become drawings. The full name of the *Tremors* project is *Photogenic Drawings: Tremors*, a title that is taken from Fox Talbot's name for his first photographs. In the course of subtracting the apparatus of photography, Breuer ends up referring to

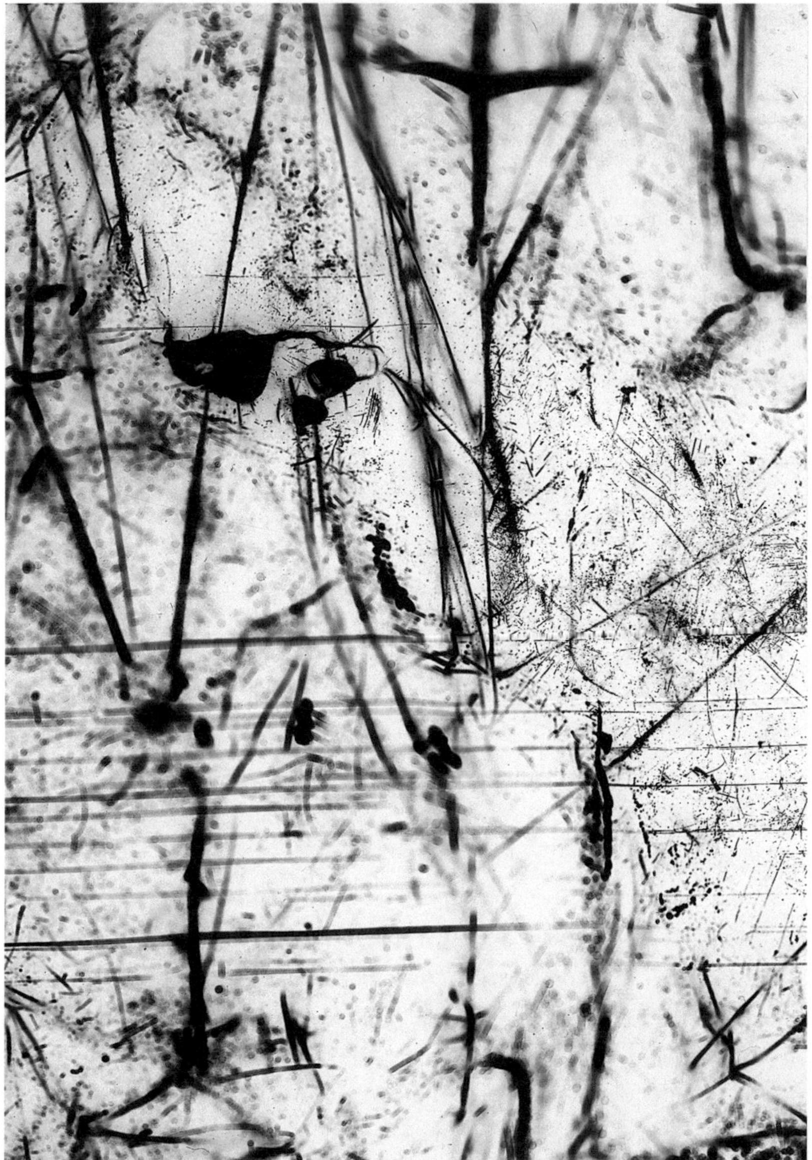

another kind of representation. The photographs in the *Photogenic Draw-ings: Tremors* series are more different one from another than similarly constructed photographs in earlier series, and that is appropriate because they conjure different kinds of drawing. One print is an open-architecture gestural abstraction, another looks like a detail of Cy Twombly or Antoni Tàpies, a third looks like parts of a De Kooning drawing.[33]

Breuer's work, like Umberg's, suggests that it is not interesting just to renounce optical representation. What counts is to try to make an image that owes as little to ordinary photographic representation as possible. The burn marks in *Tremors*, like the scratched and cut paper in the earlier work, are Breuer's acheiropoietai (images, in the Christian tradition, made without human intervention—an echo of the deeper history of the aniconic and religious traditions).[34] They are immediate evidence of the world, and criticisms of the elaborate efforts that most photographers make at representing the visible world.

15 *The Problem of Quality, Revisited*

Blur, darkness, grids, and anti-optical images are four rungs on the ladder of disorder. It matters that some of the rungs are easy to catch hold of, and others—like Bataille's *informe*—can hardly be grasped. What could be simpler than underexposing a photograph or unfocusing a camera lens? Things are essentially different in painting and printmaking, where blur has to be manufactured. Perhaps it is inevitable, then, that blurred and dark images are fairly uncommon, if prominent, in late-twentieth-century painting but ubiquitous in photography. To take five examples at random: Robert Stivers's *Red Dogs* is an unfocused image, with barely discernible animals in it; Lynn Geesaman takes photographs of misty landscapes; Roy DeCarava makes nearly black photographs; Kati Toivanen's photographs have an extremely shallow focus; Sylvie Readman's steel-gray, unfocused *Vestige* could be a panoramic view of tiny ripples.[35] Photographers like Sally Mann make nostalgic landscapes whose narrow focus recalls early photography, and Bill Jacobson and Fred Weber work in similar ways. Barbara Astman's *Frames* (1997) are black-and-white figures, smeared as if they were movie stills (they are from a series called *Scenes from a Movie for One*). Lori Newdick has been exploring blurred figures, which she juxtaposes with pulp-fiction paperback covers of femmes fatales.[36] Jennifer Ramsey has photographed overlooked places such as the space under the bathtub and bed, as well as the lint on the floor. Her photographs, printed large, are blurry and often dark, and almost hide their everyday origins (Color plate 7).

A special issue of *Photovision* (2001) explored the intersection between images of the sky and blurred art photography. Susan Coolen's *Astral Projections* is in the issue; it is a series of three photographs in which something that appears to be an asteroid is gradually blurred into a nearly black smudge. Also included are Riwan Tromeur's *Des Grands nords*, a dark empty landscape in cloudy gray and black, streaked with moonlike shapes; Aleydis Rispa's round print called *Planeta Scor*, apparently a planet like Jupiter, but with more ferocious storms; Richard Misrach's *Clouds (Orographic Stratiform)*, a bleary, dark, unfocused cloudscape. Sometimes there are strong formal parallels with pictures in this book: Rispa's foreboding, stormy roundel is a ringer for the cloud chamber photograph reproduced here as Figure 55. The special issue of *Photovision* only samples a larger practice. Pamela Bannos, for example, makes manipulated astronomical images—a practice that has been compared to Vija Celmins's painting.[37] The photographic response to darkened, out-of-focus heavenly scenes is generational and centers in the 1990s; most of the photographers in the *Photovision* special issue were born around the mid-twentieth century and have been making unfocused images for the last fifteen or twenty years.[38]

This thematic extends well beyond photographs of illegible figures, cottony out-of-focus faces, blurred stars, deserts, and skies. Even Joel-Peter Witkin's overworked faux-tintypes, Robert Gober's photo project for the 49th Venice Biennale, and Catherine Opie's photographs of Ron Athey (made using a room-size Polaroid camera) are part of the interest in blur and shallow depth of field.[39] Both blur and darkness are hallmarks of Hiroshi Sugimoto's work, whether his subject is theaters, architecture, or the ocean—also a subject of Opie's.[40] In the realm of commercial art there are the many shallow-focus photos of food, tableware, and cosmetics. (I have been told that they first appeared in the early 1990s, in Martha Stewart's magazine *Martha Stewart Living*.[41]) Digital art and design have followed suit, particularly in Daniel Brown's work.[42] Especially in the wake of Morris Louis and Cy Twombly, painters have also found easy kinds of blur; I think, for example, of Jean Degottex's dashed-off calligraphic clouds, and Martin Barré's soft, nearly empty canvases with smears at the sides.

The problem is that as it stands, much of the work is mediocre. The critical literature follows this lead, providing impressionistic commentaries on belatedness, the loss of memory, the affection for clumsiness, faint melancholy, the embrace of meaninglessness, obsolescence, the departure of the aura, sophisticated evasions, missing objects, ineffective repressions, loss of space, loss of language, hopelessness. These notions (which I have taken from some exhibition catalogues I happen to have on my desk) can all be important ideas, and they can be crucial for understanding some work: but they are disorganized and finally lost in the day-to-day criticism that gets written in the art world. It is easy to find work that plays with unfocused shapes, and it is easier still to write about it. The challenge is to find the work that is trying hardest to understand such things as formlessness, darkness, blur, or the anti-optical.

One useful criterion of value is that the work should not descend the ladder too quickly. Borgeaud's work may not quite make that grade, because he means to avoid whatever is clearly visible, not in order to say something specific about representation or vision, but just to open the door to mystery and poetry (see Figure 9). For him the cosmos is an unimaginable "wall," immuring us on the earth: "the more we know how much the universe is unknowable," he writes, "the more we can take the measure of the narrow space left to us." [43] The "narrow space" is interesting, and in terms of the bracket (as I defined it in Chapter 1), *La Nuit, la pluie* gives the viewer a glass- or paper-thin space. I wonder, however, if the darkness beyond the window is too freely offered. It is as if Borgeaud gripped the sides of the ladder with both hands and slid all the way down. If the darkness were any more perfect, it would be a photographic monochrome, like the pictures sent back from a webcam at the South Pole in the middle of Antarctic winter. (Such pictures are perfectly black except for the date and time stamp. The caption assures viewers that they are pictures of the South Pole.[44])

Borgeaud's pessimism in *La Nuit, la pluie* makes me think again of Bleckner's *Flora and the Future* (see Color plate 1). In terms of the sublime, both pictures aim to move the threshold closer, compressing the part that is understood, making more room for what is not. Again the limen is very near. Almost from the instant I encounter Bleckner's picture, I have stepped forward to the limit of my understanding. There is almost no space left for painting to work, or for painted shapes to show. Catching sight of either *La Nuit, la pluie* or *Flora and the Future* is like opening a door onto a dark room: I stop short and peer inside, seeing what there is to see. But Bleckner also shows me that I am trapped. The painting apparently represents a lunar reflection off some oily, dark surface, but the paint itself is flat and waxy, like a shop window smeared with soap so no one can see inside. These are Krauss's brackets at work. There is no sliding into reverie in Bleckner's picture: the viewer is clamped in place. In photography, a good strategy for avoiding the easy slide into nothingness is to blur the image to a certain point, and let that point matter. An excellent example is a series called *unanswered: witness* by the photographer P. Elaine Sharpe. She visited places where murders and other calamities had taken place and intentionally focused her camera on the empty foreground, where the people involved in the tragedies would once have stood. Her photographs are blurred, but that is because the plane of focus is sharply positioned on an absent subject.[45]

Another test of quality is that an artist's oeuvre should give evidence that the limits of representation cannot be reached all at once, by a single straightforward strategy or insight. (It should be difficult, perhaps even tiring, to climb down the ladder into the darkness below.) Richter's paintings based on photographs propose an analogy between inaccessible memory and unforthcoming surface. Each picture or series of pictures manages the parallel between failed memory and half-effaced image differently (see Figure 5). "One can only make bits of images," Luc Tuymans has said, in a conversation on the subject of Richter.[46] Tuymans's paintings are exploratory: he paints Morandi-style still lifes, faces blanched like Manet's *Olympia*, interiors pared down to just a corner or a ceiling, piles of pillows, shadows, black-and-white silhouettes, empty mirrors. He tends to return to several dozen subjects, and he has as many strategies for leaving each of them half painted. At the end of the 1990s it was clear that he was working, in as many ways as possible, to find ways to keep "reducing" paintings (to use his words), to something at once "clear" and enigmatic.[47] (I am close here to Lacoue-Labarthe's idea of the pallid sublime—the one that begins to flower wherever beauty is exhausted. "Contrary to appearances," Lacoue-Labarthe says, this enervated sublime "wants to provide a weak thought, that is, a thought precisely *without* grandeur."[48] That fits Tuymans's work well enough, and perhaps also Borgeaud's.)

It is also a good sign—this would be a third criterion—when the photograph itself does not reveal where focus and unfocus should be—or, to say it differently, the artist seems not to be sure where clarity meets blur. If the plane of focus misses its object, creating blur where there shouldn't be blur and focus where it isn't needed, a picture can become perplexing. Jennifer Ramsey's photographs have that effect (see Color plate 7). Here the camera's eye looks under a bathtub, into a dark space between smooth surfaces. The camera is set to miss the picture's subject, and it captures a few hexagonal tiles instead. Ramsey has also jarred the camera, creating a double image, and she has printed it in nocturnal blues and grays. It is as if the plane of clear focus is too fragile, or as if clarity itself has gone wrong: the picture doesn't say.

These three criteria are not sufficient to distinguish good work from bad, but they are a start. Each of these criteria can make it evident that the artist has looked hard and considered how to present a lack of clarity. That is what I look for: the effort of trying to see well, the failure of that effort, and the acknowledgment of the failure in the image. The criteria and possibilities are precisely the same in blurred scientific images.

16   Presence

Art photography, I have argued, has developed a characteristic set of strategies for avoiding clear representation, including blur, darkness, the ruined grid, and the anti-optical. There are other photographic strategies that could be enlisted, including the growing range of digital filters in Photoshop and other programs, motion-induced blur (which can be made to resemble painterly smearing), and experimental optics, including pinhole cameras. But this is enough to indicate the principal options. It's a large, complex field, poorly theorized, and this is only a start.[49] The concepts are diverse, and so are the examples, and each concept would require a book to itself. But I hope I've said enough to suggest that blurred and dark photographs are being made by the thousands, and they require attention commensurate with the criticism lavished on the more prominent figural photographs from Thomas Struth to Beat Strueli. I have tried not to shirk the question of the quality of this work, but that, too, is a difficult problem—dark, blurry, formless art photographs are exceptionally easy to make and do not often repay concerted attention.[50] I will return to the issue of quality at the very end of this book; I will conclude this chapter with a brief meditation on the concept of presence, which will serve as a bridge to the following chapters.

When there is talk about representation, then the question of presence cannot be far behind (or rather, it will have been ahead of us all along). Presence is a nearly talismanic word in contemporary philosophy, and for readers who come at this material from the vantage of philosophy and literary theory, presence will be a fundamental term. It is taken as one of the principal constructions that allowed Western talk about existence and truth to get under way. In the vast and disorganized project to deconstruct unified presence, truth, and certainty, there is relatively little work on the reconstruction of truth. Presence and intention are not wholly erased in post-structuralist thinking, but few writers attend to the rebuilding of the remnants of presence out of the ruins of old structures. In painting as in post-structuralism, what tends to be most compelling is the ruin of whole meaning and the concomitant fall from presence.

In this regard, scientific imaging is a tonic, because even when the objects it tries to find keep disappearing under its very eyes, it continues to search, never entirely reversing the aporia, but often discovering new sources of clarity. A recent example in astronomy is the search for distant galaxy clusters.

(And again, as in the pages on the Wolf-Rayet star, I will describe this in some detail: I would ask readers in the humanities to note that the scientific detail in the next few pages is not measurably more or less than the detail accorded to the descriptions of art photography in the preceding pages. Note that the appearance of inappropriate detail is purely a perspectival effect. If photography or art history is your field, you may well feel that this chapter was scattered, sketchy, or insufficiently historical, because you will know the full discourses that I have been sampling. If astrophysics is your field, you may feel something of the sort about the next chapter. But in a project like this, the question isn't full fidelity to a given discipline—every discipline is sampled, none is given in an unabbreviated form—but the possibility of finding ways of talking that are commensurate or even comparable across "geometric" and "literary" borders.)

In the 1990s some competing theories of the origins of elliptical galaxies turned on whether or not there were galaxy clusters, and not just galaxies, early in the universe's formation. To search for galaxy clusters beyond redshift $z = 1$ (that is, at large distances from the earth) astronomers use X-ray telescopes, because the gravitational force that attracts the galaxies heats the gas between them, causing them to emit X-rays. The problem is that X-ray telescopes don't resolve tiny details like distant galaxies—they see with a blurry eye. In March 1999, a team headed by Piero Rosati at the European Southern Observatory in Munich reported the first discovery of a cluster at $z = 1.26$, significantly farther than any previously found.[51] What the X-ray telescope actually saw is shown in the center of Figure 15, labeled "Real." It is nothing more than an asymmetric blob, ever so slightly more intense than the back-

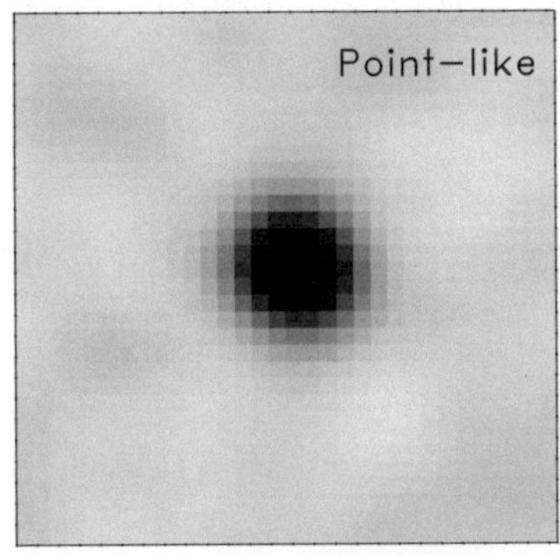

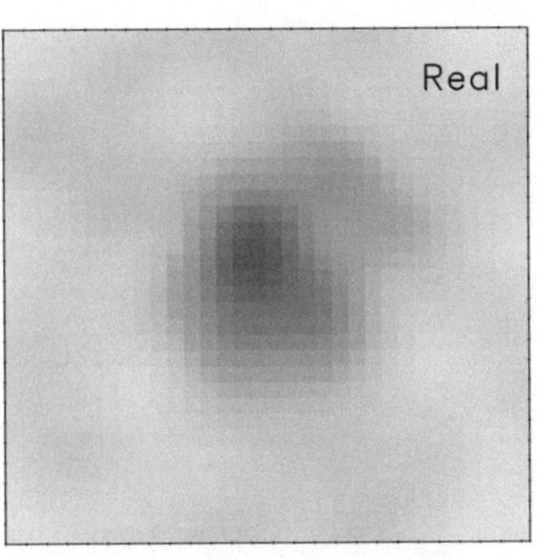

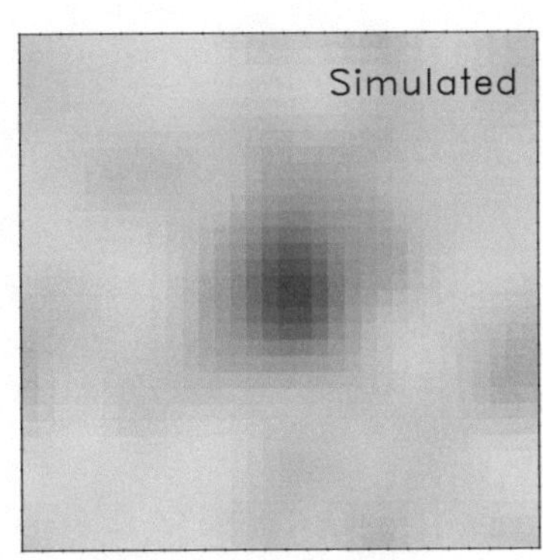

ground. The first step was to make sure it wasn't a "point source" like a star in our galaxy or a quasar in some distant galaxy, so the team imaged a nearby quasar for comparison (Figure 15, left). The "Point-like" quasar is obviously more of a perfect circle than the suspected group of galaxies. The team then computed what a galaxy cluster in that part of the sky *might* look like. They programmed a simulation, positing that the cluster would have a certain typical shape, called a King profile, and that it would be an average size (200 kiloparsecs) and have a certain plausible X-ray flux. They then redshifted the hypothetical X-ray spectrum in the computer and overlaid the result on a typical background for that part of space. The result, labeled "Simulated," shows that the signal they actually received is about the right shape and intensity.

At that point nothing more could be done with the X-ray image: it could not be sharpened or analyzed any further. They compared the image with a picture of the same part of the sky, taken with an optical telescope (Figure 16). Using the adjacent quasar as a known reference point, they overlaid the image of the X-ray contours onto the optical plate (solid and dashed contours). That made it clear that galaxies were in fact clustered in the area, and a further study (Figure 17) obtained spectra for several galaxies (those in square boxes) verifying that they are at roughly the same distance of $1.257 < z < 1.268$. Other galaxies, not yet tested at the time, are circled, and the center of the cluster is enlarged at the upper right.

Figure 15 has interesting resonance with fine-art practice: it resembles any number of enigmatic conceptual pieces that purport to give information, but don't. The image could be used as an artwork without even changing the labels. The difference would be that in the image's original context as astrophysics, each frame has a particular purpose that works *against* the inevitable aporia caused by the limits of the instrumentation

FIGURE 15

An X-ray image of a distant galaxy cluster (center), compared to a point source (left) and a computer model (right). From Piero Rosati, S. A. Stanford, Peter Eisenhardt, Richard Elston, Hyron Spinrad, Daniel Sten, and Arjun Day, "An X-ray Selected Galaxy Cluster at $z = 1.26$," preprint, arxiv.org/abs/astro-ph/9903381 (March 24, 1999). Courtesy Piero Rosati.

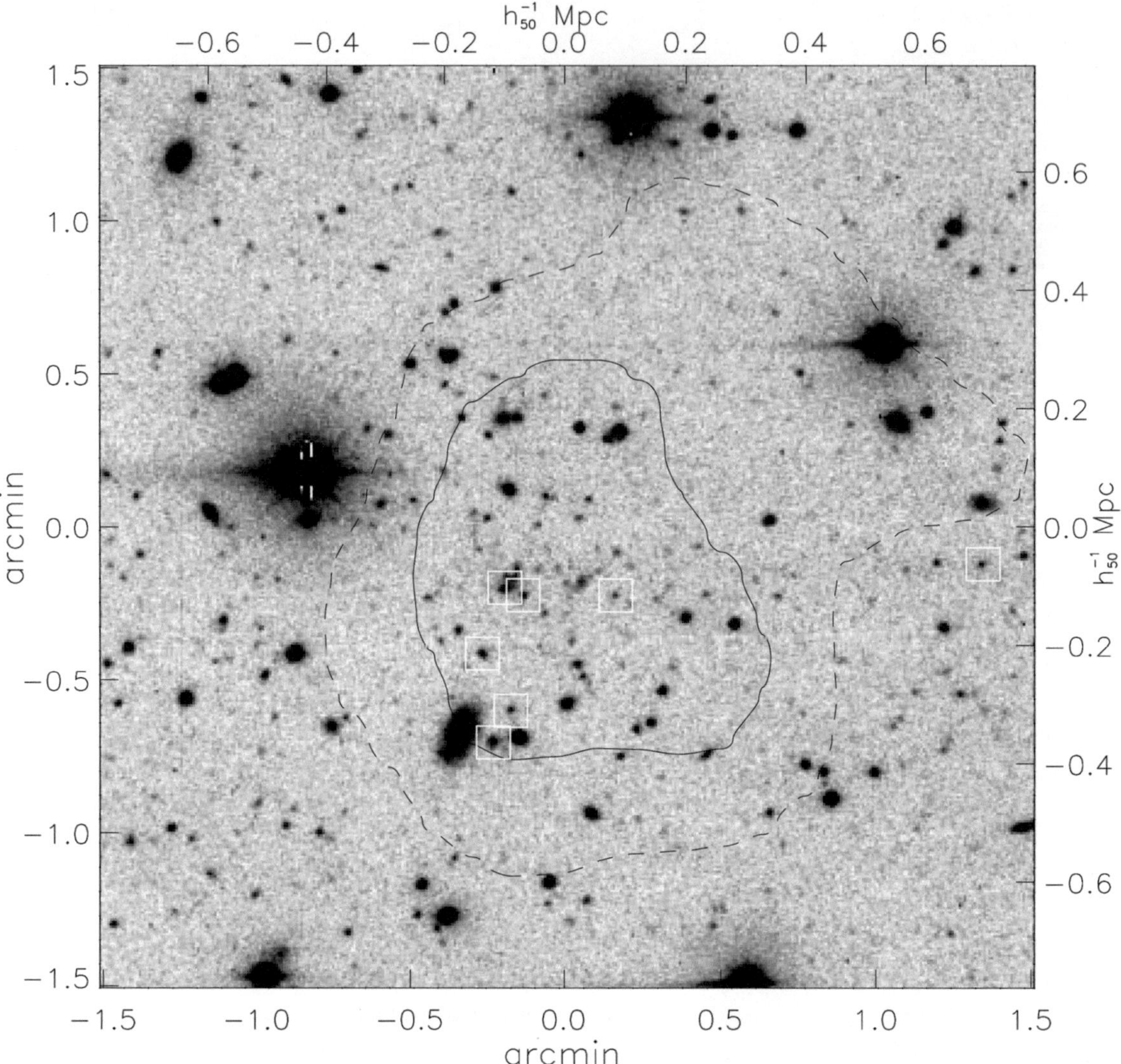

FIGURE 16

Optical image of a distant galaxy cluster, with X-ray image superimposed. From Piero Rosati, S. A. Stanford, Peter Eisenhardt, Richard Elston, Hyron Spinrad, Daniel Sten, and Arjun Day, "An X-ray Selected Galaxy Cluster at $z = 1.26$," preprint, arxiv.org/abs/astro-ph/9903381 (March 24, 1999). Courtesy Piero Rosati.

and analysis. In this case the first panel is something known, the second is something dubious, and the third is an experiment designed to simulate the second. If all telescopes were X-ray telescopes, this is where the team would have had to stop—comparing known things with half-seen things and things only visualized in mathematical simulations. It would be a kind of dead end, but one that used all its available resources to minimize uncertainty and to pull presence back into the images.[52]

Notice that the optical images, Figures 16 and 17, do not meliorate the blur of the X-ray image. They clarify the positions of individual galaxies, leaving the X-ray illumination—indicating the presence of gas between

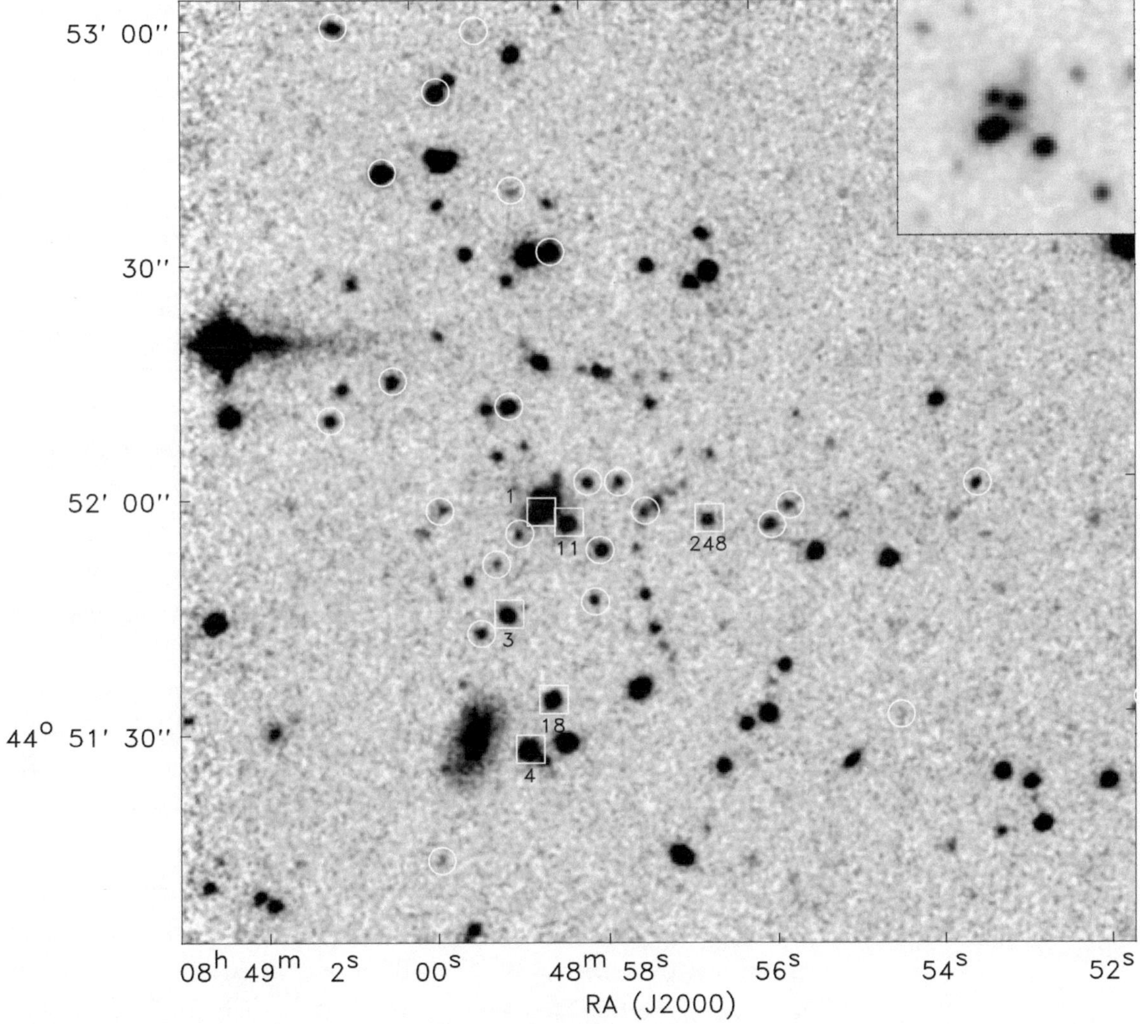

galaxies—unresolved. Certainty and presence are displaced, and they are found in another context. Stellar coordinates, marked around two of these plates, provide the link between the blurry X-ray image and the sharper optical image. Aporia and knowledge, presence and mediation, play back and forth.

Another example, more spectacular and less well understood, is the enigma of gamma ray bursts, or "bursters"—intense, brief explosions from extremely distant parts of the universe. The radiation that reaches the earth from the gamma ray bursts indicates that they are more energetic than any other energy source—more energetic, for example, than quasars, which can exceed the energy of an entire galaxy of stars. For

Detail of Figure 16, showing resolved and suspected galaxies. Inset: close-up of the center of the cluster. From Piero Rosati, S. A. Stanford, Peter Eisenhardt, Richard Elston, Hyron Spinrad, Daniel Sten, and Arjun Day, "An X-ray Selected Galaxy Cluster at $z = 1.26$," preprint, arxiv.org/abs/astro-ph/9903381 (March 24, 1999). Courtesy Piero Rosati.

several decades, gamma ray bursts were little known because they were detected at high energies, beyond the visible spectrum, and when observers trained optical telescopes to the locations of the bursts, they found nothing but empty space. The enigma was finally solved in January 23, 1999, when a burst was recorded by an orbiting detector.[53] The satellite transmitted the object's approximate coordinates to a ground-based telescope, which swiveled into position less than twenty-three seconds after the burst and recorded the first optical photograph.[54] A minute later, the image was already fading. At that point, the burst was a tiny point of light in a field of faint stars.[55] Knowing the location, astronomers trained the Hubble space telescope on the spot and recorded a detailed image that showed where the burst originated—in a very distant and strangely shaped galaxy (Color plate 8). Hubble couldn't be trained on the spot until February 8 and 9, and by that time the burst had faded to less than one-millionth its initial strength; but it was clear that the gamma ray burst was in a galaxy full of young stars, which helped narrow the theorists' explanations of what the bursts might be.

As of spring 2007, somewhat more was known about gamma ray bursts and about this particular one, known as GRB 990123, after the date it occurred. It originated a little to one side of the galaxy's center, probably in a field where many stars were being formed.[56] Gamma ray bursts are among the most powerful sources of energy in the universe—in the typically unimaginable numbers astronomy trades in, they were once thought to be 100 quadrillion times brighter than an average star. It now appears that their energy is "beamed" in one direction, making them look more powerful than they would be if their energy went equally in all directions. Still, they are only partly understood, largely because they occur at great distances from the earth and from the present.[57]

Neither the distant cluster of galaxies nor GRB 990123 were wholly recovered for analysis. They are not present—or rather, their presence is mediated by disorienting reaches of time and space, as well as by dizzying imaging technologies. Even so, properties of each have been retrieved: there is some knowledge to put against the hopeless distance between these objects and us. This is the return motion, the one so seldom encountered in the humanities, where the deconstruction of presence too often discharges the brief of interpretation.

three # Astronomy

Now I want to move on to astrophysics. In accord with the methodological agenda I set out in the Introduction, I will not try to extend themes that fit the arts. Some recent astrophysics images are remarkably similar to some recent art photography, but in the absence of a discourse that links the two fields—except, as usual, a set of simple formal parallels—I will not pursue the links explicitly. In almost all cases—except the "pretty pictures" I will discuss first—astronomical images and art photographs are spoken of in wholly different terms.

17   *Astronomy's Bad Reputation*

Recently, the astrophysics community has been giving itself some poor press with the art world by disseminating what scientists call "pretty pictures": hopped-up versions of legitimate photographs, with the colors intensified or falsified. Pretty pictures sell as calendar art and they apparently help direct public interest to unmanned space missions, but they have also worked to further alienate serious art making from serious science. The aesthetic of astronomers' pretty pictures was the subject of a study by the art historian Sam Edgerton and the sociologist of science Michael Lynch; they concluded that astrophysicists who produce such pictures are not working from any definite aesthetic.[1] Even so, the colors and compositions of astronomers' pretty pictures are clearly dependent on such things as science fiction paperback covers, Maxfield Parrish works, fantasy art, Hollywood special effects, and the gaudy colors of popular magazine illustrations starting in the 1950s—sources that have virtually nothing to do with contemporary art making, except as camp. Even committed amateur astrophotographers using expensive CCD (digital) cameras, "gas-hypered" film, and large-aperture telescopes try to inject as much color and drama as they can into their images. (Examples can be seen in several nationally distributed astronomy magazines.)

A few serious artists have tried to adapt astronomical images, but they are treacherous models. In 2006, Thomas Ruff exhibited some appropriated images of star fields, printed at an enormous scale; without their original contexts they just look decorative and perhaps a bit sublime.[2] Even Vija Celmins's sober, ambitious paintings of star fields can be ruined—they can become irredeemably hokey—when the viewer realizes that the dots do not descend from Larry Poons or Georges Seurat as much as from *Sky and Telescope*.[3] Celmins's star-field pictures are intriguing because she lets some stars fall out of focus, so her paintings begin to look like close-ups of particles suspended in a soapy fluid. But her paintings of galaxies are just that: astronomical images that happen to have been painted on canvas. (Her best work, I think, is of objects that aren't infected by astronomical pretty pictures, such as patterned ocean waves and anonymous stretches of gravel and dirt.) The same faults plague serious photographers who use astronomical images, and they also plague "art prints" made from NASA negatives, such as those done by Michael Light.[4] Some of Light's photographs are uncanny: as a reviewer put it, their "principal unearthliness" is "the blackness that seeps into the imagery from space."[5] But most are awkward transpositions of ordinary late-romantic landscape photography and painting to subjects like moon walks.

Outer space just does not mix—at least not yet—with the central themes of painting. "Painting rarely dines well on the leavings of science," to quote T. J. Clark's wonderful phrase.[6] Too much of the night sky has been co-opted for Hollywood space operas and overly dramatic public-relations efforts for it to be useful for serious painting. In the fall of 1979 Carl Sagan gave a lecture at Cornell University, with the first results of the *Voyager* mission to Jupiter. He said that he thought the photos of the Great Red Spot outdid twentieth-century abstraction. As I recall it, he said something like, "these are more beautiful than anything painters have produced"—but the pictures he showed were entirely unrelated to painting, and were therefore unusable even as models. The creamy swirls of poisonous gas were very impressive, but they have more affinity to eighteenth-century marbled endpapers or 1960s hippie cartoons than to twentieth-century abstraction.[7] As weird and desolate as some of the images are—and they wear well, in the sense that there is still nothing like them—they have no formal connections to painting.

Outside the poison well of sentiment and sensationalism there is a truly lovely desert of astronomical images that do not try to be pretty. Many have been produced during attempts to see increasingly faint objects at increasingly large distances. The shapes are so faint that they scarcely register on the film or the CCD plate—a far cry from the boisterous whirling galaxies that are presented in NASA press releases. Professional astronomical journals rarely reproduce objects in color, because color can be inexpensively coded in black and white; and they rarely bother with the familiar objects—Andromeda, the Crab Nebula, the Pleiades, the Horsehead Nebula—that still capture the public imagination. Most professional journals are stocked with images like the distant galaxy cluster I reproduced in Figures 16 and 17; they are not spectacular or eye-catching, and they are anything but obvious.

It is not easy to give a sense of the visual fragility of such objects. Even on a good night, the eye cannot penetrate the darkness enough to see even the brightest of them. They aren't just beyond normal vision— they are *far* beyond it, by several orders of magnitude. They are also too *small* to see, because they are so far away. There are many wonderful faint lights in the night sky that can be seen with the unaided eye—including the "night glow" and the Zodiacal light—but the objects in this book are not among them.[8] The images that interest astrophysicists are captured by expensive instruments and are given to us—much as Kant got his images of volcanoes and other natural wonders—in books.

18   Mathematics and Infinity

When Kant called one of his species of sublime "mathematical," he did not mean that it involves algebra, but rather that the sheer scale of objects like the starry sky conjures mathematical concepts that are inaccessible to intuition, such as nothingness and infinity.[9] The essential difference between Kant's examples and actual images is that pictures—even scientific illustrations—never get near infinity. For the most part, scientific images attempt to encompass unencompassable objects by means of real-world mathematics. Things like stars do have a "sort of infinity," in Burke's wonderful phrase, but it is measurably short of full infinity.[10] That is a sufficient reason to leave the sublime behind, even if it were widely used outside of postmodern art.

Gravitational lenses will be my first astronomical example: they occur, for example, when one very distant galaxy lies directly behind a second, closer galaxy. According to Einstein's equations, the nearer galaxy will act as a lens, bending the light from the farther galaxy around it, ideally fragmenting the image of the distant galaxy into four parts (the "Einstein cross"), or, more often, into many irregular fragments. The configurations of the lens can be studied using decidedly finite mathematics.

Color plate 9 (top) is a small detail from a deep-sky image called the Hubble Deep Field South (I will introduce it formally later in the chapter). In this detail, an orange elliptical galaxy is in the same line of sight as a much more distant bluish galaxy. The farther galaxy appears around the edges of the nearer one, like an image bent by a piece of bottle glass. The large pixels indicate that the image is near the limit of the telescope's resolution. Rennan Barkana, Roger Blandford, and David Hogg, at the Institute for Advanced Study, Princeton, have analyzed this image by simplifying it still more—a typical mathematical strategy. They assume the foreground galaxy is a perfect sphere, and that the bluish streak at the upper right can be divided into four components. Reading in clockwise order, they label them A, B, C (the fainter, elongated strip) and D.[11] They ignore the blue dot at the upper left, positing that it could be included in a future study. With those simplifications, they test several mathematical models to see which one might produce the best fit with the data. In other words, they begin again from scratch, feeding numbers into Einstein's equations to try to produce a picture similar to the one from the telescope.

Gravitational lenses involve exact relativistic equations applied to inexact lenses (the lensing galaxies are made of stars, so they are "grainy" and do not perfectly follow mathematical models) and incomplete observations (the galaxy in Color plate 9 is known only as a couple of hundred pixels, not as the stars and dust particles that actually comprise it). Within those limits, the equations predict that a gravitational lens will produce multiple images separated by angular distances proportional to the square root of the lens mass.[12] The more accurately the lens is modeled, the better the estimate of the number and positions of the images.

In the case of Color plate 9 (bottom), astronomers' models fits the phenomena very closely. Here the mathematical model of the lens galaxy is visualized, in part, as a tangential caustic (the small star shape in the middle) and a surrounding critical curve (the dotted oval). The equations predict that there will be five fractured images of the distant galaxy (drawn as small, open circles). The one in the center would be too faint to see because it would be directly in line with the center of the lensing galaxy, and they cannot detect the one at the lower left because it is below the threshold for the image.[13] The remaining three images, at the top right, fit perfectly: the small x's show the actual positions of images A, B, and D. Barkana, Blandford, and Hogg are even able to predict that other galaxies, not seen in this photograph, are pulling at the lens galaxy, distorting it.[14]

In this fashion an object so small and blurry that it almost can't be analyzed at all is reduced still further, until it fits with the available mathematics.[15] What is at work here is not an abstract Kantian infinity, but a set of particular observations and equations. The scientists would be the first to allow that with better resolution and the ability to reach to fainter magnitudes, mathematics could model the objects more closely; but in this realm—two objects partway across the visible universe—no mathematics will be fully adequate. The yellow galaxy, a nearly unimaginable object, is only barely seen, and it masks the blue galaxy, even more mysterious, less well seen, and even farther off in the "infinite" reaches of space. The clean diagram is a map of human inadequacy, because it gives us the best that the mathematics can provide.

This is my non-Kantian point: this image is not interesting because it is an encounter with infinity, but because it relies on several particular equations and on the particular conditions of observation—this blur, this faintness, these relative distances. The image and the graph fail poignantly and precisely, in two different ways. The color image is just barely sharp enough to provide something to see, to prove that there is something out there to be seen. Two little x's hit their mark, and one just barely misses. Two open circles go astray with no observations to match them. The pixels and the little circles and stars are predicted by specific routines of image analysis and computation: a real mathematical regime of interpretation, without the sublime and without infinity.[16]

19   Magnitude

Gamma ray bursts and other evanescent lights live in the middle of vast expanses of darkness. In astronomy, as in the particle physics that is the subject of Chapter 5, unimaginable size is intimately related to virtual nothingness.

Kant says that size (*Größe*) is what sparks the mathematical sublime. He was thinking of large objects, so it is fortuitous that the Latin *magnitudo* became the scientist's *magnitude*, which applies equally to very small things and to very large things. In Kant's time there were no submicroscopic objects, but it is likely he would have been interested in the scientist's ability to work indifferently up and down the scale of magnitudes, from the incomprehensible string theory lengths (c. 10^{-25} meters) and Fermi lengths (10^{-15} meters) to the equally incomprehensible width of the observable universe ($> 10^{25}$ meters). Burke was interested in microscopical creatures, and he talks about large and small magnitudes. Either extreme, he says, is sublime, so even "the last extreme of littleness is in some measure sublime." Thinking of tiny things like animalcules, "we become amazed and confounded at the wonders of minuteness; nor can we distinguish in its effect this extreme of littleness from the vast itself."[17] I don't know why Kant was not impressed by microscopic phenomena, which were a sensation in the eighteenth century; but Burke's formulation articulates what is implicit in Kant's theory.[18]

The word *magnitude* has the double virtue of not confining discussion to the unproductive difference between "natural" numbers and unimaginable infinity, and not equating sublimity with largeness. What matters in perception is the boundary between magnitudes that can be grasped and those that can only be written in symbols. There are, in fact, practical limits to each person's intuition, and practical ways of extending that intuition. The wonderful book *Powers of Ten* demonstrates this by providing a series of pictures that zoom inward from a person's hand down toward subatomic particles, and then another series that zoom up and out to encompass the universe.[19] The picture on each page is ten times smaller or larger than the one before. If the picture that represents one square meter is taken as a starting point, then the imagination can follow at least three orders of magnitude in either direction (that is from 10^0 to 10^{-3} or 10^3). *Powers of Ten* has been published in several languages and also exists as at least two films; one version of the book begins in Chicago, on a small stretch of the shore of Lake Michigan (the actual place has since been renovated, so it is no longer possible to take a copy of the book and sit in the exact spot that the book depicts). The first image depicts a man asleep on a picnic blanket. I can easily follow as the scale increases, zooming up into the air, looking down on Chicago. The pictures make intuitive sense—I can imagine making the trip in a helicopter—up to about 10^5, where the view is really from space. I can follow the downward sequence as well, zooming onto the man's hand, until the pictures reach the scale labeled 10^{-3} (one square millimeter), which is about the limit of my vision.

Of course I can *read* the book and interpret what I am seeing, but Kant would say my comprehension (*Zusammenfassung*) gives out long before my apprehension (*Auffassung*). I can keep reading about powers of ten, and powers of those powers, but I quickly have to abandon any hope of bringing it all together in intuition.[20] The first few pages of *Powers of Ten* are easy to assimilate, and then when I try to picture the next leap into outer space, or the next jump down toward the molecules in the man's hand, my thought slows down: it is as if I were swimming through molasses. I can't comprehend the picture of the earth from space. I know how it looks from television and magazines, but I don't quite see how it is connected with the aerial views I know from first-hand experience. In Kant's terms, if the change of magnitude takes too long to interpret, it cannot be used to help comprehension.[21]

Where my intuition finally gives out, analogies can help: I can imagine I am looking through a microscope or a telescope. Inevitably, there are limits, which are different for each observer. An astronaut might feel intuitively at home as far up as 10^9, where the view includes the earth and the moon. A microbiologist might comprehend everything down to 10^{-9}, the scale of small molecules. *Powers of Ten* shows that there is more to magnitude than the opposition of numbers and infinity, and more to orders of magnitude than the mechanical addition of exponents. "Intuition" in this sense is similar to Kant's sense of "comprehension": it's the point—slightly different for each viewer—when things in the image stop working the way things in the familiar world work. From that point onward it is necessary to trust analogies or to rely on the mathematics alone. (The opposite of intuition in this book is "calculation," rather than Kant's "apprehension." I take *intuition* and *intuitive understanding* as synonyms.)

It is too bad that the various incarnations of *Powers of Ten* have all been edited by scientists, because the book has a great deal to teach about the limits of intuition—a subject that working scientists don't always care about. From a scientific standpoint it probably doesn't make sense to try stretching your intuition to grasp something tiny or gigantic, because intuitive understanding rarely helps and can often be misleading. It is a constant refrain in physics: intuition won't work—it is irrelevant or useless. In astronomy, intuitive understanding has severe limits: people talk about the behavior of gases in nebulae by analogy with jets and bubbles in fluids, but there is nothing in human experience analogous to a neutron star or a black hole.

(In passing, I will mention that time, as well as space, lends itself to encounters between intuition and calculation.[22] A lovely example is the studies of the future evolution of our solar system. It is possible to put many of the laws that govern planetary motion into a computer and let the computer predict where the planets will be hundreds of millions of years in the future. The equations are not simple; they start with Kepler's laws, but they include corrections for general relativity. In the time frame of a human lifetime, or even all of history, the planets revolve placidly, closely following what Kepler would have predicted. Over a longer span of time it turns out that the solar system exhibits chaotic behavior: if its initial conditions were slightly different—for example, if a planet moved ever so slightly out of its present location—the solar system would end in a very different configuration. Figure 18 is the result of comparing two predictions for Jupiter. One image shows Jupiter as it actually is, and another shows Jupiter moved 7.5 millimeters to one side. The two futures diverge exponentially for about 20 million years and then level off after 250 million years. This is an "unsettling" result by mathematical standards, but it is also beyond intuitive understanding.[23] What could it possibly mean, in terms of human intuition, to move Jupiter such a tiny amount and then follow its changes for so long? By neglecting such effects of time in this book, I am not implying that they are less interesting—only that they are less connected to twentieth-century images.)

Those who try to push their intuition, like the successive editors of *Powers of Ten*, are hoping to control the unrepresentable by the powers of human intuition. Those who disparage or ignore intuition need (or want) to work in the inhuman domain of objects that can never be comprehended. The physicist Paul Dirac is known for this attitude: "one can tinker with one's physical or philosophical ideas to adapt them to fit the mathematics," he said, "but the mathematics cannot be tinkered with. It

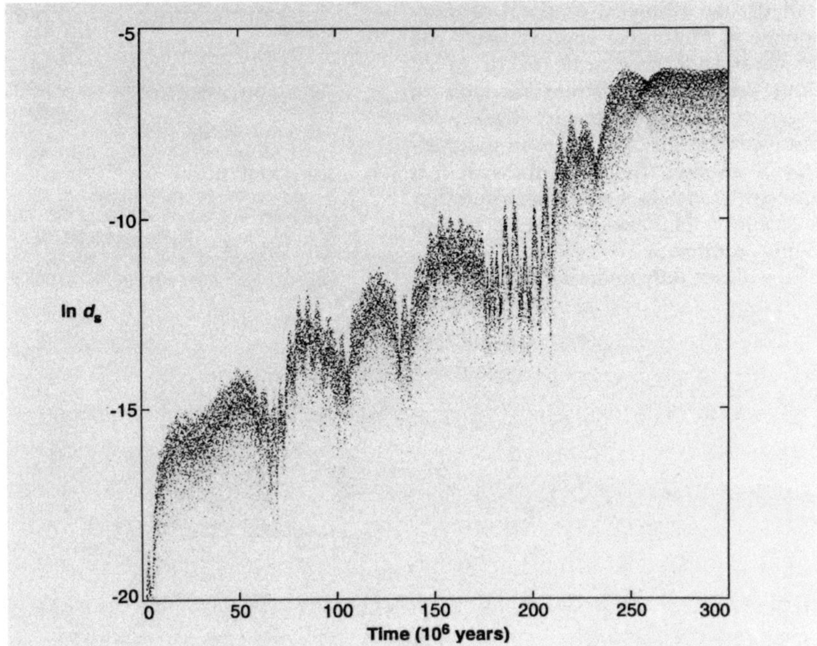

FIGURE 18

The log of the secular phase space divergence of the trajectories of Jupiter in the twenty-eight-day Stormer integrations, with one initial point perturbed by 7.5 mm. From Gerald Sussman and Jack Wisdom, "Chaotic Evolution of the Solar System," *Science* 257 (July 3, 1992): 56–62, fig. 5, p. 61. Courtesy Gerald Sussman.

is subject to completely rigid rules and is harshly restricted by strict logic."[24] Mathematics and logic become the blind man's guide to the regions beyond intuitive understanding.

20   Three Tiny Galaxies

It is fascinating to negotiate these failures of intuition, watching as whatever is counted as unrepresentable begins to appear around the edges of comprehension. Let me take as an example three minuscule images of galaxies.

The first images are on the small glass plates that were used to record the redshifts of galaxies, before digital technology. The earliest are kept at Lowell Observatory in Flagstaff, Arizona. They are unexpectedly tiny— only two inches by three inches—given their enormous subjects. When I saw them I was amazed by how fragile they looked: the images are faint and monochrome grey, easily lost in the reflection of a table lamp. It's odd to consider how such small images could signify the vast spaces they opened to astronomy.

The second example of a tiny galaxy image is even smaller. The artist and geneticist Joe Davis—to my mind the most interesting living artist, but that's a subject for another book—has been working on a way to put an image of a galaxy into a mouse, where it would become part of the mouse's genetic code and be passed on to future generations.[25] Davis began with an infrared picture of the Milky Way galaxy (Color plate 10).[26] He displayed it in digital form, using hexadecimal notation, so that it became, in part:

Joe Davis's hexadecimal printout of the COBE satellite image of the Milky Way, detail. Courtesy Joe Davis.

```
02DE0000000000070050001102FF0C00FFFFFFFF0000000000000000005
0000000070000000000000001E0001000A00000000000070050820000000222
0000000100000000000000000000000000000000100000000000000000000
0000000400000000000000000000000000000000040000000000007005000
00030000000000000000005672707A61000000000000000000000100016170706
C000000000000003000050000700480000004800000000000187000105566964
656F0000000000000000000000000000000000000000000000000000000000010F
FFFE10001872507842100C0505010838401400056A728C384000000ABE
B3D4684000000AAAB4D648800000056FF04000400040004000040004000
4210421310439451A0A45EA6ACE66496A8A73326ACE84010000AAFF6
ACD88210005AAFF6F1094A4015AAFFF7775B9C810BBFFBF3E0B3A0
B2D89294873766B335A8D4E2B7B97775372ED6668772F773177546ECB
0C62042104010400318720E51CC4106362275A28520949656A686ACD6
F1066AD5A4A88210051BFFF5E4A88000010FAFF562984000000AAFE41
6584000000FEFF41A984000004AEFA394584000044FEBE188384000040E
AEA0C8384000000B58604218401200D6060108384007102040408418400
EA0100000C428400AF0000001CA48400BE01000035A98400AB0000003
E0C8421FF5500004A4D8821FF5500005AAE8C83BF5B01017398A148FF
AA10106B5490A4FEA900003DC98401FA40000018A48400FA04000018A
48400EB00000018C58400E900000008428000FA00000008418400E900000
004218000EE0000001083840000380000A0040000000700AE00780007000
100010003000300040000000D000C00100050005000BE00BE0028FFFE00
1010517569636B54696D65AA20616E6420610000280003000E125669646
56F206465636F6D70726573736F7200002800080000210D20617265206E6
56564656420746F20736565207468697320706963747572650D000000FF
```

He then devised a series of codes that change the sixteen decimals of hexadecimal notation into DNA base pairs. The challenge was to create a code that would accurately transcribe the image and still be viable as DNA in a living cell. His final "supercode," which I will not describe in detail (it reads amino acids from a base-20 code) yields this sequence of thymine, guanine, adenine, and cytosine DNA bases:

T**GGATCC**CCC**GAAGA**CCCTGACCCAGTGTAAAAAGGATCACC
CTTTTGAGCCCCTTCCCCCCCTCTCCCAGGGGCTGACTGAGC
CCCCGTA**GACGTC**TAATATGGGTAGTTAAGGCTGTCCCTCCT
TTTGTCCCTTTGGGGGCAGACCCTTTGGGAAAACACCCTTTG
TGCCCAGCAAATA**GACCGGT**TAAGGCTCACCCTTTTGAGCCC
CTTCCACCTAAAAATATCCCTGAACACTAAAAATAATATCCCT
TTGCCGCACCCTTTTA**GGGCCC**TAATCGTGTCCCTTTGCAAC
ACCCTTTGGTCCACCCTTTTGAGCCCCTTTA**GTGCAC**TAAAA
CCCCGGGG**GTATCC**CCTGATTT**ATGCATGC**TAGGTGACTGTA
AAAATAGTGAATACTTAAGGTTGGCCCTTTACGCCCTGATTAC
TTGCCTGCCTATAAGGGGGCAAGCCCGGGTAG**CCTAGG**TAAA
AGCCCTGATTTAAAAGCCCTTTTGAGCCCCTCATAAAATCCCT
GATCATAAGGACTGCCCTGATACTGTAAACTCCCT**AGATCT**T
GATCCTTTTTATTAAAGCAAATTTT**GATATC**TATTTATAATCT
GGGTAACCAAGACCCTGATCCTAAAGAGGGTGAACTCCCCCC
TAG**TCTAGA**TAGTGACTACTGTGACATTCCTGTAAAAATGACT
CCACTCCCCGCCCTTCCTTCCCTCCACCGACTCCCCTTCCCC
CCCTTTAG**TCATGA**TAGGGTAATTTAGCAAATAGT**GAATGC**T
AGGGTGAAACGCCGACTTAAAATCCCTGAAAAGGAAGCGGTT
CTTAAAGCAAACCCTAG**CCGCGG**TGATTAAAATCCCACGAAA
GGGTGATCGTTATCACATAAAATCCCGCATTTTGAAGGGGCCT
TAAACGCCCTTTACACCCTTTTA**GTCGAC**TAAACTCCCTTTAC
GCCCTTTACCCCCTTTACGCCCTTTTGACCACTCCTCCACTC
GCTCCTCCGATTCTTTCAACCAATCTTGTA**GCTTAGC**TAAAA
ATAGTGAAATTAAGCGAAAGGGTGACGATATATCATTT**AGTAC**
T**TAAGCGAAATAGTGAA**CAATG**A**GCGCGC**TGAATTAG**CGGC**
CGTAAGCCAAAGGGTGACGAACTCCCCTTAAAAGCCCTGAAA
AAGGGGTTAAGCGAAATAGTGAGCGTACACCACTCCCCCCTT
TA**GGTCTC**TAAGCGAAATAGTGAAGGGGTATAATCTGGGTGA
CTCCATTCAATCCCCTTTTAAGAGAAAGTCGGGTTTTGAGTGT
GTTTA**GGCGCC**TAAAAAGGGTGAATTAAGGGTGACACCTCCA
GAGGGGGAGGGCGGATAG**GGATC**TGACCAGCGAACCAGCAGT
ACATCAATT**CACTGCGTG**TGATATTAAAGCAAAGGGTG**ACGC**
GTTTAAGCCAAACCCTGAGTTCGACAAGTTAAGGGAAAGGGT
GAATTGTGTGTTC**GTGCAGA**GTTATATATAAAAATAGTGACTG
TTGAGCAGGTGTGCGCTTGTGTTTCTTAAAAATAGTGAGA**GCA**
GCCGCTACACCTCCACTCCTCCCCTCCTCCCCCGCTACTTA
AGGGTGACACCGATTCTGCGCTCCTCC**TACGTA**C**ACATGT**TA
TAGTAAAAATAGTGACAACTTCACCATTCATTATTTTAAGCGA
AATAGTGATATAGTGAACTATAAAGCAAATAGTGAGCGTTTA

Joe Davis's base-20 supercode, en-
coding the hexadecimal code into
DNA base pairs, detail. Courtesy
Joe Davis.

G**CCAAGG**TAAAAATAGTGAGGCTCCTATATAAAGCAAATAGTG
AGTTTATAAAAATAGTGATCAAACACCACTCCCCTTGATCTTA
AAAATAG**TCCGGA**TGAGGGGGGGGTTGTAAAAATAGTGATCAA
ACTAAAAAGATCCCTGATCCGTGAGAAGGGGTTTACAATACTT
AAAATCCCTAG**CCTCAGC**TGAAAAAGGGTGAATCCTTGATATT
ACTTAAAATCCCTGAGGGAGGGGTCCTATAAAGCAAATAGTGA
TACTTAACAACCCTAGTA**GGCCGGCC**TGATCAAGAGGAACGA
TTCTTACTTAACGACCCTTTTGACTCGGGAAGGACTACACCGA
CTTAACGACCCTGATCCCGAAAGAAATAG**GCTAGC**TGACCGC
ACCGACTTAAAATCCCTGAAGTTACTACCTCCACTACTCCCT
CACCCCGTTACCTACCCTCCACCGACTCCCCCTTAG**AAGCTT**
T**GAGCTC**CCACCTCCCTCCCACTCCTACTCCCCCGAAACCCT
TAAGATCCCGGGTGACTCCAACTCCCCCAATAATCAGGGTAG
CGGTCCGTAAAATCCCTTTTGAGCAATCACTCCCCCAGGATA
AGCTCCCTTTCGACCCTGAGTTATAATCGAAATAGTGATACTC
CCCCAAAGTAG**ATCGAT**TAAAATCCCTGAGGACCGCACTCC**A**
CTGGGGTTTTTAAAAGCCCTGATCAATCGTACACC**ACTGGG**G
TTTTTAAAAGCCCTAG**CTCTTC**TGATTATAAAAATAGTGAAAG
AACGCACCGAGGGTTTAAAAATAGTGAGCCCTCCCTTGCGATA
CAACTTCACGGGGAAAACTCCCTAG**CTCGTG**TGATCCTATAG
CATATGTAAAAATAGTGAGTTTCATTAAAGTCCCTGAAATCGG
GAAAATTAAAGACCCTGAGGTGCATACTCCTA**GACTAGGTC**T
GACCTGGAATTAAATACCCTGATACAATCACTCCCCCGGAACC
TTAAGATCCCTGATACAATCTGAACTCCCCCGAAGTAG**CCAT**
AGAATGGTAAAATCCCTGATACGCTTACTCCTAATAGTGACC
CGAATTAATTCCCCTGAACTCCAACCCCCCCGGAATAATTCCC
CTAG**TGGCCA**TGAACTCCTACTCCCCCGAATTAATTCCCCTG
ATCCACTACCCCCCCGAGATAAAATCCCTGATCCACCGACTTA
AGATCCCTAG**CTCGAG**TGAGATAAAGACCCTGAAACCCCTTA
ATATCCCTAATTTTGAGCCCCAAGACCCCTGTAAAAAACGCCC
TGATTGAGTAAACGCCCTAG**ACTAGT**TAATTTACGCCCTTTA
CGCCC**GGGAC**ACCCGGGTAAGTCCCCTTTTACCCCTGAGTTA
AGTTCCCGGGGTTCCCTAATTTGTTCCCTAG**CAATTG**TAATC
GTTTGTACCCTCGTTTGTACCCTGAAGGACCCCAGGACCCCC
ATAAAAACCCACAGGGTGAACCCCCTCCCTCCTTTAG**ACGCG**
TTGACTTGTTTATAAAAATAGTGATTATAACCCTGAGTATAAA
AATAGTGAGTTTGATCTATAAAAATCGTTTTAAAAATAG**AATA**
TTTGAGTTATTTAAGAGAAATAGTGACACCTACTGATTTAAAA
ATAGTGAGATTGAATCCTGAACCTACTTAAAGACCCTGAAATA
G**CACGTG**TAAAAGCCCGGGGGTTCCCGGGTGAACTCATTTATA

TAAAAATAGTGATTATCTATTTTAAAAATAG**CCATGG**TAGGGT
AGTGAGGCACCTATCTATTTACGTTAAAAATAGTGAGGTTAAA
AATAGTGAGTTGC<u>**CTGCA**</u>TATTTGCGTGCGTTAAAAATAG**TG
CGCA**TAGTGAGGTGCATAACGACCCTGAAATAAACACCCAAA
GATCCCTGAACTCCGTCACCTACTTGCATT**GAATTC**ACCTTA
AAAATAG**ACCTGGT**TGAGATTGAATTTATTTATCTATTTATCC
ACCTGTCTTAAAAATAGTGAGGCACCTGATGCGTATTTATTTG
ACACCTGTCTTAG**ACCGGT**TAAAGCAAATAGTGACTATAAAAA
TAGTGATTGCTAAGGGTGACACCTAG**CGTACG**TGATGCCTAT
AAAAATAGTGATTACTGAGTGTCTGTTTAG<u>**TCGCGA**</u>TAGGTGA
TTGAGCATATTCCGTTAAATACCCGAAGGGTAGG**GTCTTC**GG
<u>**GCTGCA**</u>G

This supercode is a biologically viable and stable form of the digitized image of the Milky Way. Once it is sequenced and inserted into a mouse's ear, it can be passed on from generation to generation. It could conceivably even be inherited by species descended from the mice, and barring significant random mutations Davis expects it to remain stable for "geological time."[27] He thinks of his project as a variation on the "pictures" that animals and plants already have encoded in their genetic material, and he says his DNA-encoded galaxy picture may be "redundant," because "mice and other living organisms already inherently possess subtle 'maps' of the local cosmos" in the form of responses to diurnal and annual rhythms.[28] There are "pictures" hidden in many microscopic forms, and—who knows?—perhaps mice even have an awareness of the Milky Way.[29]

The third tiny galaxy image is the best known of the three: it is of the galaxy "in Orion's belt" in the movie *Men in Black*. In the film, Orion is a cat, and he wears a collar with a small, transparent plastic ball. In the little ball there is an entire galaxy that preserves any number of cultures and living beings. In one scene a woman, petting the cat, comes to realize that the bauble contains an actual galaxy, slowly rotating before her eyes—a near-infinity in nearly nothing.

These three small galaxy images capture an enormous magnitude in a small compass. All three are negotiations of the end of representation: in each, something that can only be apprehended through calculation is abruptly brought down to the size of intuition. Orion's galaxy is the simplest and the most simpleminded: it gets its effect by short-circuiting the realities of both science and representation. The Lowell Observatory plates do not possess the impossible level of detail of Orion's belt, because they are bound by their medium (so are the reproductions of galaxies in *Powers of Ten*). Davis's work is remarkably complex, and in describing it I have condensed a hundred-odd pages of calculations into a couple of sentences. It is especially hard to see, in Davis's project, where intuition ends and calculation begins. What exactly is accessible to intuition? Certainly not the Milky Way itself; and not the DNA or its unforeseeable multiplications in the mouse's innumerable future descendents. Davis finds himself putting the words *map* and *picture* in quotation marks; and indeed, what kind of "picture" is he making—something shrunken, supercoded, distorted, and metamorphosed into a molecule, and then multiplied and preserved in the cells of a mouse's ear? As always, what is interesting is the unexpected threshold between what can be intuitively understood and what cannot. In Davis's work, the threshold is mobile and multiple: a signal of what would, in visual art, be a postmodern sensibility. The movie, the book, and the Lowell Observatory plates are simpler, but they all operate where intuition meets calculation.

21   *The Deep Fields*

If the history of late-twentieth-century images were not biased toward fine art, one of the central works of the period would be the collection of photographs jointly known as the Hubble Deep Field. The Hubble Deep Field is a picture—or rather, an aggregate of pictures—of a tiny patch of sky near the Big Dipper. The idea was to point the Hubble Space telescope at a part of the sky far from any bright star and see what it would record if its cameras were left open for a maximal amount of time. For over 100 hours the telescope peered into space, and it brought back an astonishing image, full of galaxies fainter than any that had ever been seen (Color plate 11).[30]

The Hubble Deep Field was made in December 1995. In October 1998, the same procedure was followed for a small area of sky in the Southern Hemisphere, producing the Deep Field South.[31] Again, an apparently blank stretch of sky yielded thousands of faint galaxies. The deep fields are extremely small parts of the sky, almost at the limit of what I can imagine. Deep Field South was centered on the constellation

Tucana, the Toucan (Color plate 12, top, in the box). This view corresponds, more or less, to what you might see if you looked up at a cloudless night sky from a place south of the equator. Color plate 12 (bottom) is a close-up of Tucana, with a small box showing the general area of the HDF-S. Within that box the Hubble focused on three tiny rectilinear areas, too small to be reproduced this illustration. One was shaped like a stairway, and the other two were square. Together they were the parts of the sky imaged by the HDF-S. The minuscule square area in the dead center of the box in Color plate 12 (bottom) yielded the picture shown as Figure 19.[32] Each plate found thousands of galaxies where none had been known before.

The inhabitants of this distant realm are decidedly strange. Brighter objects, like nearby galaxies with magnitudes of 19 or 20, tend to be elliptical, spiral, or barred: that is, they have predictable, orderly shapes. After magnitude 25, almost one-third of the objects are "peculiar."[33] There are "head–tail galaxies resembling tadpoles" and all kinds of asymmetric objects like the one in Color plate 13, which has a "red nucleus" ringed by "many blue knots."[34] This galaxy is visible in Color plate 11, toward the top center, where it appears among many other unclassifiable objects. There are only a few known "tadpole" galaxies in the nearer universe, but several dozen can be seen in Figure 19 and Color plate 13.[35] At first, the astronomers were perplexed; they weren't sure if these objects are proto-galaxies or older galaxies being torn by collisions and tidal forces.

The two Deep Fields were an astonishing achievement. In terms of the information they yield, and entirely aside from their amazing subject matter, they are among the most complex images made in any century. Dozens of scientists continue to analyze the data in the images, which continues to help sort cosmological models.[36] The faintest galaxies remain elusive—hard to analyze, and even hard to count.[37]

22   At the End of the Visible Universe

The HDF-N's most wonderful visual act was the attempt to see *beyond* the faintest galaxies on the plate—to see something in the black regions between the faint bright spots, at the very end of the visible universe. It seemed as if the Deep Field images reached up to a region named "the wall of dust extinction," where visibility drops off.[38] What if it were possible

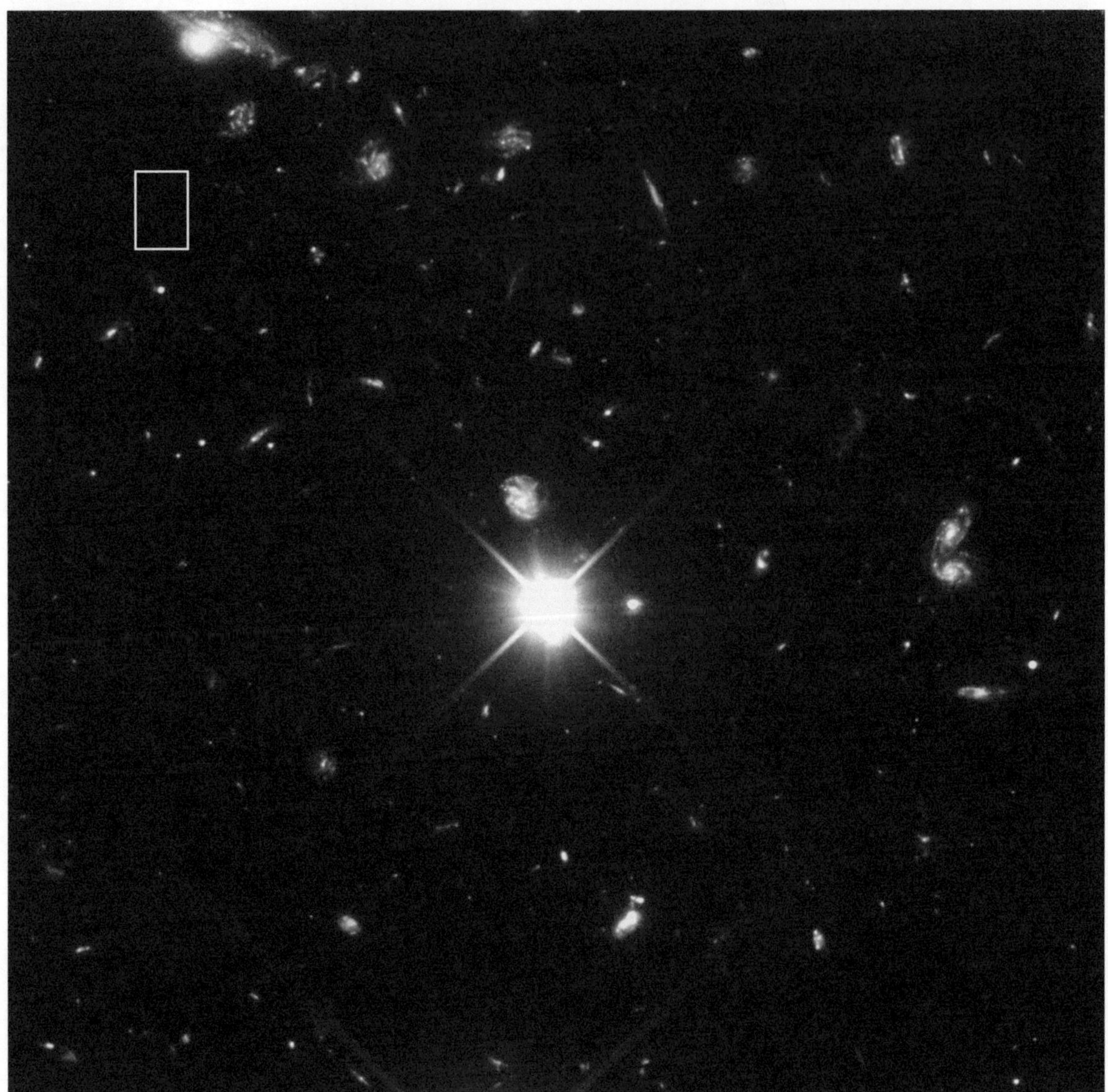

One of the HDF-S images, taken with the Hubble telescope's STIS camera. Courtesy NASA.

to discern something there, among the dark pixels? The telescope was at its limit, but if a better telescope were to be built (and one was already being planned), would it see farther? Would the black voids resolve into little lights and fill with even more distant galaxies? Did the visible universe continue, or was the HDF–N a view of the very end of the visible universe?

It was clear that the space between the faintest lights in the HDF–N is not a perfect void, because it is not a uniform painter's black. Rather, it is a set of pixels with slightly different colors and luminosities. If you enlarge part of the HDF-N far enough and avoid anything that looks bright, you find yourself looking at a shifting ocean of pixels: blue–black, black, green–black, red–black—there is no uniform darkness between the farthest galaxies. Could the faint glitter be resolved into farther galaxies? Astronomers had tried to answer that question before the advent of the Deep Field images by looking at the "black" background between stars, which was always comprised of undetected faint galaxies blurred into their surroundings. In 1974, Stephen Schectman studied photographic plates covering large portions of the sky and found evidence that there were entire clusters of galaxies hidden in the apparent flat, black fields.[39] In 1988, another study found evidence that an apparently perfectly black background had "residual surface brightness" that might indicate unseen galaxy populations beyond the pictures' limits.[40] The Hubble telescope then verified that there *were* galaxies beyond what had been visible before. The question was whether the HDF-N also hinted at things to come, or whether its background really is a formless void.

In a 1997 paper, Michael Vogeley of the Princeton University Observatory called the imperfect blackness Extragalactic Background Light (EBL), and he studied it by subtracting every possible light that might come to earth from nearer than the limits of the visible universe.[41] To do that, he first deleted all the visible galaxies in the HDF-N, erasing the areas around each one that had been used to estimate the galaxies' magnitudes. Those areas were already twice as large as the galaxies' apparent sizes. Then, for good measure, he deleted additional ellipses twice as large, so that he erased pixels far out into the surrounding darkness. In all, he subtracted fully 30 percent of the pixels in the image. Then he defined an "angular autocorrelation function," essentially comparing discrete areas with areas next to them to see how the brightness fluctuates. Taking $\mu(\mathbf{x})$ as surface brightness, the autocorrelation function is

$$C(\Theta) = \langle \mu(\mathbf{x})\mu(\mathbf{x} + \theta)\rangle - \overline{\mu}^2$$

where $\overline{\mu} = \langle \mu(\mathbf{x})\rangle$ is the mean surface brightness of the sky, and Θ is an angular measure of distance on the plate. If there are objects that lie "below the detection limits of deep optical imaging," Vogeley writes, they "might be inferred" from their statistical signatures in the "object-masked 'sky.'"[42]

He concluded that if the HDF-N is a fair sample of the "sky" between galaxies, then it includes the majority of all galaxies. Fainter galaxies could well be hidden in the darkness, but not many. The only things that could be beyond the faintest galaxies in the HDF-N would have "extremely low surface brightness" and would be large and diffuse. At the resolution of the HDF-N, those elusive objects are "confusion-limited": their images overlap with one another, making it difficult to tease them apart. If the resolution of the picture could be improved, then those evanescent objects might be detectable. Paradoxically, Vogeley pointed out that telescopes on the ground have a better chance of searching for very large, faint objects because their cameras see more of the sky. Objects "much larger than several arcseconds" would be easier to see with other kinds of telescopes.[43] The Hubble looked right to the end of the universe of galaxies. Beyond it there could be large, faint lights, but no more galaxies.

Still, it is hard to quell the thought that shapes are latent in that darkness. The "skies" of images like the HDF-N are anything but placid, painted backdrops. They shimmer and boil with colors: they seem alive, even moving. There is nothing, to me, that is more uncanny than these "summery" scenes from the very end of the visible universe, especially when the image is in color, as it is in Color plate 14. (Color plate 14 is a detail of the upper-right portion of Color plate 11.) The image looks for all the world like a deep azure sky, dark like the sky is near the zenith. It even looks as if the "sky" has scudding clouds. The color is false but plausible because the redshifted wavelengths recorded by the telescope are color-corrected and stacked back into the familiar blue, green, and red channels used in computer-assisted display. Scientists sometimes call these images "true-color" (in quotation marks) or "pseudo-color."[44]

One of the deepest views into space is the ancillary photograph reproduced in Figure 19. If you look closely at it—say at the upper-left corner—you see the same patchy clouds, creating the same weird sense that this is really a summertime sky with wisps of cumulus and cirrus. Figure 20 enlarges part of the upper-left corner of Figure 19, revealing a "nocturnal sky," the way it looks lit by a full moon. In the black areas between the "clouds," the "sky" looks quiet; Figure 21 enlarges the area in the box. But with the contrast and brightness turned up, the pattern looks anything but random (Figure 22). Black, wormlike forms meander around small bright spots, and it seems there must be much more waiting to be seen.[45]

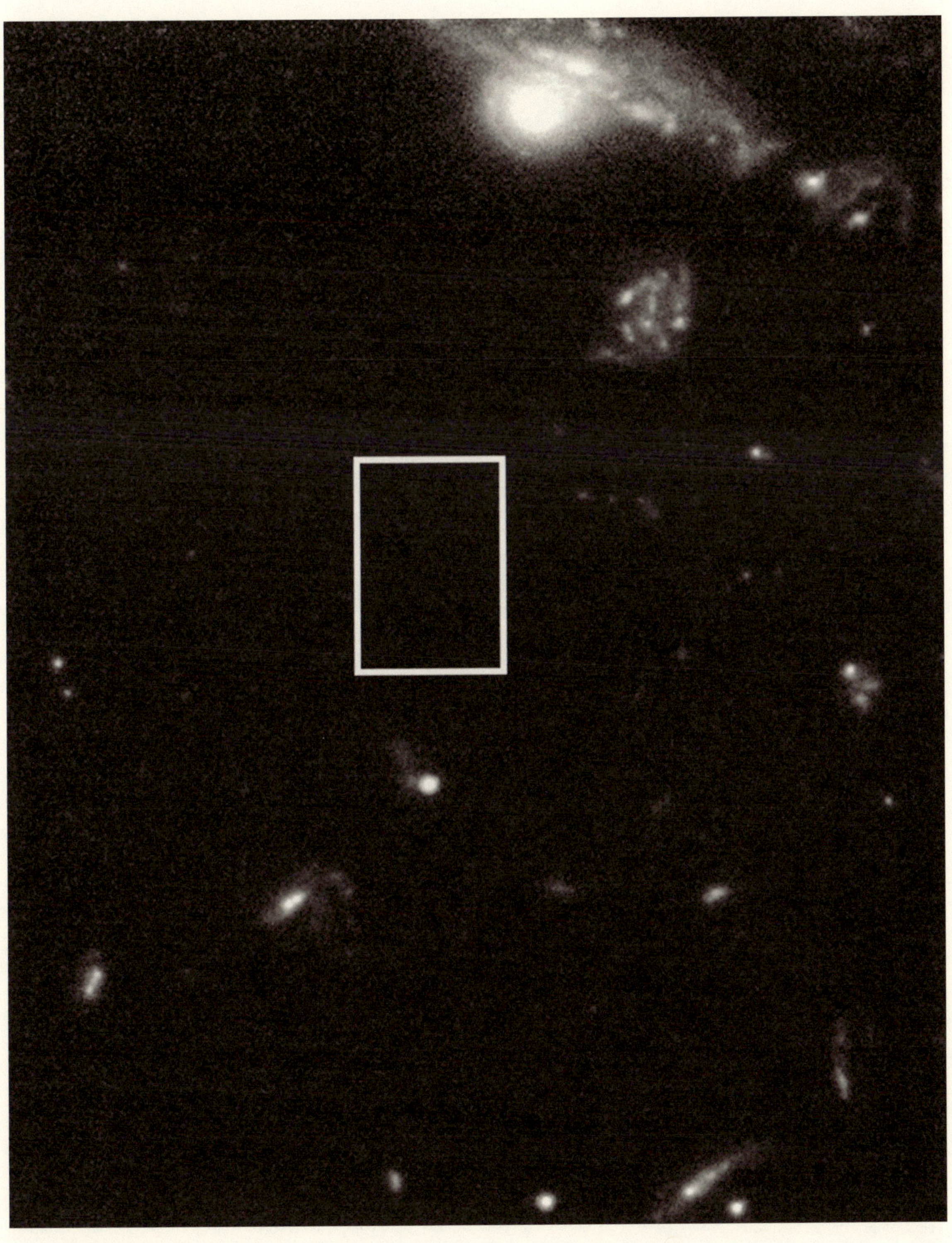

There have been several attempts to see into that hazy darkness. Among them is the paper "Diffuse Dark and Bright Objects in the Hubble Deep Field," written in late 1997 by Changbom Park and Juhan Kim, both in the Department of Astronomy in the National University, Seoul.[46] They found "tiny spots" of light "embedded in extended backgrounds." The shapes they claim to see are highly variable—not at all like the verified galaxies; Park and Kim call them "galaxies in the process of active star formation and merging," "primordial galaxies," or "proto-galactic objects." In addition to such objects, they also found smooth, dark shapes, where the "sky" is darker than in surrounding

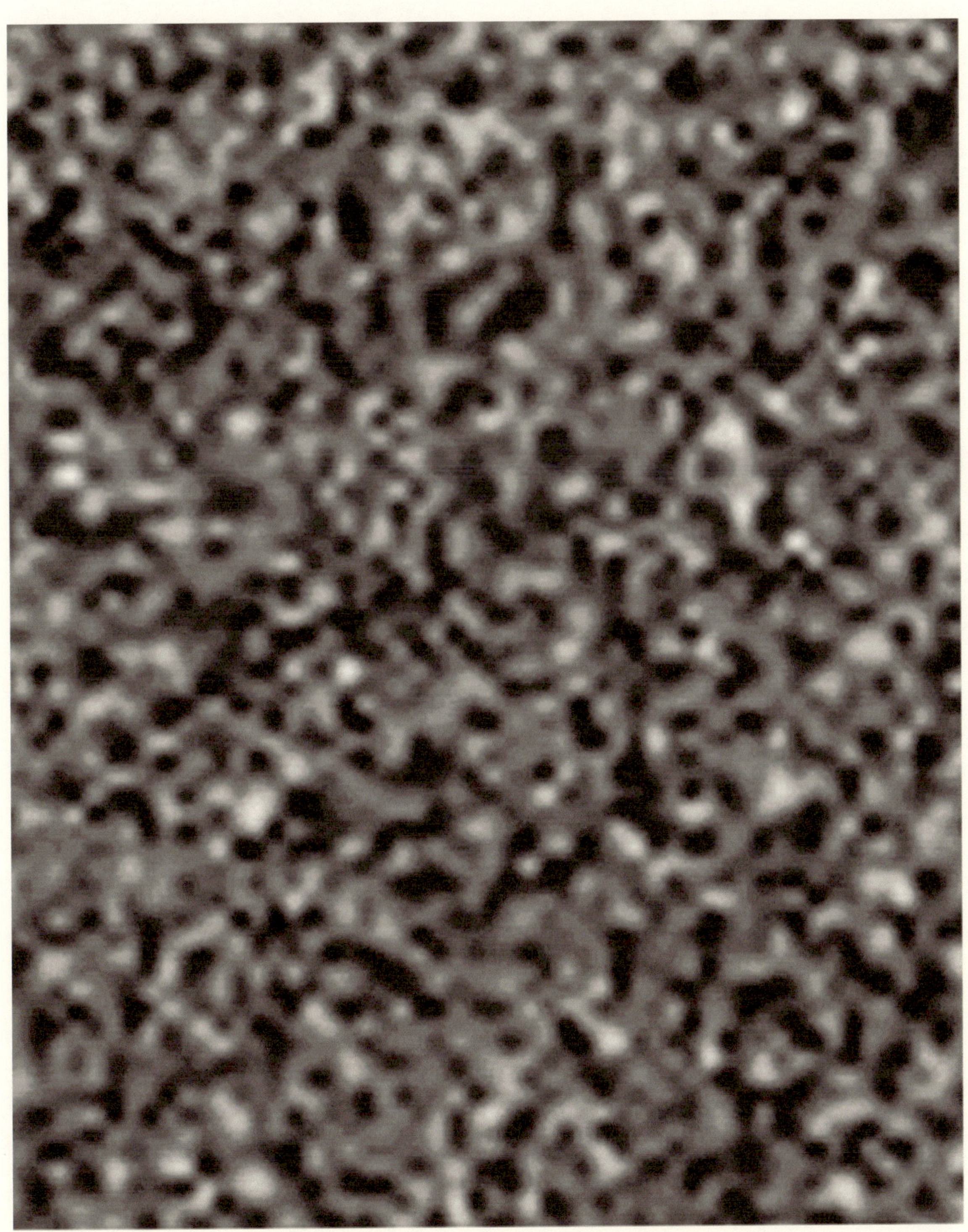

areas; they interpret those as "dark clouds" between the "proto-galactic objects."[47]

They found the objects by masking the sky, as Vogeley had done, and then smoothing the pixels and seeing what emerged. Smoothing routines work by smearing pixels over a certain area, and Park and Kim chose a twenty-pixel-wide blur, which corresponds to 0.8 seconds of arc. That way they "avoid[ed] picking up objects that belong to a single galaxy," and also increased the signal-to-noise ratio. Their initial survey found 5,359 objects, which they narrowed by looking only at those that show high redshift. They calculate that at $z = 3.6$, these objects have an absolute magnitude (M_V) of $-14 \sim -16$, much fainter than the most distant galaxies.[48]

These dark clouds and proto-galaxies are so faint and so entangled in the background light that they cannot be seen by simply looking at the images. Park and Kim provide an amazing illustration that purports to show the objects, though to my eye it shows nothing (Figure 23). Each row is a single object, seen in three different wavelengths. The pictures are difference maps, showing the appearance of the objects when they are smoothed over 4 pixels (0.16 arcseconds) and 100 pixels (4.0 arcseconds).

The graphs along the right side plot the surface brightness of each object across the field, measured in arcseconds. The top three objects are considered bright, and the bottom three are considered dark (notice the negative surface brightness, −SB). The top three rows are Park and Kim's eighth-, thirteenth-, and eighteenth-brightest objects, each of which is said to be about one-quarter the width of the image. But are they really there? I do not have trouble believing that a mathematical analysis can discover objects that my eye cannot, but the authors also offer a phenomenological description. "Most of these objects," they say, "are sprinkled with many noise-like glares and show extended backgrounds. Some of them are highly elongated with emission of connections and seem to be undergoing merging." It is clear that the objects—if they are perceptible at all—do not have bright cores, and that they are "diverse," irregular, and multicolored (because their pictures in the three wavelengths don't match up).[49] Certainly, if these are discrete objects, then they are irregular and multicolored; but nothing else about them, I think, is visible without mathematical help.

Inevitably, future images will make these clouds less hazy. The next generation of instruments, including the Next Generation Space Tele-

Six bright objects (rows 1–3) and six dark objects (rows 4–6) in the HDF-S. The first three columns show the objects at different wavelengths. Courtesy Changbom Park and Juhan Kim.

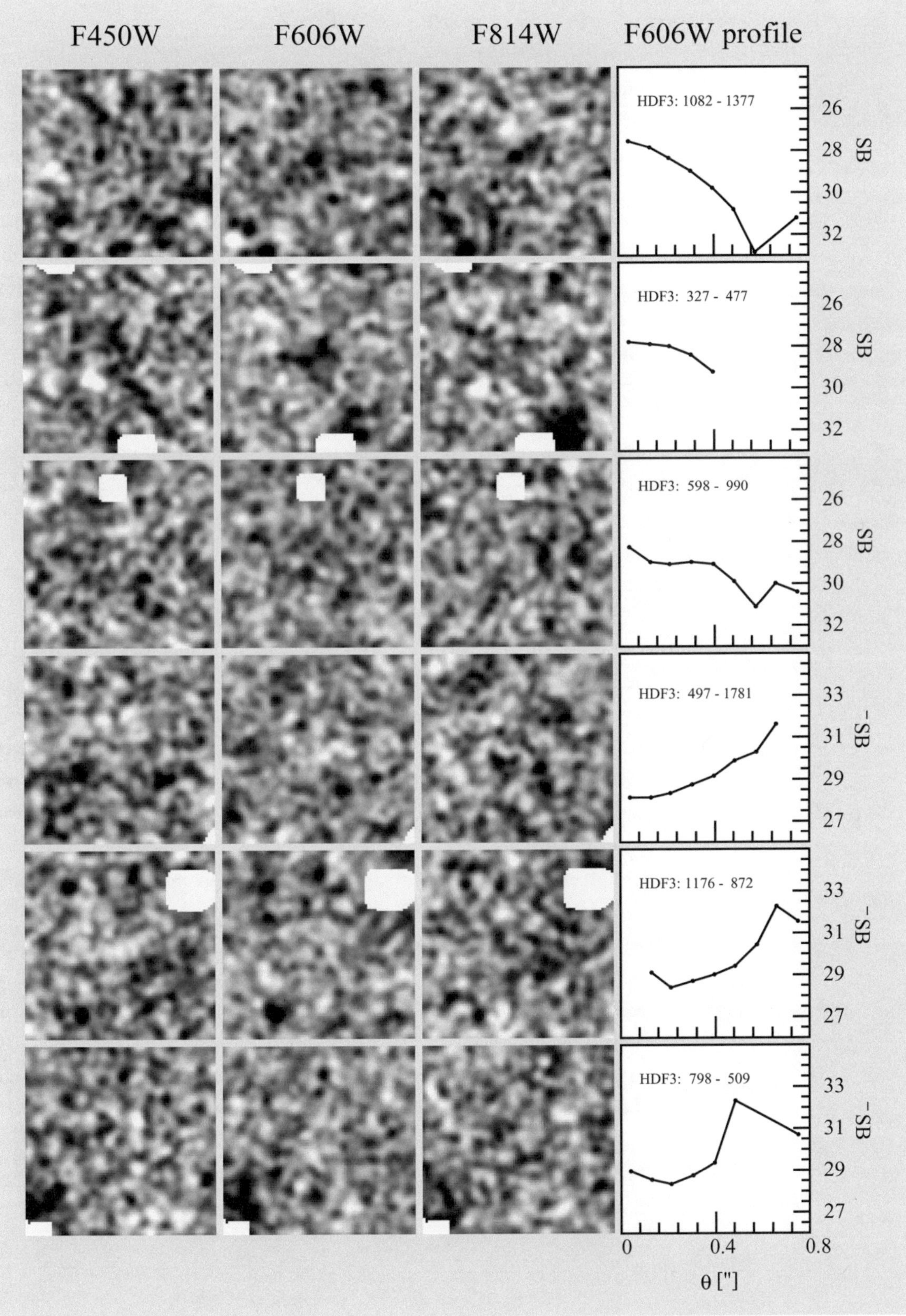

F450W
F606W
F814W
F606W profile
HDF3: 1082 - 1377
HDF3: 327 - 477
HDF3: 598 - 990
HDF3: 497 - 1781
HDF3: 1176 - 872
HDF3: 798 - 509
SB
26
28
30
32
33
31
29
27
θ ["]
0
0.4
0.8

scope, is designed to see vague clouds, halos, "residual unresolved sources,"[50] and perhaps even "primordial fog particles" weighing up to 10^{26} kilograms.[51] If the analysis in papers like Park and Kim's is correct, then many such objects will remain "confusion-limited," blurred beyond repair, simply because they are inherently nondiscrete.[52] With the HDF-N, "we are now looking at the edge of the universe of galaxies," into a region that does not possess neatly bounded objects in neatly namable shapes, such as the ordinary round, elliptical, and barred galaxies. Even "irregular"—the catch-all final category of galaxy classification—does not describe these kinds of objects, because they are not discrete in the way that irregular galaxies are, separated from each other by many times their own lengths. They are, in fact, nearly formless, and they come in turn from a universe even older and even more without form—a universe just at the border of what the astrophysicist Piero Madau calls the "dark age" of the universe, before it ionized and lit up with stars.[53]

23   Uninteresting "Infinity"

It might seem as if this is the end of visible or representable objects. But the ill-defined "clouds" in the HDF images are not the farthest things that have been seen.[54] In other parts of the night sky there are quasars that are significantly more distant. Even though most of the starlight in the universe was formed in the epoch $1 \le z \le 2$ (z is the measure of redshift), objects are known from much greater distances. Several are known from $z > 5$, and records are being broken continuously. Those objects are exceptionally interesting for cosmology, but they are oddly uninteresting as images. Their unimaginable distances ensure that most are "pointlike"—that is, they cannot be resolved into shapes.

By coordinating radio telescopes from different parts of the world, astronomers have been able to make detailed images of a few of those very distant objects. In early 1999, astronomers using Very Long Baseline Interferometry published an image of what was then the most distant known radio-loud quasar, with a redshift of $z = 4.72$ (Figure 24, left). This quasar, the authors say, "may be the most luminous steady source in the Universe," but it appears "almost unresolved"—it looks slightly elongated, but so close to a perfect circle that it might as well be a point source.[55]

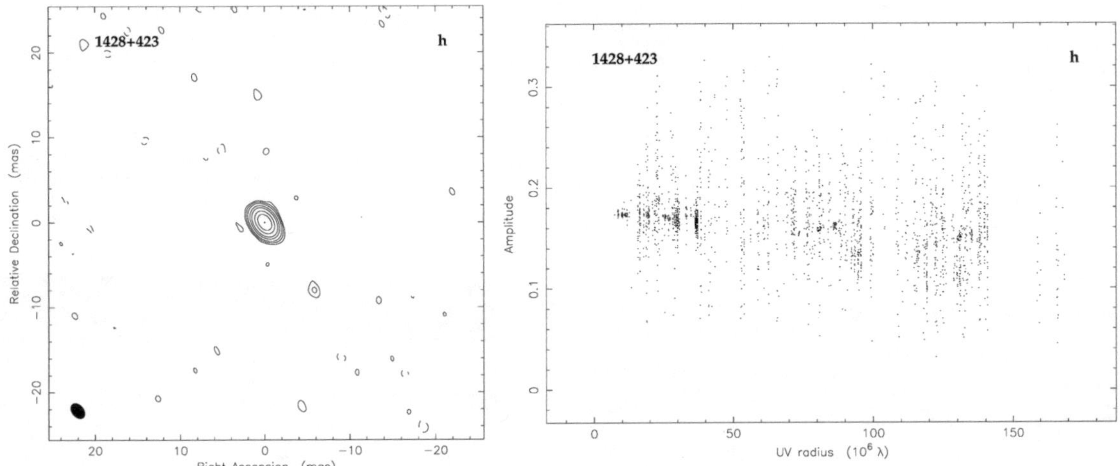

FIGURE 24

A radio-loud quasar at $z = 4.72$.
Left: Very Long Baseline Interferometry graphic. Right: baseline length plot. Courtesy Zsolt
Paragi.

It is a physical fact that the most distant object will be the least well seen and may make the least interesting, or at least the simplest, picture. And yet it is strangely compelling, this flat, colorless image of an object farther than almost anything else that can be seen. It is orders of magnitude beyond unaided vision, and it is nearly beyond optical wavelengths as well (Figure 24 is a radio image, mapped so it looks like an optical image). It is also beyond the capacity of any one seeing device (it took eight radio antennas scattered around the world to see this quasar). The right-hand side of Figure 24 is a plot of the information gathered from the eight radio telescopes. It compares the strength of the signal (the amplitude) with the "baseline length" (the distances between the antennas measured in wavelengths of radio emission). The slow descent of the plot, left to right, is a sign that the quasar is resolved—just barely. It also demonstrates how counterintuitive the process is that results in the image at the left. What could be farther from anything that might be called vision or ordinary image making?

In 1999, one of the most distant objects in the universe was a galaxy at $z = 6.68$, recorded by a team headed by the astronomer Hsiao-Wen Chen.[56] That is a record that has been broken several times over since then, but images of such objects are very self-consistent: the objects are both visible (in the sense that they are unambiguously recorded) and invisible (in the sense that there is almost nothing to see). The strip along the top of Figure 25 is an ordinary photograph (reproduced in

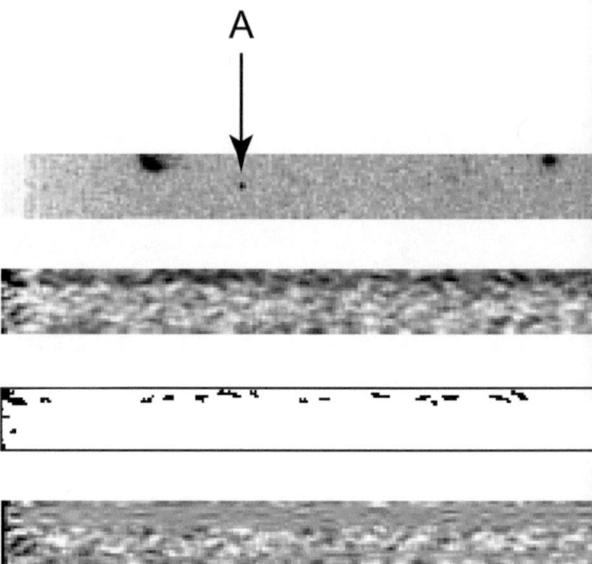

FIGURE 25

A galaxy at $z = 6.68$. Top row: optical plate; second row: open-slit spectrograph; third row: segmentation map, with the emission line of galaxy A indicated; bottom row: open-slit spectrograph, with emissions lines except galaxy A's removed. Courtesy Hsiao-Wen Chen.

negative) of the object, which is labeled A. It is two pixels wide and two pixels high. (B is another, closer galaxy.) What could be closer to invisibility than this set of four fragile pixels? They are the vanishing objects at the end of calculation, the last terms in the sequence of magnitudes.

The top row of Figure 25 is cropped in an elongated format to match the three images below it, which are actually spectra of the same region. The Hubble telescope sometimes makes images in "slitless spectroscopy," in which the spectra of the objects are smeared across the field, left to right. The second image has the spectra of all the objects in the top image, "dispersed" according to wavelength. (The image does not have to be in color because the horizontal position of a pixel corresponds to a range of wavelengths: colors can be read off a scale, like any other property.) The arrow in the second panel points to the principal emission line for galaxy A: it is also two pixels high, and the entire two-pixel strip to the left and right of the arrow also belongs to galaxy A's spectrum. Hsiao-Wen Chen and her colleagues then use an image analysis package to analyze the smoothed dispersion image into discrete emission lines. The third strip from the top of Figure 25 is the resulting "segmentation map." There galaxy A's emission line is picked out of the rest of its spectrum. Just below it and to the right is one of the emission

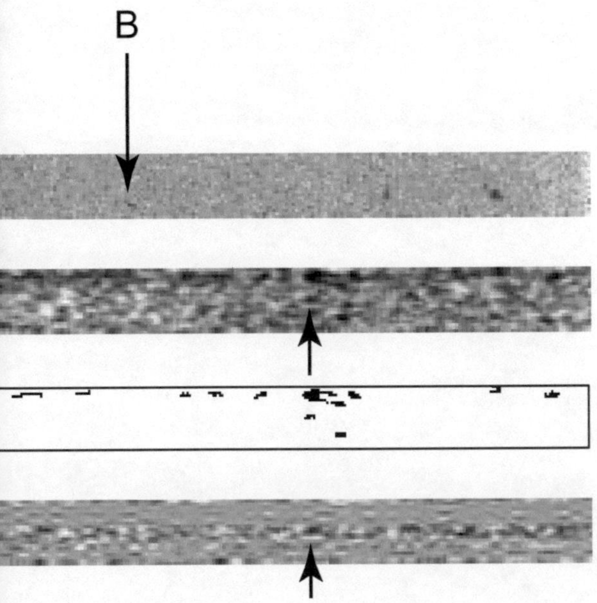

lines of galaxy *B*, and another is visible halfway across the segmentation map, at the same level. The brighter galaxies at the top of the original image appear as scattered marks along the top. The bottom strip is the same as the second, but with all the sources except galaxy *A* subtracted. The authors say that this result is "robust against image processing," meaning that their various procedures, including the segmentation program strangely named SExtractor, can be modified and still produce the same results. From these tiny bits of evidence, the authors produce a graph of a well-differentiated "continuum spectrum" (the area to the left of the emission line), a single strong emission, and a "statistically insignificant but suggestive 'dip'" toward the end of the spectrum.[57] Those three minimal pieces of evidence were sufficient to assign the galaxy its high redshift, and they made it—as of April 15, 1999—the most distant object in the universe.

It is fascinating, staring at these minimal images and wondering about the object they half conceal. The authors interpret the single emission line at $\lambda = 9334$ as Lyman alpha, abbreviated Lyα. If this were an ordinary object, near at hand, Lyα would be at $\lambda = 1216$—evidence of an enormous redshift, from the ultraviolet all the way past the visible spectrum and into the infrared, putting the object at an unimaginable distance in space and time. Even more remarkable in this context is the

Vertically taken aerial photograph of the northern shear margin of an Antarctic ice stream on the Ross Ice Shelf. From R. B. Alley and I. M. Willians, "Changes in the West Antarctic Ice Sheet," *Science* 254 (November 15, 1991): 959–63, fig. 3, p. 961, photo 12-116. Used with permission.

simple fact that the emission line is the signature of a particular atomic state: the spectrograms not only bring us one of the most distant objects known, but they have something to say about the *atoms* within that object. The "dip" at the end of the spectrum brings still more ethereal information: it tells something about the scattered atoms in space between the galaxy and us. The dimming of the spectrum is interpreted as light

absorbed by the "Lyαforest" of neutral hydrogen atoms in interstellar space.

How strange these representations become when they are at their limits. Their dullness is fascinating in its own right, as if transcendence were secretly boring. It seems that when the last and most elusive object is finally captured in an image, it turns out to be gray, dull, and nearly dead.[58]

24 *Vertigo*

Intuition gives out when magnitude passes a certain point. That moment can cause vertigo, as comprehension gives way to apprehension. A simple example is the scene in *Men in Black* when the woman stares transfixed at the galaxy in Orion's collar. Being in possession of a range of magnitudes can quell the vertigo: the tiny image of the distant galaxy can be disorienting, but it is shored up by the many measurements that determine its actual distance.

In other instances the vertigo is more acute. What, for example, is shown in Figure 26? It is not a picture of space—that much is clear from the way it twists out of focus at the top. It could be something extremely small—the shallow depth of field implies as much—but there is also a strange little row of scratches along the bottom, in perfect focus. The image includes a little scale, like a microscopist's or an archaeologist's scale, at the lower left. To me the image seems small, and if I had seen it out of context, I would have guessed that the scale measures a millimeter or so in length. Actually, it is very large. The image is an aerial view of part of the Ross Ice Shelf in Antarctica. The little scale is a kilometer long, and the little hook-shaped scratches along the bottom are crevasses. The picture was taken to show how a chaotic region of ice fractures formed between two smooth areas—two of the region's five ice streams.[59] So it is not a tiny, textured surface seen up close: it is a large, cold surface, far below the camera lens.[60]

A number of artists toy with uncertainties of scale. Pollock's all-over paintings have struck many viewers that way, and the same pertains to paintings by Rothko, early Stella, Martin, and Umberg. When I first considered these issues, I thought that vertigo might be a radical version of uncertainty over threshold, because it seems to produce an especially intense experience of the sublime. I also considered that vertigo might be a kind of disease of the sublime, because it brings on the experience too quickly, with too much force. But the image in Figure 26 is better taken as a final reason why the sublime is not appropriate for analyses of scientific images—and the reason is the little scale at the lower left. The images in this chapter, and in the following ones, are marked by an *incremental and measured* laying-down of intuition, not a sudden influx of unencompassable experience.

plates

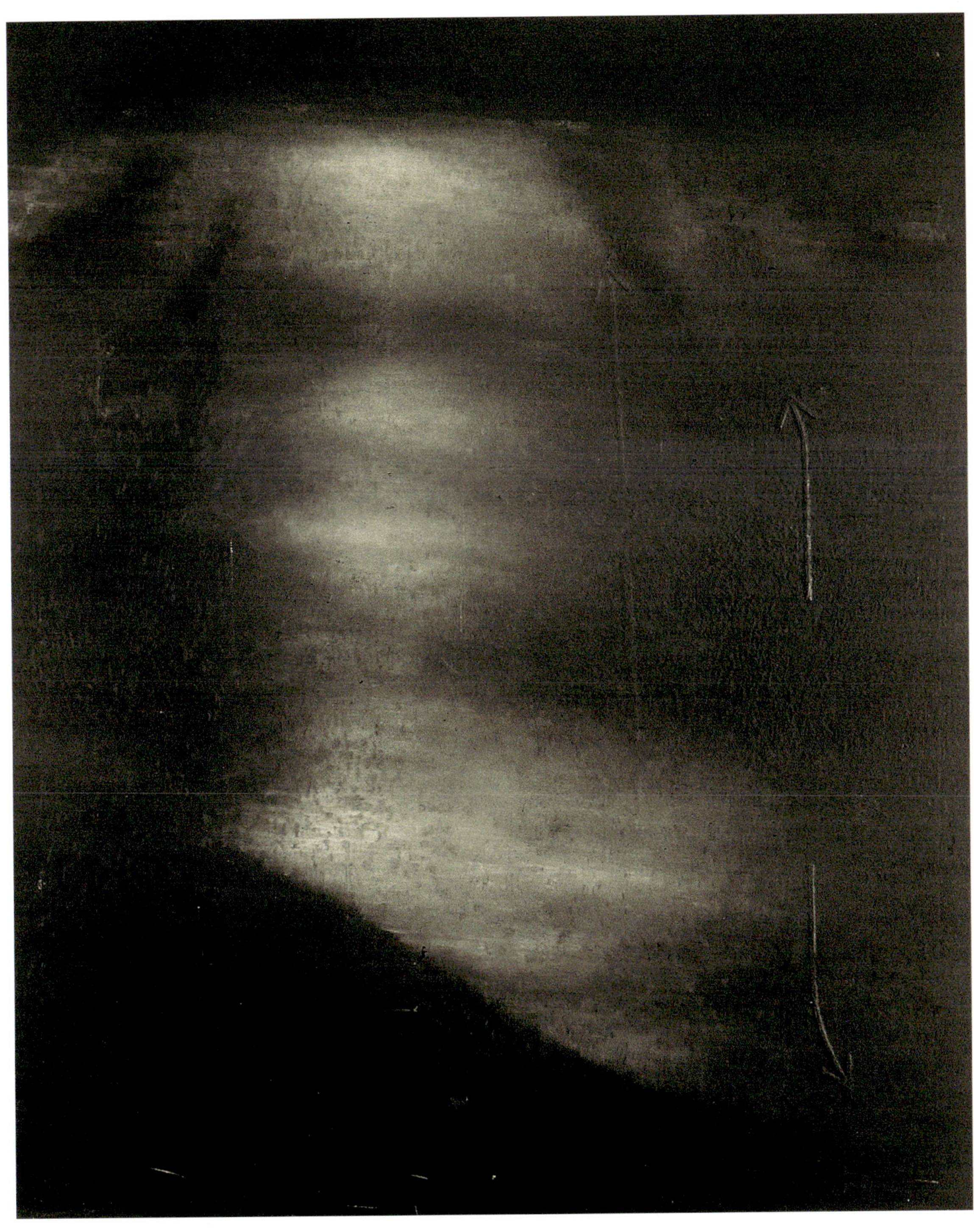

Ross Bleckner, *Flora and the Future*, 1983. Oil on linen. Private collection. Courtesy Mary Boone Gallery.

PLATE 2

Günter Umberg, *Untitled*, detail
of surface, 2000. Courtesy
Günter Umberg.

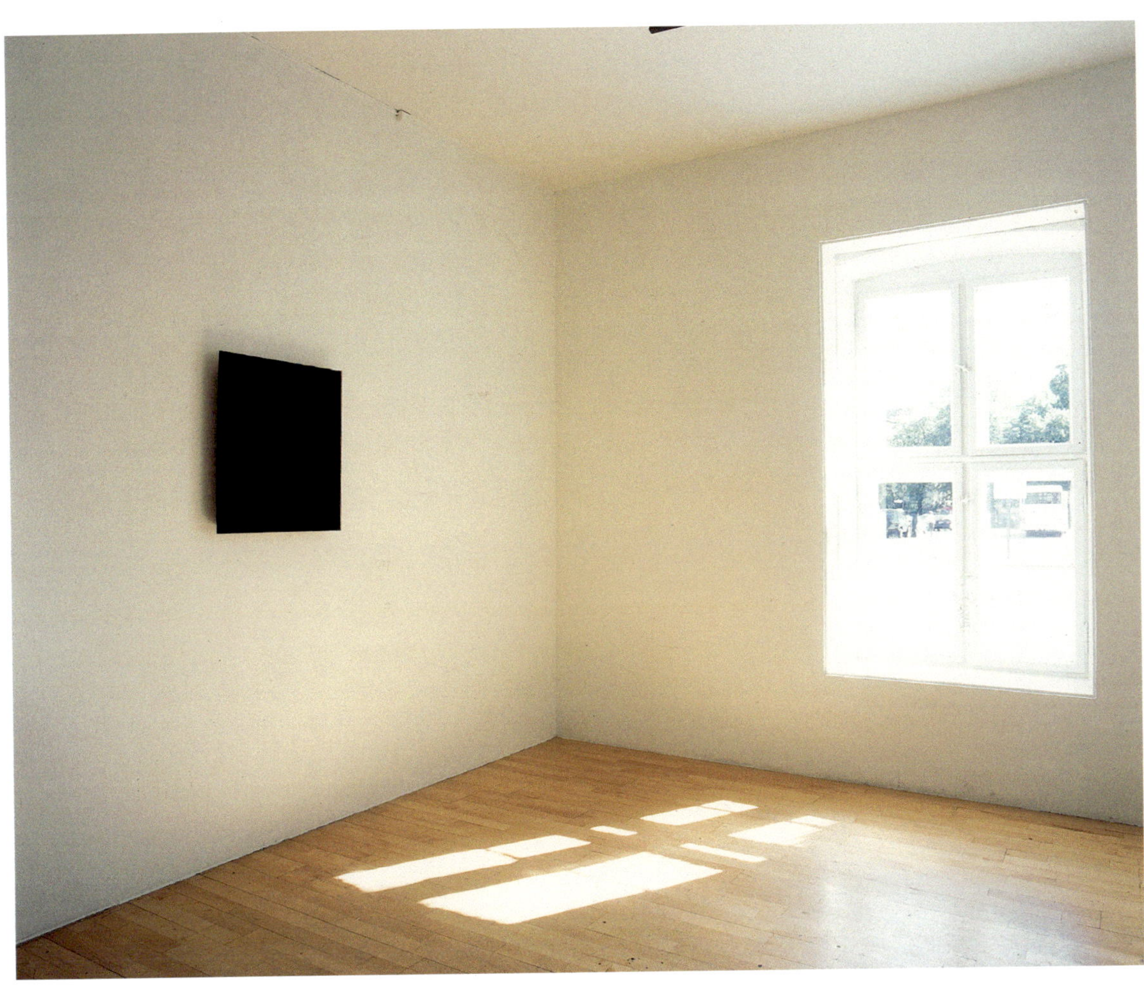

PLATE 3

Günter Umberg's installation
view of a painting in the Galleri
Riis, Oslo, 1998. Courtesy
Günter Umberg.

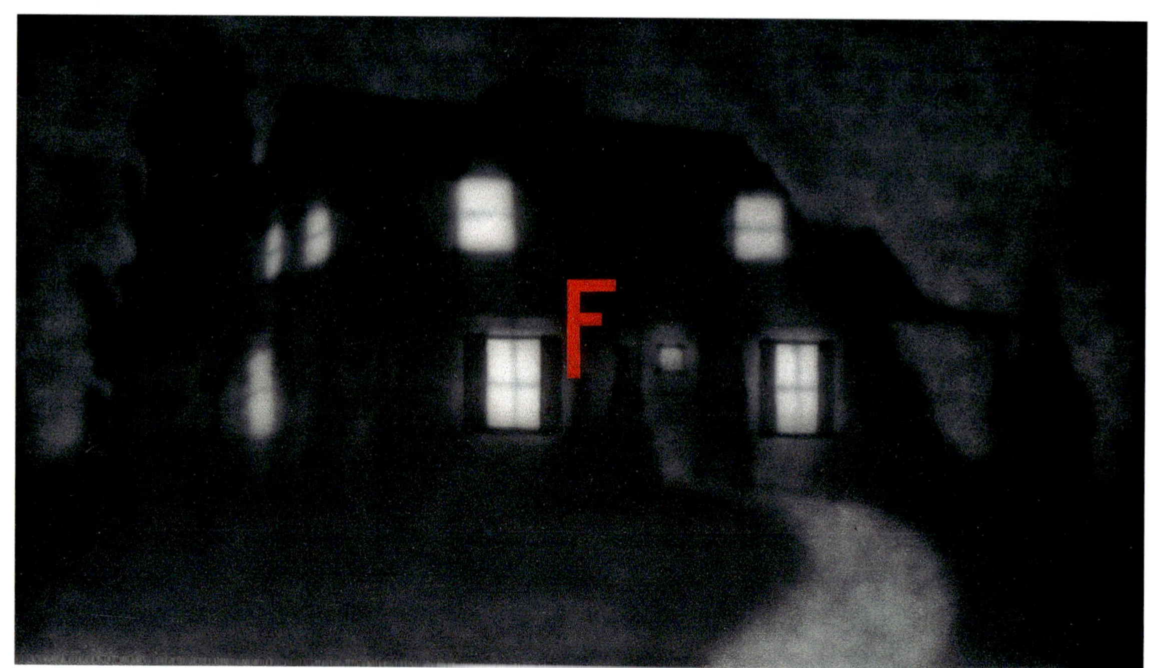

PLATE 4

Edward Ruscha, *F House*, 1987.
Acrylic on canvas. 46 × 80
inches. Chicago, Art Institute.
Gift of the Lannan Foundation,
1997.153.

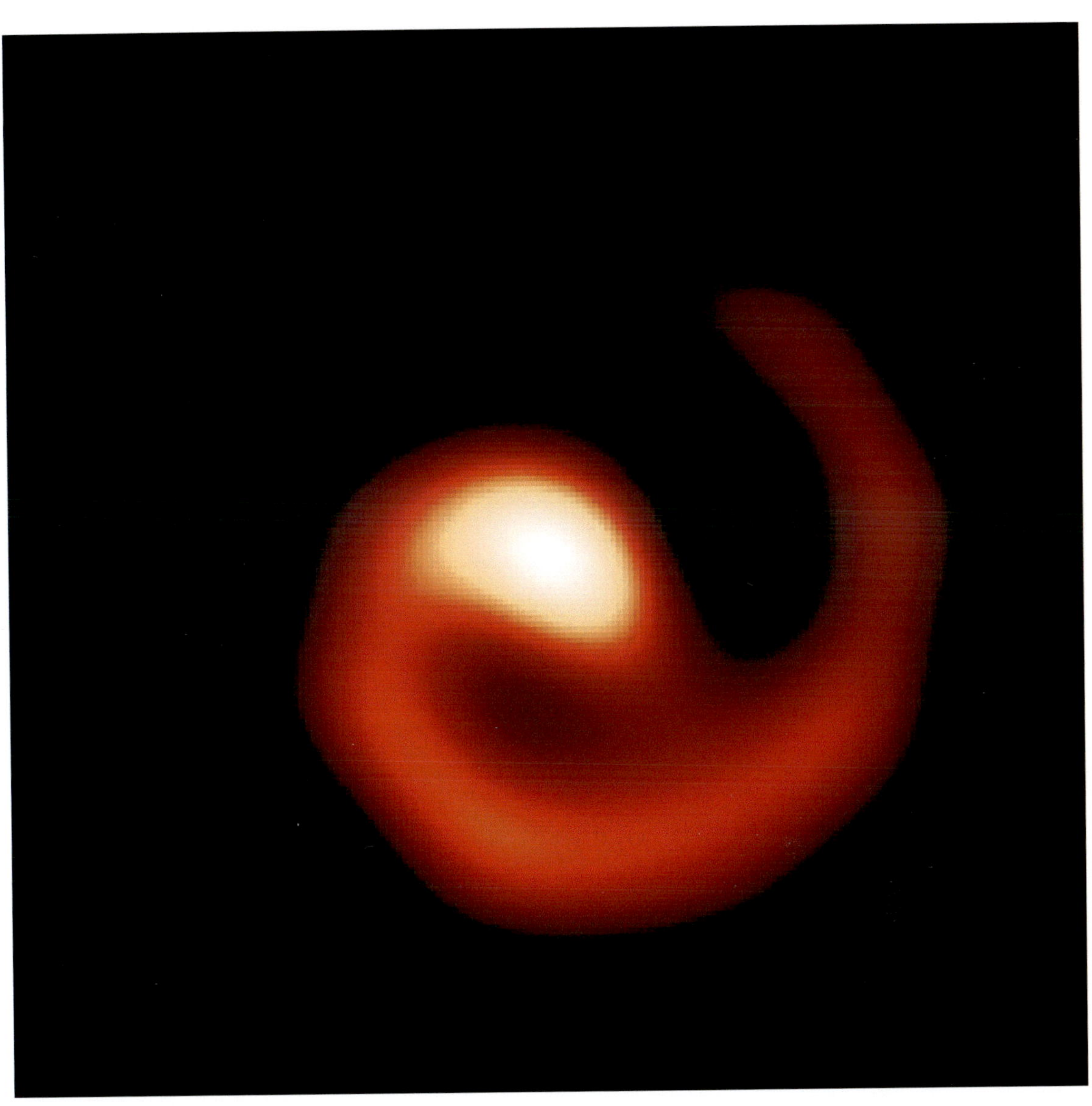

PLATE 5

Wolf-Rayet star 104. Photo
courtesy Peter Tuthill, John
Monnier, and William Danchi.

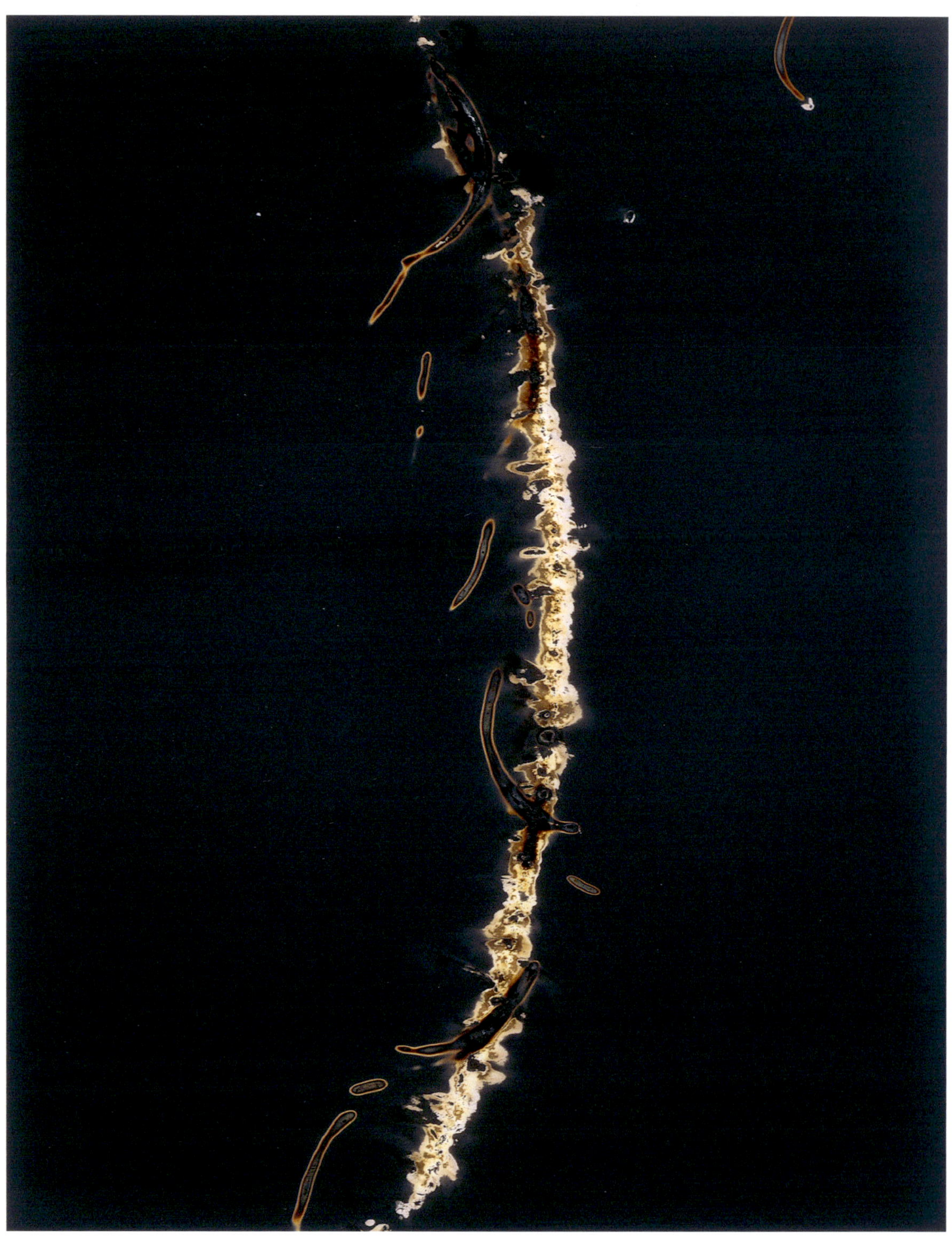

Marco Breuer, *Untitled
(Fuse/Day)*, 1999. Silver gelatin
print, burned. 18 × 14 inches.
Courtesy Marco Breuer.

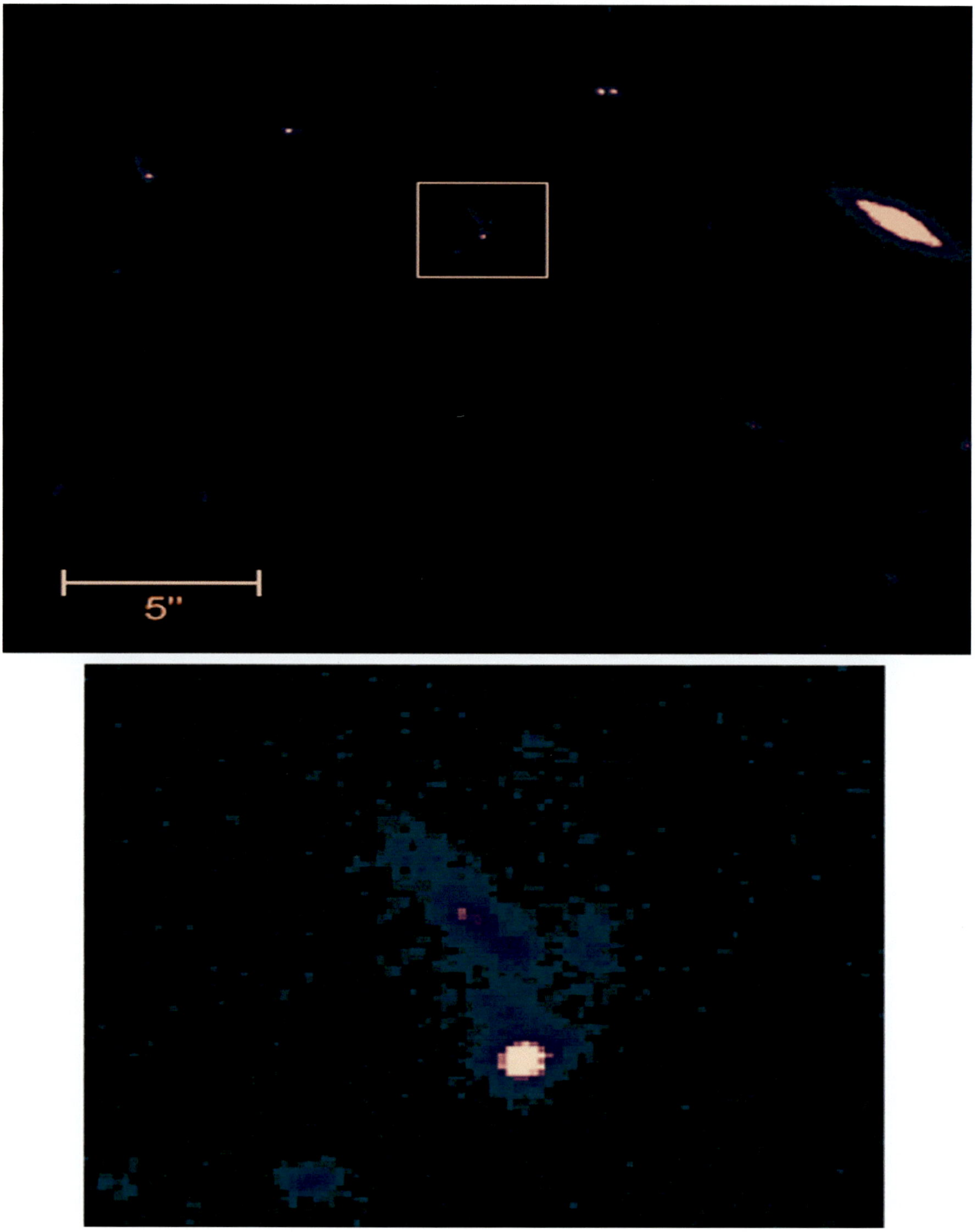

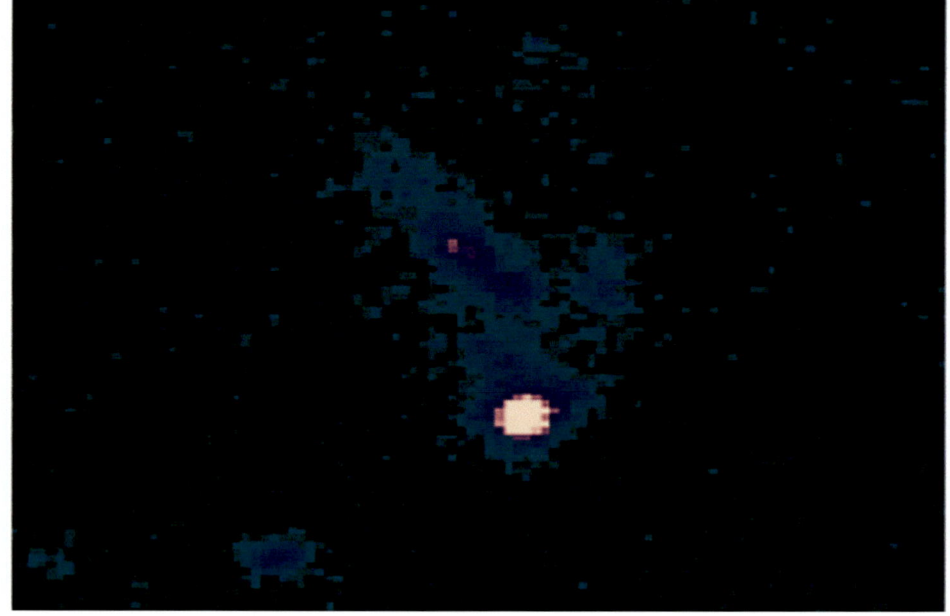

PLATE 8

Gamma ray burst GRB 990123
in its host galaxy (top) and with
detail (bottom).
Courtesy NASA.

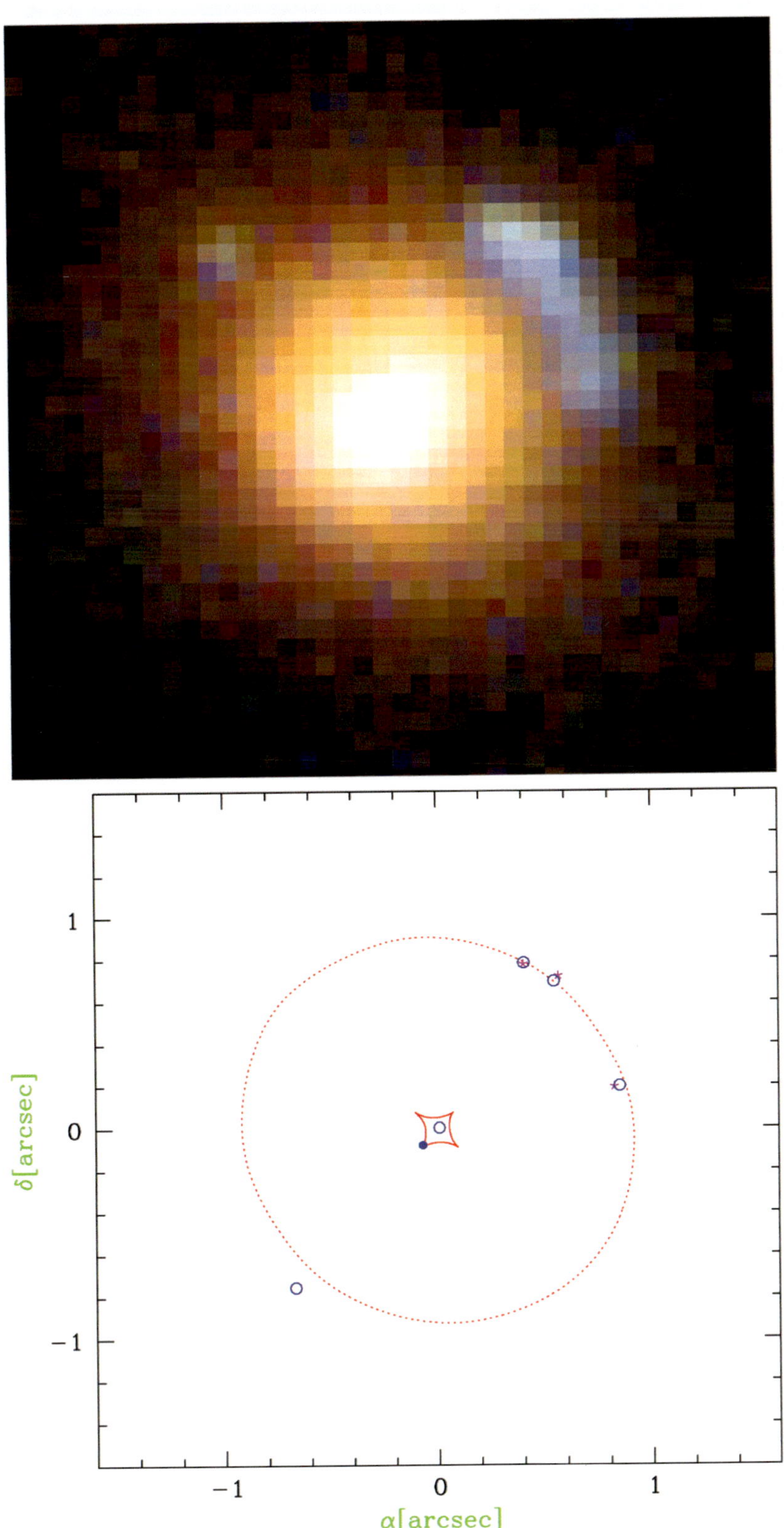

PLATE 9

Top: a gravitational lens in the
Hubble Deep Field South,
HDFS 2232509-603243.
Bottom: a mathematical model
of the gravitational lens.
Top: courtesy NASA; bottom:
courtesy Rennan Barkana.

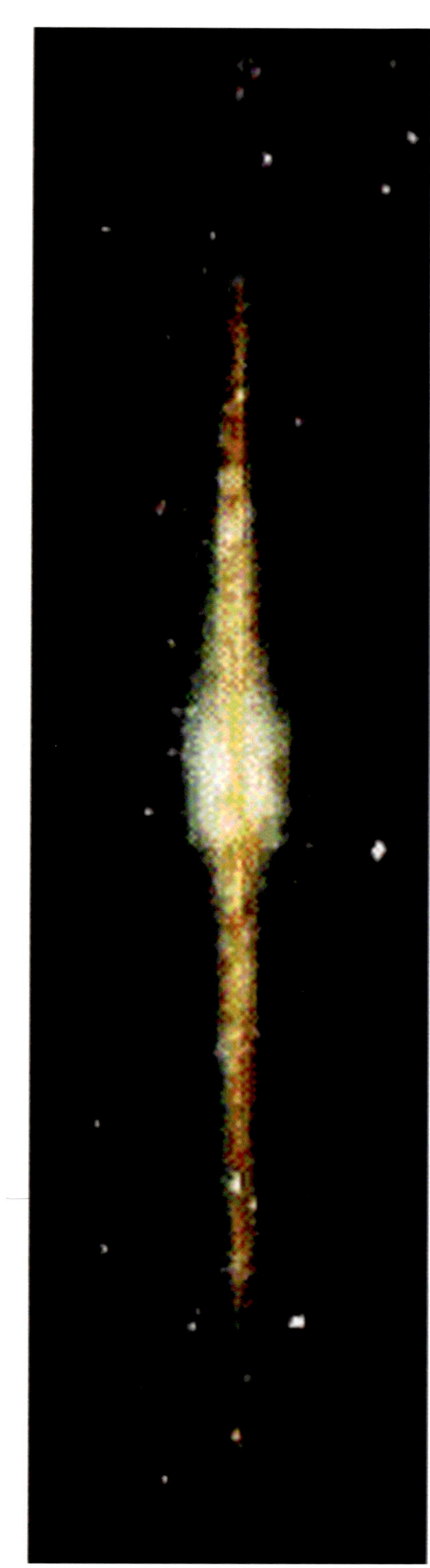

PLATE 10

The Milky Way galaxy, imaged by the COBE satellite. COBE/DIRBE infrared image of the Milky Way ©Edward L. Wright. Used with permission.

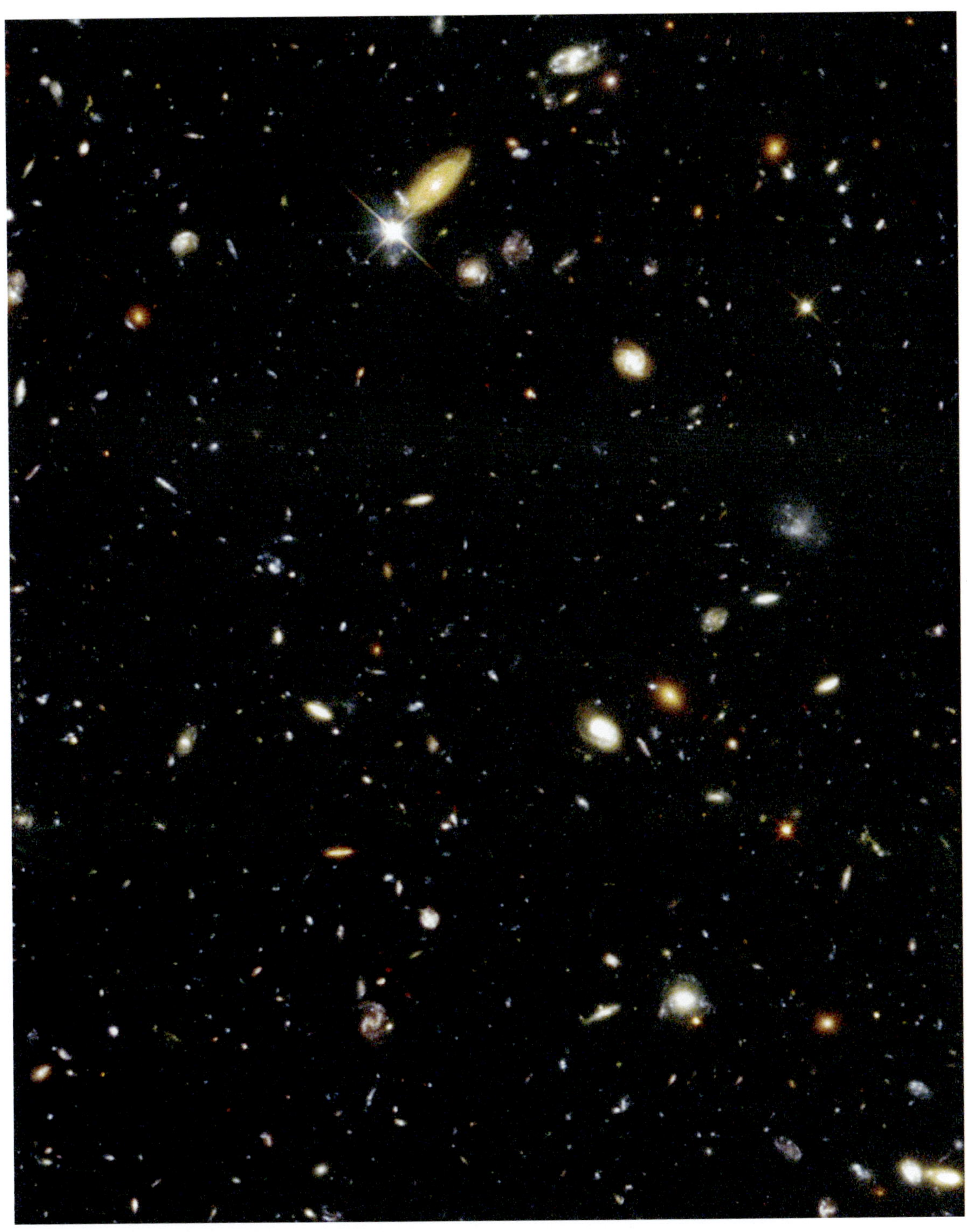

PLATE II

The Hubble Deep Field North
(HDF-N), detail. Courtesy
NASA.

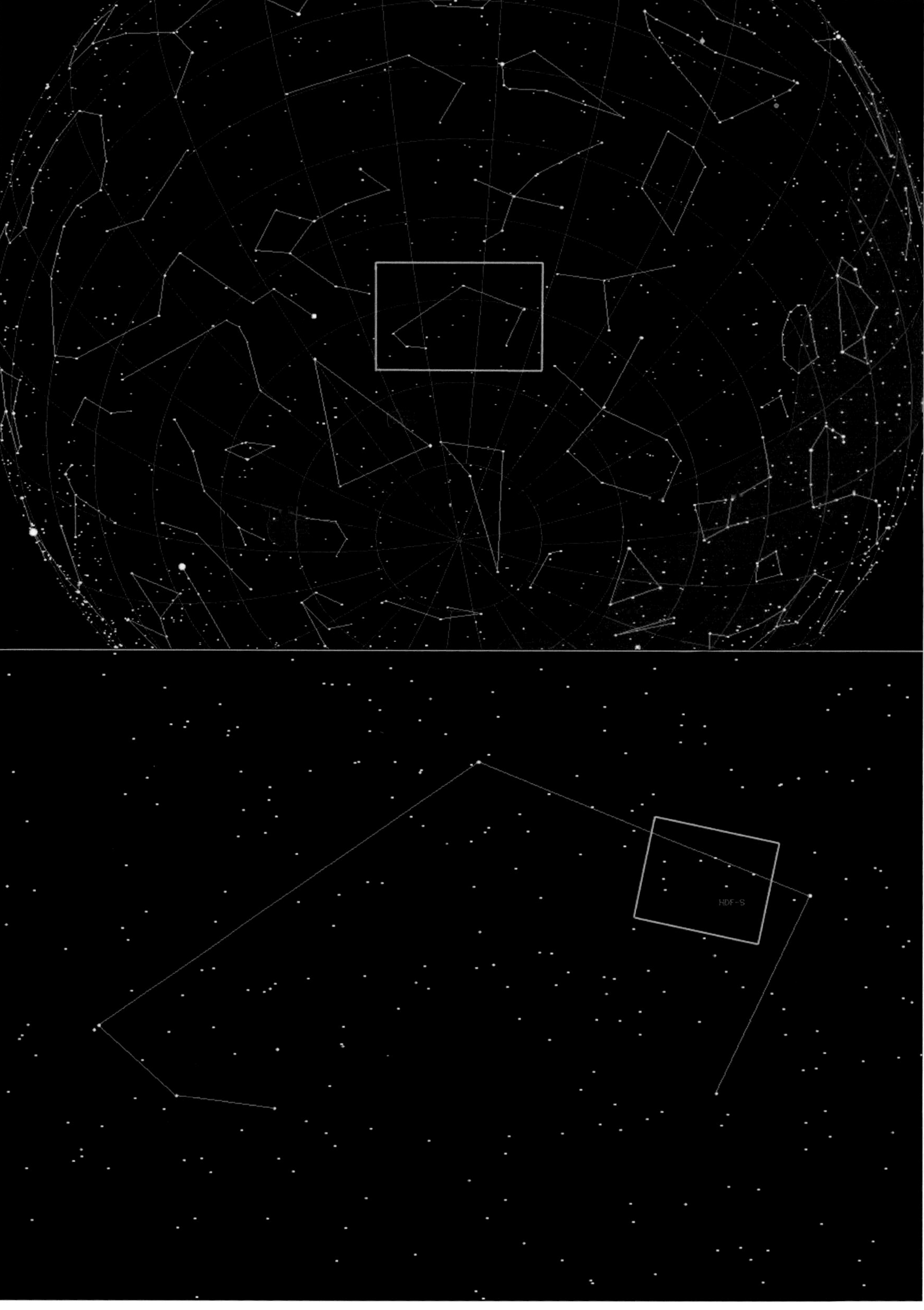
HDF-S

Opposite top: constellations
in the Southern Hemisphere,
centered on Tucana (box).
Opposite bottom: the
constellation Tucana, with the
general area of the HDF-S
(box). Below: the specific areas
of the HDF-S images.
Graphics by the author
(opposite top, bottom)
and NASA (below, inset).

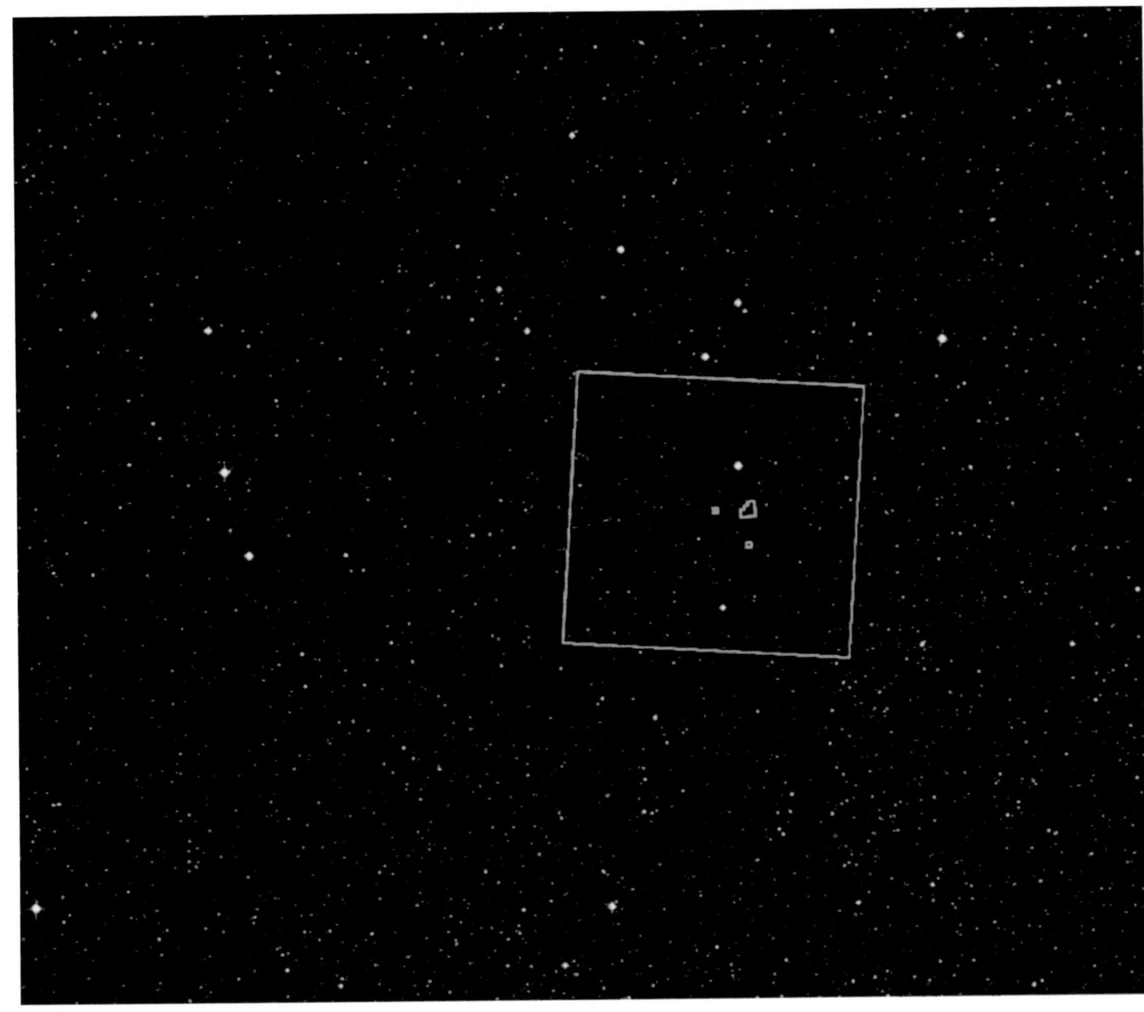

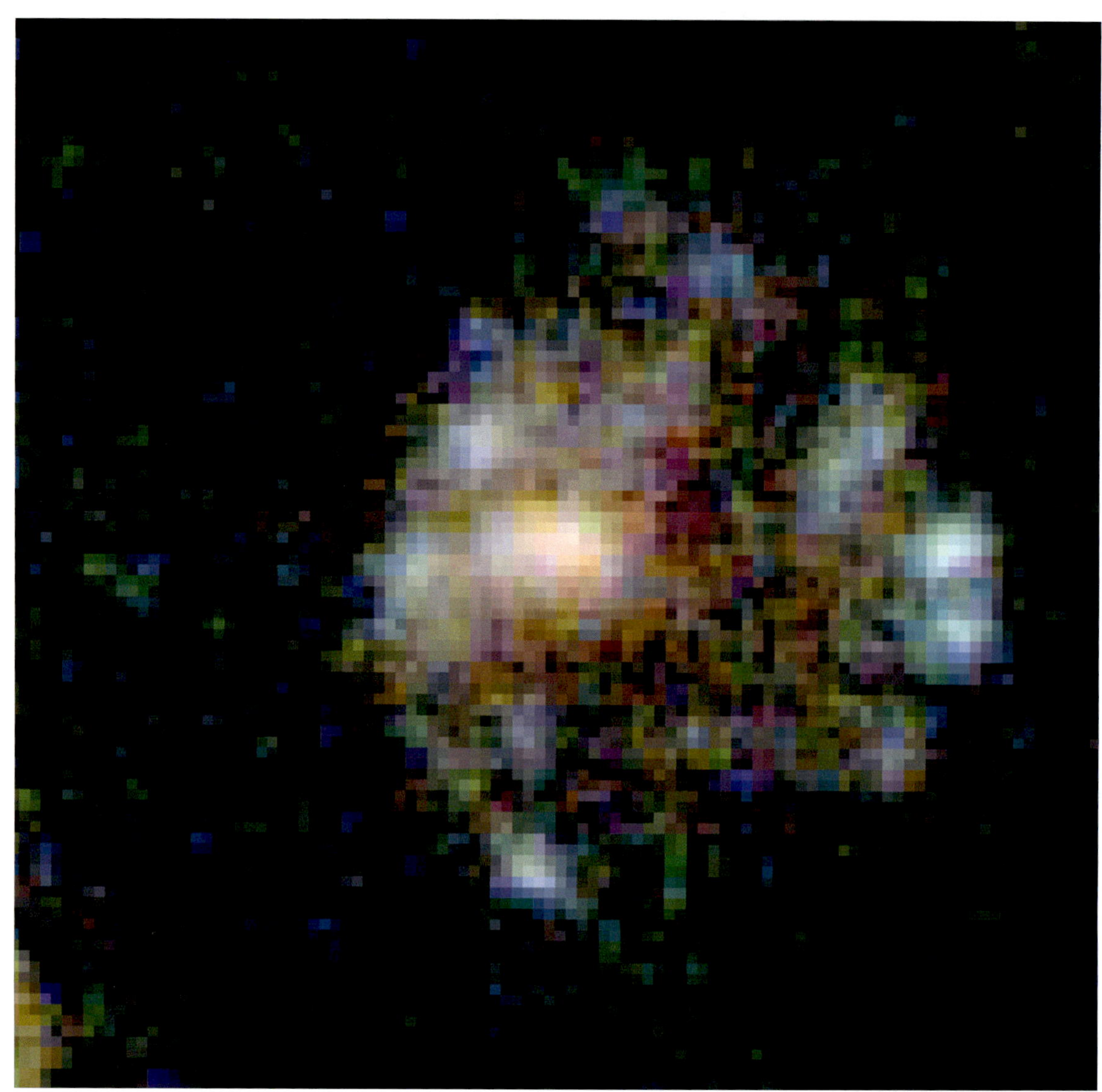

HDF-N 3-312, a possible
proto-spiral with
asymmetrically located reddish
nucleus embedded in a
structure containing blue knots.
Courtesy NASA.

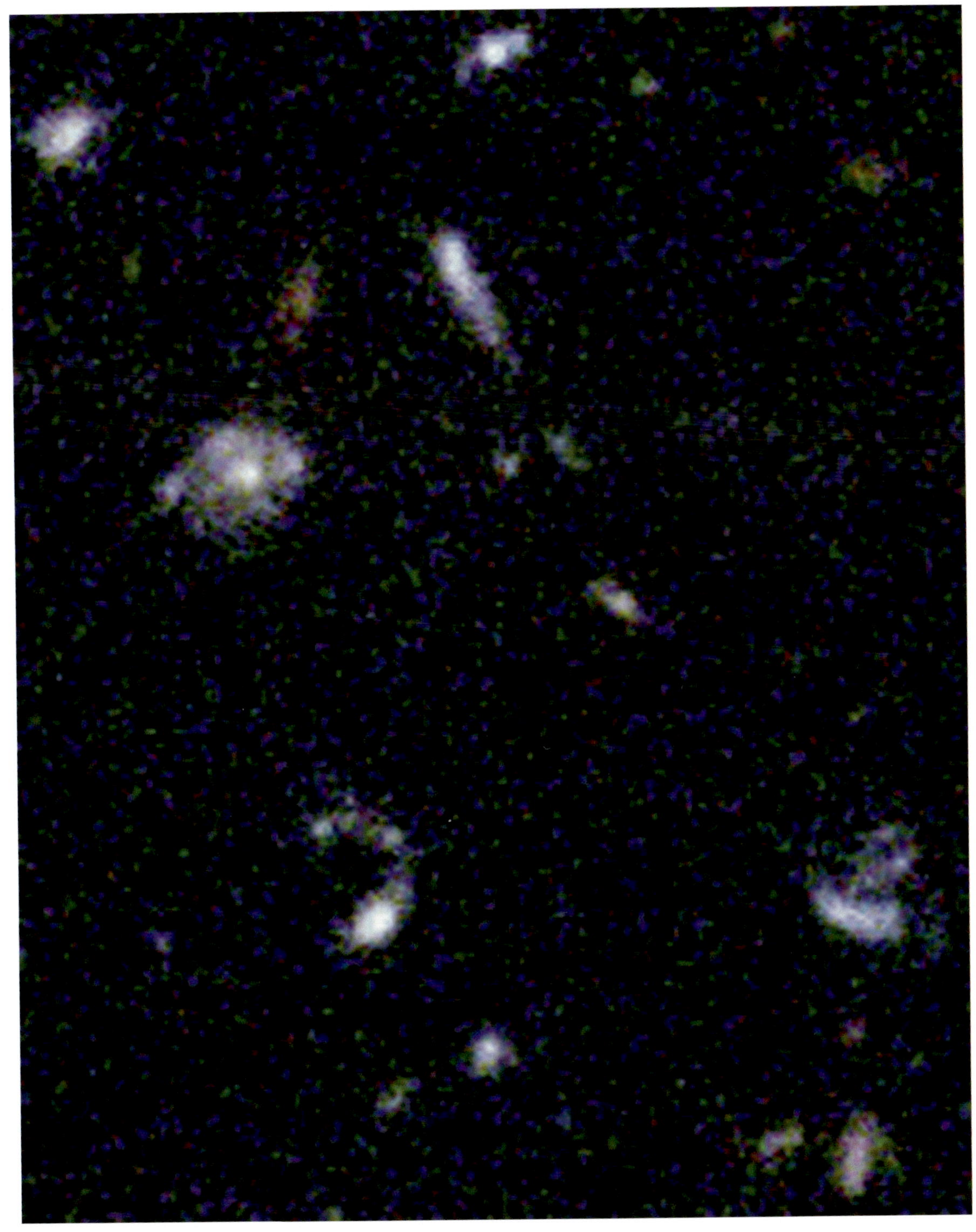

PLATE 14

"Clouds" in the HDF-N.
Courtesy NASA.

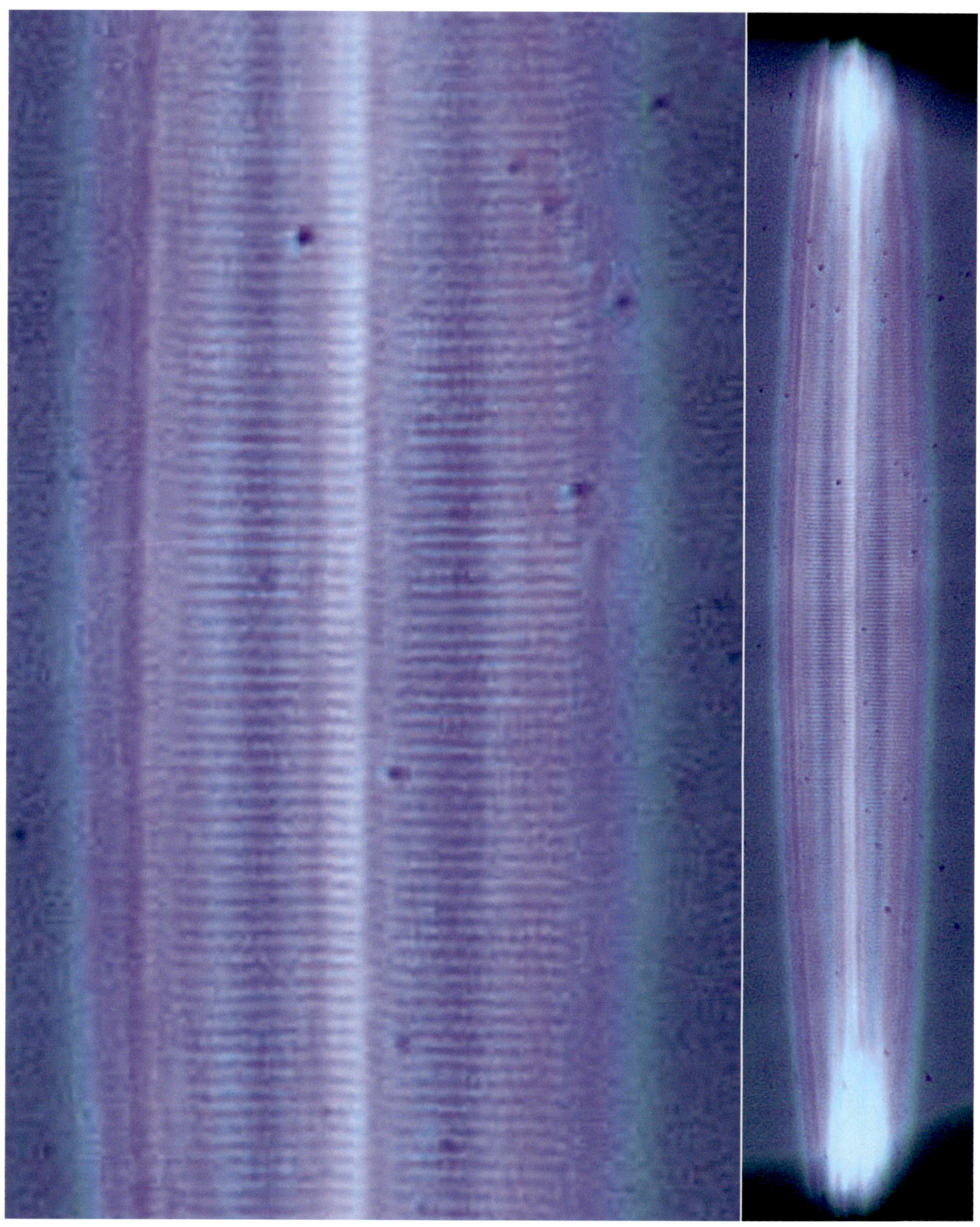

The diatom *Amphipleura pellucida*. Digital differential interference contrast image by William Varnell, M.D. Left: detail. Right: entire. Courtesy William Varnell, M.D.

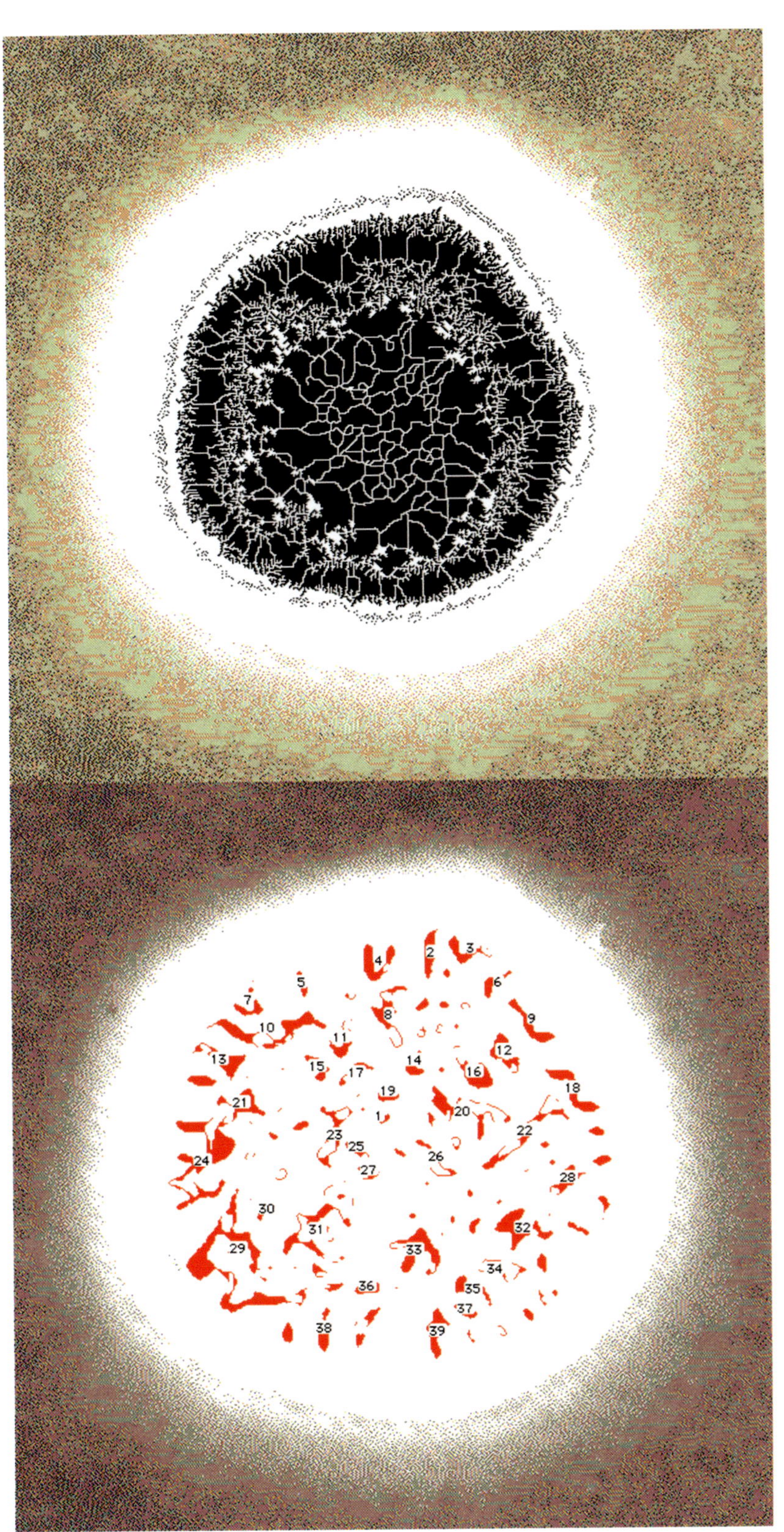

PLATE 16

Images derived from Figure
29 (a), analyzed using NIH
Image to find "watershed areas"
(top) and "particles" (bottom).
Diagram: author.

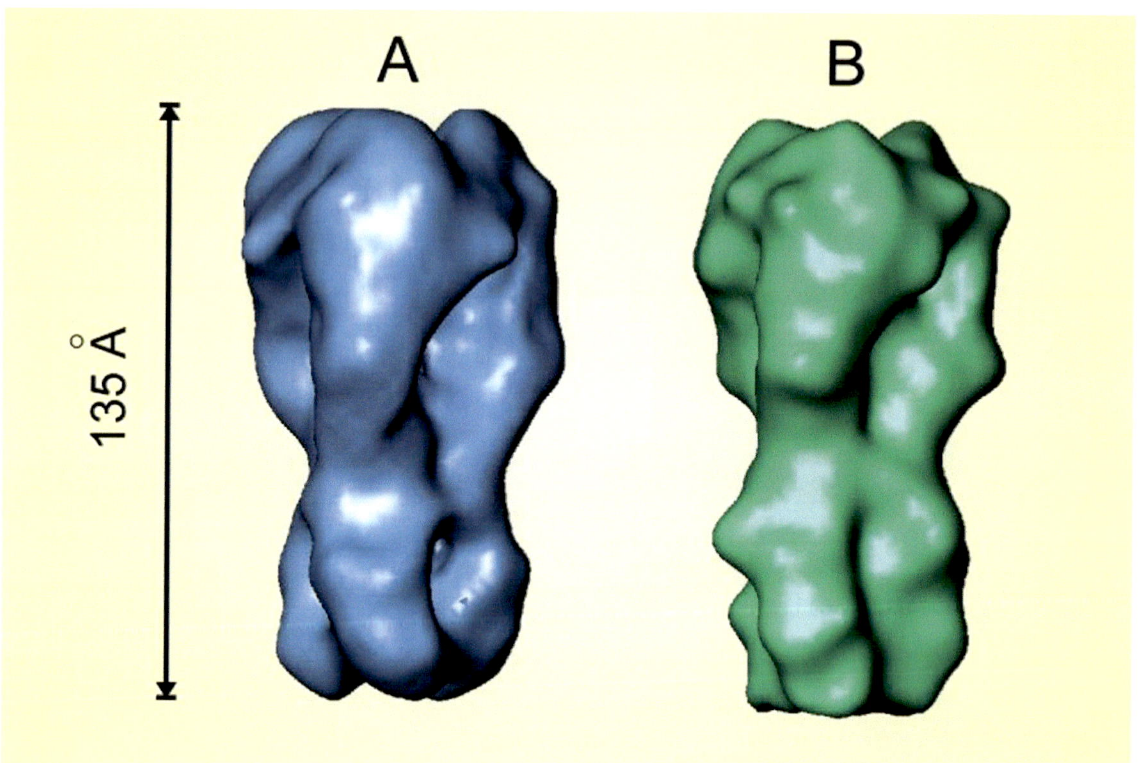

PLATE 17

3-D reconstruction of
influenza haemagglutinin
molecules under neutral pH.
From "Structure of Influenza
Haemagglutinin at Neutral and
at Fusogenic pH by Electron
Cryo-Microscopy," 257, fig. 2.
Courtesy Christoph Böttcher.

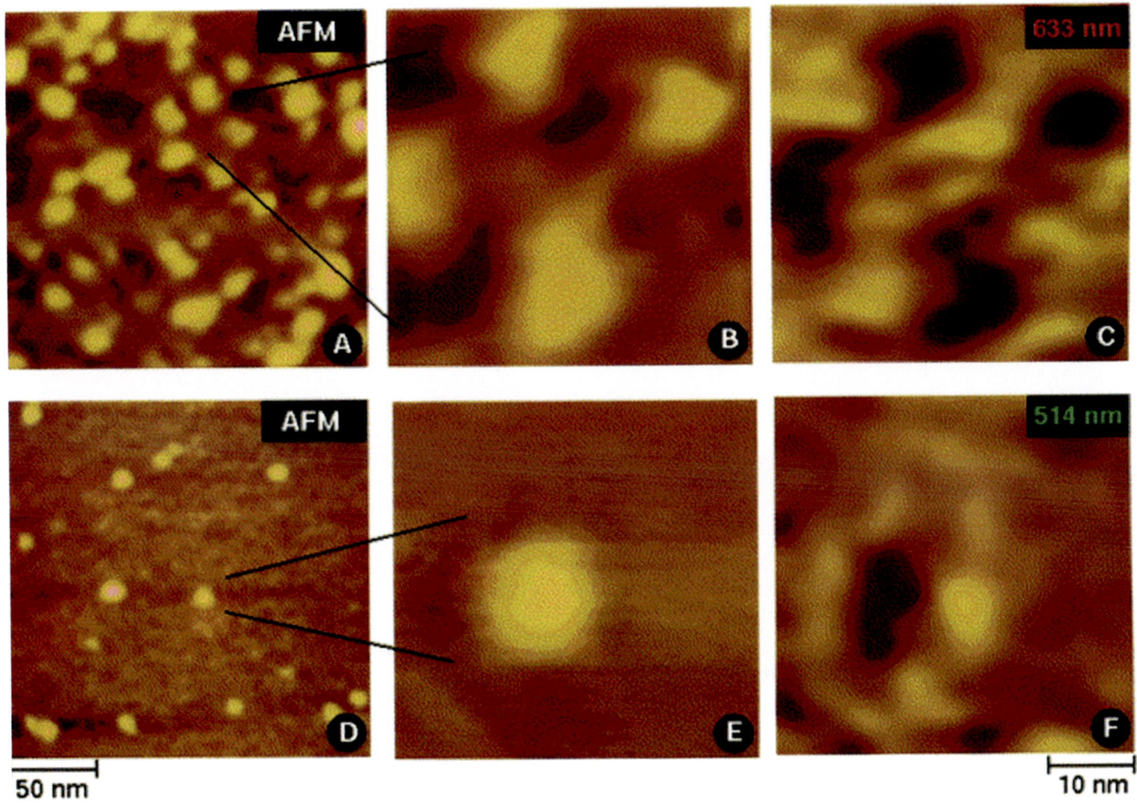

PLATE 18

Absorption images of individual molecules of coomassie blue (top) and rhodamine (bottom). A, B, D, E: AFM images; C, F: NSOM images. The wavelength in C was 633 nm, and in F 514 nm. From H. Kumar Wickramasinghe, "Progress in Scanning Probe Microscopy," *Acta Materialia* 48 (2000): 347–58, Fig. 12.

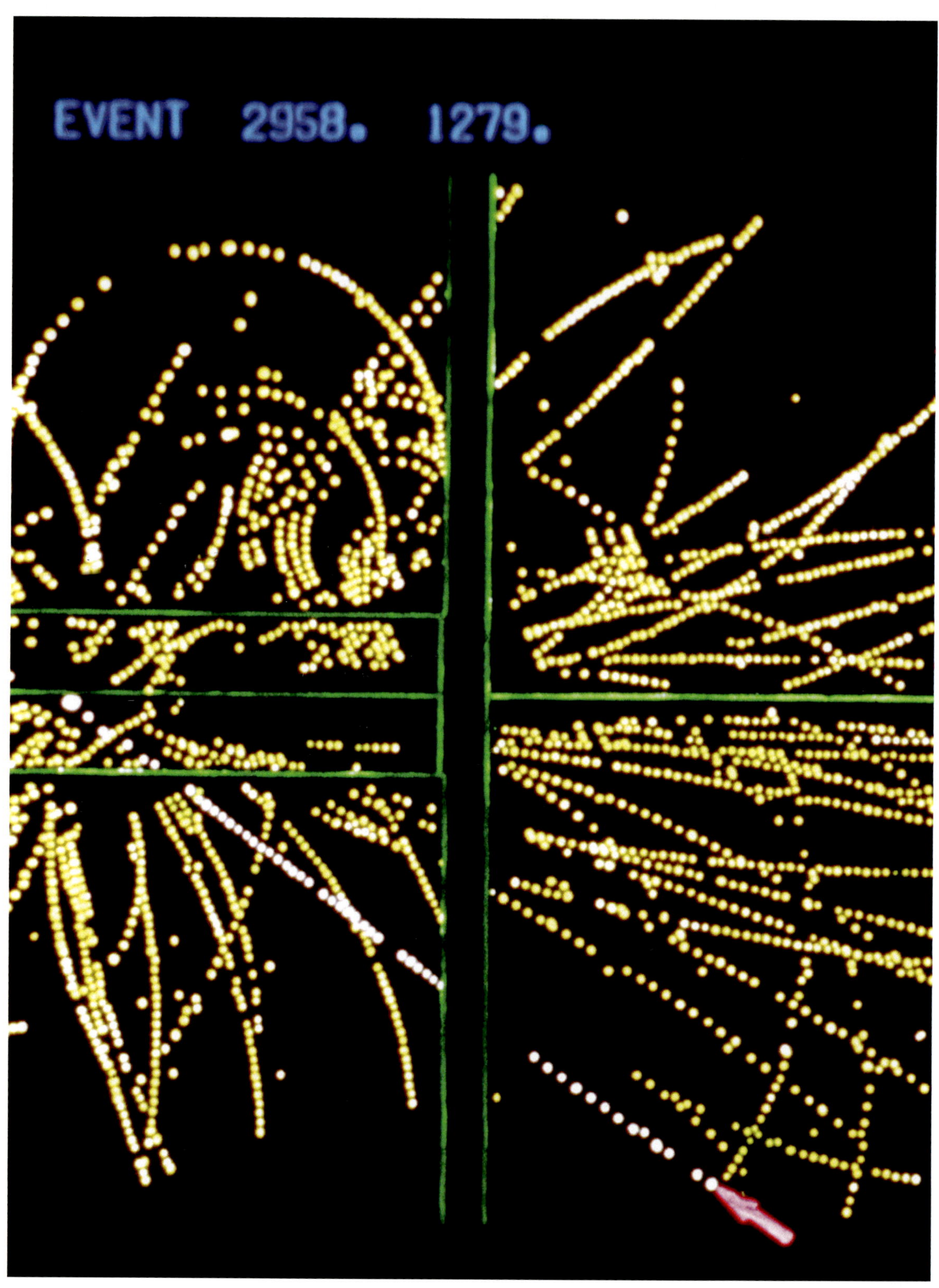

PLATE 19

Discovery of the W particle, detail, 1982. Courtesy CERN,
photo 69576. The event number has been moved into the frame
of this detail.

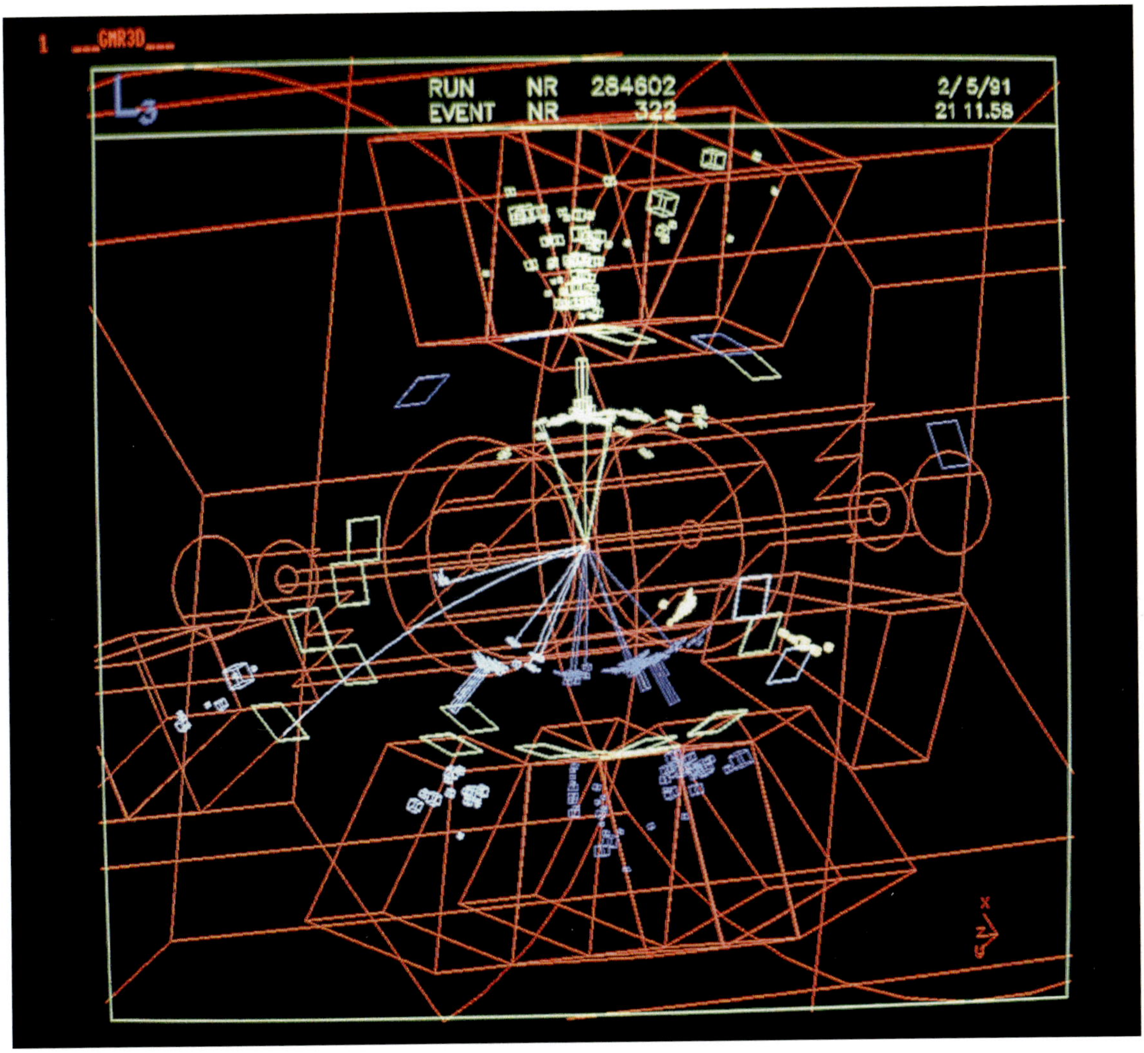

PLATE 20

Decay products of a Z^0 in the CERN L3 detector, May 2, 1991. Courtesy CERN, photo 9106038_07.

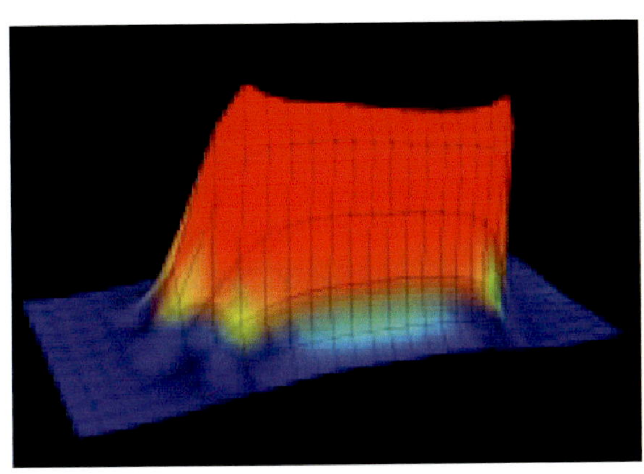

PLATE 21

Monopole contribution to the action density distribution in Maximal Abelian Projection of SU(2) measured on a lattice $V = 32^4$ at $\beta = 2.5115$. The physical quark-antiquark distance is 1.2 fm. From Nora Brambilla and Antonio Vairo, "Quark Confinement and the Hadron Spectrum," arxiv.org/abs/hep-ph/9904330 (April 14, 1999), p. 56, fig. 15. Courtesy Nora Brambilla.

Bernd Thaller's two-slit experiment, six frames from an animation. Courtesy Bernd Thaller. Image made using the CD that accompanies *Visual Quantum Mechanics: Selected Topics with Computer-Generated Animations of Quantum-Mechanical Phenomena* (New York: Springer, 2000).

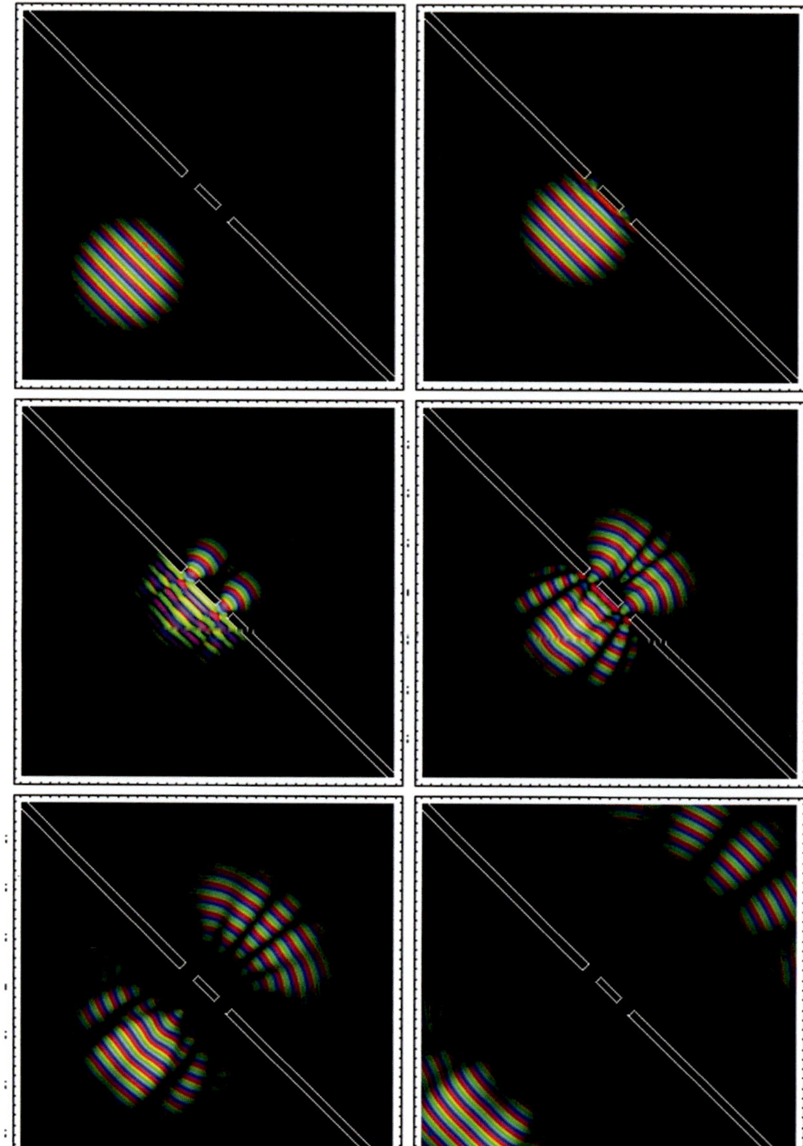

four Microscopy

And now on to a field as different from astronomy as astronomy is from painting. I won't say the objects in this chapter are the microscopic correlates of the enormous structures in the previous chapter: they aren't, and the problems involved in seeing them are different. Optical microscopes continued to develop throughout the twentieth century, moving further from representation in the ordinary sense, to the point where light and shadow themselves had to be interpreted as if they were abstract symbols instead of universal markers of solidity and relief. At the same time, electron microscopes necessitated a new and even more complex theory of resolution.

Those two issues—the reinterpretation of light and shade, and the limits of resolution—affect both light and electron microscopy. The end of the twentieth century saw a revolution in instrumentation, allowing entirely new kinds of objects to be visualized—well beyond what intuition would suggest could ever be made visible or captured on photographic film or CCD plates. The newest microscopy involves no lenses at all. Taken together, the developments in late-twentieth-century microscopy constitute an astonishing new domain of image making, unexpectedly radical even in relation to the limits of astronomical imaging.

25   *The Blinding White Arena*

The view through an ordinary optical microscope, as Vladimir Nabokov said, is of a blinding white arena. Telescopes do not usually have binocular eyepieces (although they are becoming popular among amateur astronomers) because the two eyepieces divide the already dim starlight, and because a view of the stars can never be three-dimensional. Microscopes tend to have binocular eyepieces, even when the image isn't in stereo, because the eyes are naturally suited to binocular vision. When looking with two eyes, even the flat microscopic arena appears uncertainly but distinctly raised in relief. (Astronomers report the same phenomenon: stars appear to arrange themselves in space. In both non-stereo microscopy and astronomy, the effect is an illusion.)

A standard light microscope, of the kind in general use since the mid-nineteenth century, will produce an image that has little sense of relief or contrast. Small objects such as bacteria, individual cells, and microscopic algae are generally completely translucent. A small protozoan may well disappear into the fluid that surrounds it, and bacteria have to be carefully lit to make them visible at all. In the white arena, microorganisms appear white, trembling nearly invisibly against a white, bluish, or greenish background. It helps to manipulate the optics so their watery bodies appear thicker, denser, and less flat.

The first recourse is to close the microscope's iris diaphragm, which raises the contrast and reveals the creature. The problem with that solution is that the dark outlines that are produced do not belong to the creature itself: they are artifacts of the stopped-down optical system. The creature itself remains largely invisible—what appears to be its own edges may not belong to it at all.

The *manufacture of contrast* is a problem specific to microscopy. In one sense it has been solved: in optical microscopy it is possible to achieve convincing relief, apparent shadows, apparent bulk, and density. But with the enhanced contrast comes a trade-off, which makes this particular limit of image making so interesting: as the contrast goes up, the reliability of the image sinks down. What appear to be shadows may not be; what seems to be a light source may be an artifact of the optics; what looks solid may be empty medium. Contrast-enhancing techniques turn the universal elements of all images—light, shadow, relief—into unfamiliar signs that have to be interpreted in ways that are deeply counterintuitive, as judged by contemporaneous image-making protocols in adjacent sciences.

(I put it that way because *intuitive* in this chapter doesn't denote a transhistorical or neurophysiological concept, but one that floats along with interpretive communities. That is not to say that those communities aren't often so stable as to appear wholly natural. When something is lit from the upper left, and casts a shadow to the right, I am inclined to agree with writers like E. H. Gombrich that there's a hardwired element in our response to them. The images produced by some microscopic techniques would present perceptual challenges to a very wide range of observers, in a wide range of times and places; but luckily, the relative historical stability of intuitions of three-dimensionality is not my topic here.)

A second problem specific to microscopy is the attempt to see ever-smaller details. The *limits of resolution* are entirely different issues in microscopy and in astronomy. Telescopes are limited by a number of factors, including their size, the coating and design of the optics, the atmosphere they have to look through, and the software used for image analysis. In microscopy the resolution can be limited by the wavelength of the light itself, because the objects are so small they are commensurate with the periodicity of the photons that illuminate them.[1] In the last decades of the twentieth century the resolution of microscopes was pushed to limits set by light, magnetism, and electricity themselves—and in some cases, by sleight of hand, resolution was even pushed beyond the "theoretical" limits set by quantum mechanics. Microscopy is the domain where light looks at objects that are light–size, and the result is a reciprocal blur of object and illumination.

26 *Lines of Pearls*

In the nineteenth century a microscope's limits of resolution were often tested by focusing on diatoms, which are algae whose shells (called frustules), each comprised of two pieces (called valves), set into one another like a box into its top. Each valve is decorated with tiny corrugations, holes, and bumps (called areolae, puncta, striae, raphes, and many other things). Almost all diatoms are microscopic, and most are boat-shaped. Some loom large under the microscope, and others are wisps, nearly impossible to distingush from the water in which they float. One species in particular, *Amphipleura pellucida*, was taken as the ultimate test object. In an ordinary microscope, it looks like a transparent canoe (Figure 27). Only the most protracted efforts, and the finest adjustments of illumination and optics, reveal that its surface is scored with tiny ridges, nearly 100,000 per inch (Color plate 15). *A. pellucida*'s ridges were first resolved in 1868, and then, by pushing the microscope to its limits, it was discovered that the ridges are composed of tiny "pearls" (Figure 28).[2]

As far as I have been able to find, twentieth-century optical microscopists

FIGURE 27

The diatom *Amphipleura pellu-cida*. Photo: Martin Mach. Achromatic objective, *N.A.* 1.3, Hyrax medium; slide prepared by K. D. Kemp. Courtesy Martin Mach.

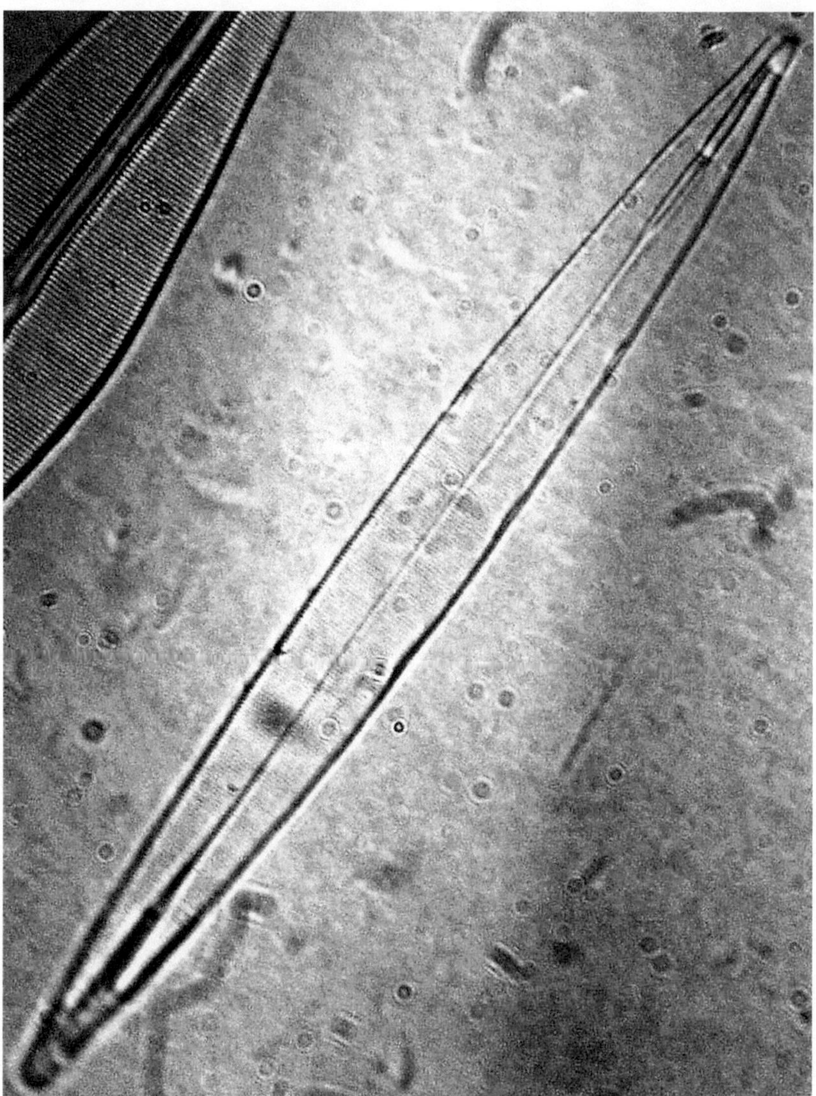

did not better the nineteenth-century result recorded in Figure 28 (left) and they rarely equaled the less impressive result in Figure 28 (right). (Compare Figure 28, right, with Color plate 15, which represents the best of late-twentieth-century amateur work.) Despite the invention of various optical tests, diatoms have never entirely been replaced as test objects, and several biological supply companies still offer test slides with different species of diatoms. At the very end of the century, in 1999–2000, amateur microscopists were still working on *A. pellucida*, producing beautiful images, but none better than what microscopists had done over 130 years

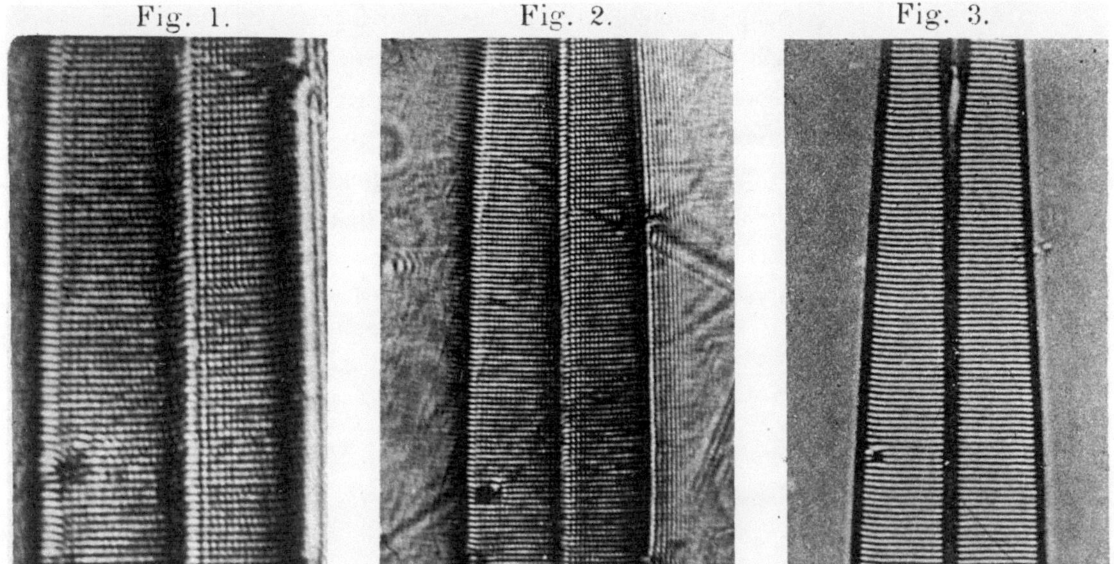

FIGURE 28

The diatom *Amphipleura pellu-cida*, the valve resolved into "pearls." Photo from William Carpenter, *The Microscope and Its Revelations*, 7th ed. (London: J. & A. Churchill, 1891), plate XI, fig. 1, between pp. 524 and 525.

before. (I myself have test slides of *A. pellucida*, but I have been unable to equal the results given in Color plate 15).

At the limits of optical microscope resolution it becomes difficult to distinguish real contours from illusions produced by the interaction of light with objects. Tiny droplets look as if they have bright centers or lines. That is the Becke effect, produced when the objects act as microscopic lenses, concentrating light that passes through them.[3] At high magnifications images are also limited by diffraction patterns: the objects are so small that the light waves bend around them or spread out as they pass through transparent areas. The Becke effect, diffraction patterns, chromatic aberration, astigmatism, and spherical aberration were all issues in nineteenth-century microscopy. Combined, such problems made it nearly impossible to determine the nature of the smallest objects.

The diatom *Pleurosigma angulatum* was another object of discussion. Leopold Dippel, an early authority on microscopes and diatoms, agreed with Albert Nachet, a French microscope builder, that *P. angulatum*'s smallest features were tiny disks, not hexagons as they appeared to be.[4] Ernst Abbé, author of the diffraction theory of microscope imaging that is still in use, reported lines within the round disks or pits. Despite their differing observations, these nineteenth-century reports were unexcelled until electron microscopy. It turns out that *P. angulatum*'s shell has two layers: one with rounded pits, and another with small slits.[5] The topology of the shell is enough to confuse even the best microscopists using optical instruments. Usually the electron microscope has the definitive answer, but not always—the optical microscope "sections" the material differently, revealing inner structures not visible in scanning electron microscope images.

In theory, optical microscope resolution is limited by the wavelength of the light and the angle at which the light diverges into the objective lens. In mathematical terms, an optical microscope should be able to resolve two objects separated by a distance d, where

$$\frac{\lambda}{2N.A.} \leq d \leq \frac{\lambda}{N.A.}$$

Here λ is the wavelength of the light, and $N.A.$ is the numerical aperture of the objective lens, a measure of how wide a cone of light the lens admits. In a typical case the light is a monochromatic green, where λ is about 0.55μ (microns; 10^{-6} m); and the numerical aperture is 1.25. That means a microscopist using an optical microscope would expect to resolve objects separated by about 0.22–0.44μ. In practice, the limit of resolution is 1–4μ, because real objects are more complex than the theoretical limits allow. As recently as 1981 there were debates about this limit and about whether diatom features could be seen more accurately by optical or electron microscopes.[6]

In the last two decades of the nineteenth century, it appeared that the way to resolve these issues was to avoid the complex topographies of diatom shells and look only at perfect geometric forms. Friedrich Nobert (1806–81) invented a machine that produced lines on glass slides, with distances as close as 0.1μ. Nobert's slides were test objects throughout the century. The extremely fine rulings were widely sold and cited, even though no one could resolve the lines; it was assumed, correctly, that his ruling machine could produce the intervals he claimed.[7] Nobert's finest rulings were not resolved until 1966, using electron microscopes. The substitution of ideal geometry for natural objects would have settled the problem of resolution, were it not for the fact that "ordinary" objects like *A. pellucida* and *P. angulatum* do not behave as if they are simply repetitions of ruled lines. Thus the history of resolution in microscopy is not linear: some nineteenth-century microscopists achieved results no twentieth-century light microscope has bettered, and some twentieth-century electron microscope studies have failed to resolve issues raised by light microscopists. The nonlinear history is partly a consequence of the objects that have been studied—no single test pattern or form, it turns out, is sufficient to measure all resolution issues.

27   Flint Glass, Oblique Illumination, Darkfield

There are many ways to enhance the contrast of nearly transparent objects such as diatom shells and test lines.[8] It is helpful, for example, to mount biological specimens in a medium that has a different refractive index. (Watery creatures in water don't provide much contrast.) Diatoms, which have refractive index 1.44 (their shells are a form of opal) show up poorly in water (index 1.33), but they show up well in a resin called "syrax" (index 1.62). The exceptional resolution in Figure 28 was made possible partly by floating the diatom in an unusual viscous medium of refractive index 2.42.

The difference in refractive indices aids the production of contrast, and it also creates an opportunity to improve the resolution. Slides and coverslips were made of flint, which has refractive index of 1.72, permitting light rays to proceed nearly straight through the slide, specimen, and coverslip without significant refraction. When the same viscous medium is used to coat the slide, it can be "fused" with the lenses of the microscope, creating a single uninterrupted optical pathway of refractive index at least 1.72. That, in turn, raises the *N.A.* because the cone of light can be very wide. Ordinarily the limit for optical microscope objectives is around *N.A.* 1.3; the photograph in Figure 28 was made with a special objective of *N.A.* 1.6.[9] The light used to illuminate such specimens was typically a monochromatic blue, to take advantage of the relation between short wavelength and high resolution. Late in the nineteenth century some microscopes were even made for ultraviolet light, pushing the resolution a little higher.[10]

Contrast (but not resolution) can also be raised by rearranging the light source. In the nineteenth century, that meant moving the light from side to side, producing oblique illumination, or else blocking all direct light and letting only refracted light enter the objective lens, creating darkfield illumination. Each method has its strengths: oblique illumination can be used to calculate refractive indices, and darkfield is useful for counting objects like red blood cells.[11] Each method also introduces its own typical patterns of light and dark. If an object is illuminated by oblique light, then it will appear bright on the side from which the illumination comes, as you might expect—but only if the object has a higher refractive index than the medium surrounding it. Otherwise, the light will appear to come from the opposite direction. Such a disruption of the normal order of light and shade is, I think, impossible to comprehend intuitively: no matter how long I look at such an object, it seems that the light is coming from the side that is brighter; but depending on the refractive indices involved, it may not be.

28 Phase Contrast

Oblique illumination, slides made of flint, and darkfield can all be modelled using Abbé's theory of image formation.[12] His theory, together with Köhler's theory of optimal microscope illumination, set the limits for optical microscope performance.[13] To push beyond those limits it is necessary to make use of properties of light that ordinary microscopy does not use. Historically, phase contrast was the first technique to do so.[14] It works by making the phase of the light waves visible (as opposed to their wavelength). In ordinary light microscopy, the phase of the light is irrelevant because it cannot be sensed by the eye. In physical terms, then, phase contrast microscopes make use of information about the object that was discarded in ordinary light microscopy.

A typical phase microscope has two disks in it: one placed in the stream of light before it reaches the slide and the other placed so it intersects the light after it has passed through the subject. Both plates have annuli (concentric rings like targets). Light that goes straight through the object on the slide, without being bent, is filtered out so that very little of it remains. Some light that is bent as it travels through the object is held up just one-half wavelength by a "phase plate." The result is an image in which the object's surroundings are relatively dark, and the object itself is a mixture of very bright and very dark areas—bright where the light is in phase with the light that was not bent, dark where it is one-half wavelength out of phase with the unbent light.

Phase contrast images raise another problem of interpretation. Ordinary microscopes register changes in the object's opacity or thickness: where the object absorbs more light, the image is dark. Light can also be dampened (lowered in amplitude). Both absorption and dampening are common effects in everyday life: sunglasses dampen illumination, and the sky is darker when particles absorb the light. Phase microscopes record absorption and lowered amplitude, but they also register how the object *slows* the light that passes through it; so a phase contrast microscope produces images where light and shade do not mean what they do in unaided vision. A photograph taken through a phase contrast microscope may look as if it shows a translucent object with a certain thickness or density, but those are just guesses, and they may be far off the mark.

Strong transitions from light to dark may not be caused by portions of the object that are more or less opaque They might be caused by phase transitions. A brilliant area might be caused by the two streams of light coming through in phase, and a dark area might be caused by the two streams coming through one-half wavelength out of phase. As the thickness of an object increases, the waves of the two streams will first be in phase, then out by one-fourth, then one-half, then three-fourths, then a whole wavelength, and so on—producing stripes where there is really only a continuous slope. (A pyramidal object, for example, may appear like a stepped ziggurat.) These effects have to be anticipated, because they exist side by side with ordinary light-dark differences.

In addition, there will often be a "phase halo" around the object—a brilliant ring of light that may or may not belong to the object itself (see Figure 29, center). The halo is generated when optical gradients shift the image of the object's edges away from the phase annulus so that it passes through the nonabsorbing portion of the phase plate. Ordinarily the phase halo is "added" to the object, whose actual contour has to be inferred. Even when the phase halo is part of the object, it can be hard to see it that way, because most objects don't have brilliant halos.

Phase contrast works most effectively where the optical path changes most quickly—from the surrounding water to the cell wall of an amoeba, or from the valve of a diatom to one of its ridges. Phase contrast effects flatten out when the optical path becomes more level—for instance, within a cell or across a fairly flat surface. In microscopist's terms, edges show up better than "wedges," places where the object slopes or changes only gradually. As a result, "the centre of one region will appear the same shade of grey as the centre of another region of different

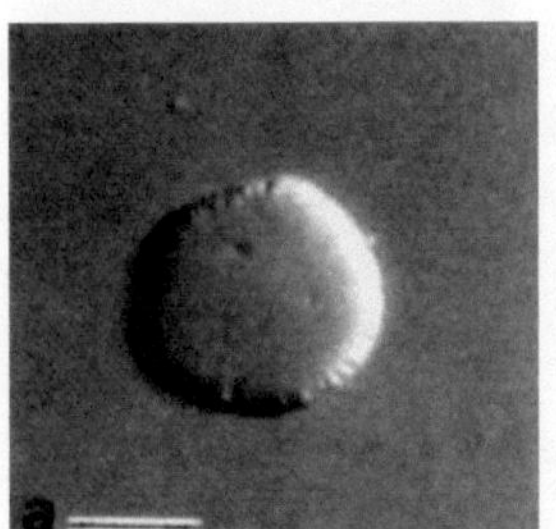
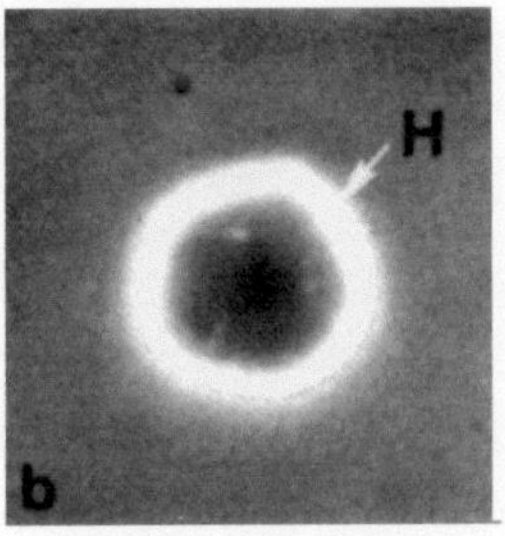

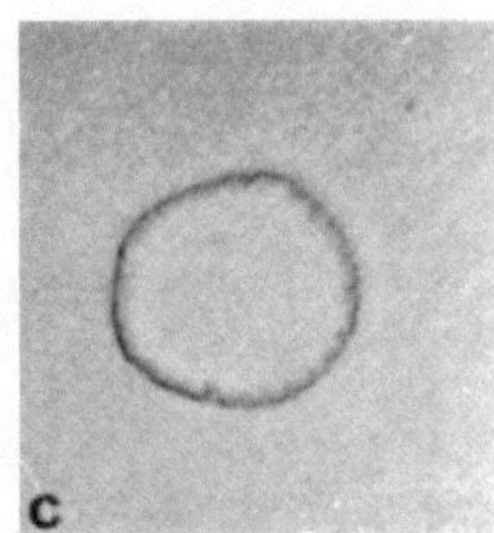

FIGURE 29

Comparison of Hoffman modulation contrast (left), phase contrast (middle), and ordinary bright field (right). Bar = 11 μ. From Robert Hoffman, "The Modulation Contrast Microscope: Principles and Performance," *Journal of Microscopy* 110, no. 3 (1977): 205–22, fig. 14.

refractive index," making a protozoan's cell look the same as the water that surrounds it, except for the coruscating halo.[15]

Microscopists can work around these problems by taking photographs using various lighting techniques to help decide what counts as an "artifact" and what is a genuine property of the object. But even when it comes to a method as common as phase contrast, the difference between the behavior of light under the microscope and in unaided vision is so great that it can seem as if the microscopic image is an entirely new phenomenon— as different from ordinary seeing as a stock-market graph is from a snapshot. The philosopher of science Bas Van Fraassen argues that it does not make sense to say that we see through a microscope: in his view the instrument creates images by diffraction, not by ordinary image formation as in the human lens; in addition, the microscope presents objects that can never otherwise be seen or related to medium-sized physics objects. In Van Fraassen's view, the problems of interpretation I have been describing are at the beginning of insuperable epistemological barriers to interpreting microscope images on the analogy of unaided vision.[16] Yet in the face of actual practice, such arguments become counterintuitive. It is one of the themes of this chapter that even the most outlandish ways of generating light and shade result in images that are almost effortlessly interpreted as displaying variants of ordinary lighting conditions. As the philosopher Ian Hacking has argued, it is rarely difficult to make decisions about the reality and even the realism of a microscopic image.[17] On the other hand, it would be treacherous to underestimate the strangeness of phase contrast images. It takes practice to read phase contrast images, just as it takes practice to interpret graphs and charts; but phase contrast images are perceived and interpreted as realistic images, and graphs aren't. It is unusual in the history of images to find a case in which an image that is taken to correspond with the world in the way a snapshot or video does is interpreted with the help of counterintuitive and apparently arbitrary rules. Hacking's answer to Van Fraassen is convincing, but it does not cover counterintuitive directions of illumination, phase transitions, phase halos, and slowed optical paths that masquerade as absorption or dampening.

29   *Differential Interference Contrast, Hoffman Contrast*

In the years 1952–55, Georges Nomarski developed a system called differential interference contrast, which combines elements of phase contrast with polarization.[18] (From a physicist's perspective, interference contrast makes use of yet another property of light, its polarization, which was ignored by phase contrast and ordinary light microscopy.) Interference contrast images can be spectacular—full of spectral colors, with a strong sense of raking illumination. They turn nearly colorless objects like paramecia into garish multicolored objects that look solid, as if they were cast in iridescent plastic. (Differential interference contrast does not produce bright color, but creates solid, sculptural relief.) The technique is widely used but also strongly misleading, requiring even more circumspection than phase contrast images. What appear as hard sloping contours are really gradients of phase shifts, which can be caused by changes in refractive index or density just as well as by changes in thickness; and what appear as brilliant colors are nearly always artifacts. The optical system that generates interference contrast is more elaborate than phase contrast, and full interpretation of the images also requires viewers to take polarization effects into account.

Hoffman modulation contrast, invented in 1975, was intended as an improvement on Nomarski's differential interference contrast. In Hoffman contrast, light is only allowed to hit the object obliquely, and then it is filtered in three ways. In a typical arrangement, light that passes straight through is cut to 15 percent of its full strength; light that is straightened by the object is cut to 1 percent; and light that is bent even further by the object passes through at full strength. The result is a strong illusion of oblique illumination (Figure 29, left). As in ordinary oblique illumination, the sense of raking light is an artifact of the technique and has to be interpreted as a map of the slope of the object rather than the shade or shadow produced by an ordinary light source.

Hoffman contrast answers several shortcomings of other kinds of contrast enhancement. Unlike phase contrast, it preserves the sign of the object gradient (whether it slopes right or left), substituting a light and dark division for the uniform glow of the phase halo.[19] Hoffman contrast also has a greater depth of field than interference contrast systems, as well as an equivalent resolution; and because interference contrast systems use polarized light, they have trouble with birefringent materials. (The fact that Nomarski interference contrast is nearly universal, while Hoffman has only a small market share, is due in part to the companies that promote the two technologies; but it is not irrelevant that interference contrast systems can produce brilliant colors, which makes viewing more entertaining.) Both methods can work near the resolution limits of light microscopy, and both work effectively to enhance and produce contrast. Yet both present formidable challenges to ingrained—and possibly even hardwired—ways of seeing the world. In my experience it is never possible to look at a shape that seems to be illuminated from the right and imagine it is illuminated from the left, and it is not possible to look at an object that seems to be dark and imagine it is light.

30   The Rise of Image Analysis

Phase contrast, Hoffman contrast, and interference contrast are products of the middle third of the twentieth century. Among more recent developments, confocal microscopy and two-photon or multi-photon absorption microscopy (TPA or MPA) are methods of delivering a precise point of laser light to a fluorescing molecule in an object, thus picking a single molecule from among millions.[20] The technique has been used to "cheat" the resolution limit by allowing single molecules to identify themselves by fluorescing.[21] Quantitative polarization microscopy, invented by Shinya Inoué, is unsurpassable—and sometimes irreproducible—on certain objects such as microtubules.

But the greatest advance in light microscopy in the last thirty years has been image analysis, starting with the discovery that a video camera could be used to augment contrast.[22] Today the emphasis in light microscopy is on image-analysis software, including deconvolution, image simulation, and image combination—all techniques of image altering that are largely unknown in the humanities.[23] In biology the preponderance of work in the United States is done on the platform known as NIH Image, which provides a standard toolbox for contrast enhancement, color replacement, and other functions.[24] In my experience biologists in particular tend to use NIH Image's built-in options so that basic threshold analyses and color palettes such as "Fire 1," "Fire 2," "Ice," and "Spectrum" can be easily recognized in journals such as *Nature* and *Cell*. (The built-in palettes are quite garish: think of fluorescent purples and oranges, cool blues and pinks.) With a program such as NIH Image, a nearly featureless image such as Figure 29 (center) can be transformed into images such as those in Color plate 16. At the top, the image is analyzed into "watershed areas," and at the bottom into "particles," which are automatically counted by the software. There are two dangers in such practices: first, that the changes are made without being documented in the publication; and second, that the changes are made by scientists who do not fully understand the mathematical routines involved. The difference between a real feature of the object and one extracted by the software can easily be blurred. In particular it is common practice to make a Fourier transform of an image, edit the transform, and then do an inverse transformation. The operation cleans up the image, but when the modification is undocumented—as it often is—it becomes impossible to deduce the image's original properties.

Image analysis has reached the point where electron microscopists routinely begin with images that are essentially invisible—they look like the snow on a TV screen or like digitized pictures of gray walls. Computers are then used to extract and analyze the objects. In a typical case, a group in Berlin studied a portion of an influenza virus called haemagglutinin. When it is isolated from partly dissolved ("solubilised") viruses, haemagglutinin naturally bonds into starlike shapes ("trimer rosettes"). When rosettes are photographed at the highest resolution, they are virtually invisible against a background of gray static (Figure 30). In order to publish a picture of the rosettes, the investigators deliberately unfocus the microscope so that it produces an image of higher contrast and lower resolution

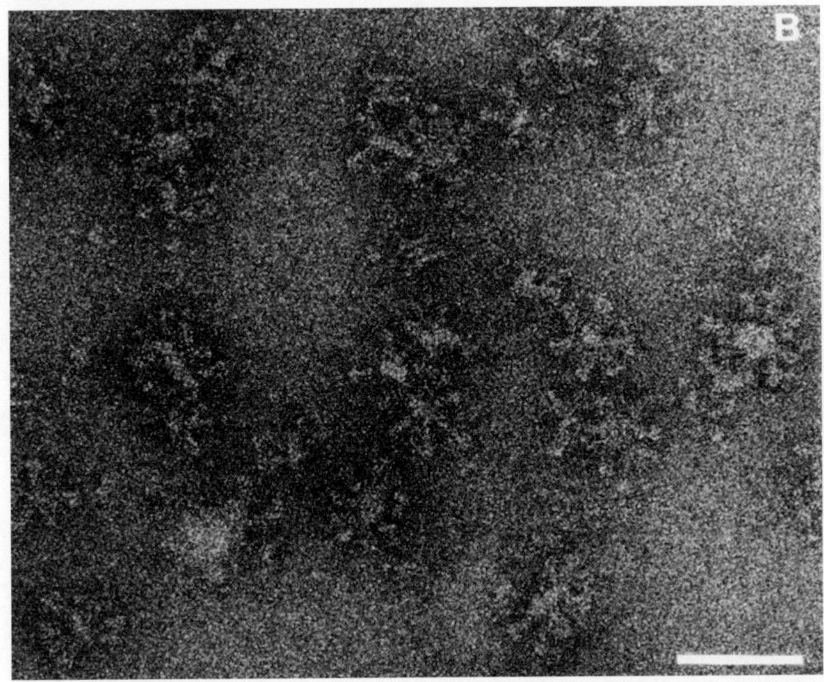

FIGURE 30

TEM micrographs of influenza haemagglutinin rosettes, highest resolution. Courtesy Christoph Böttcher.

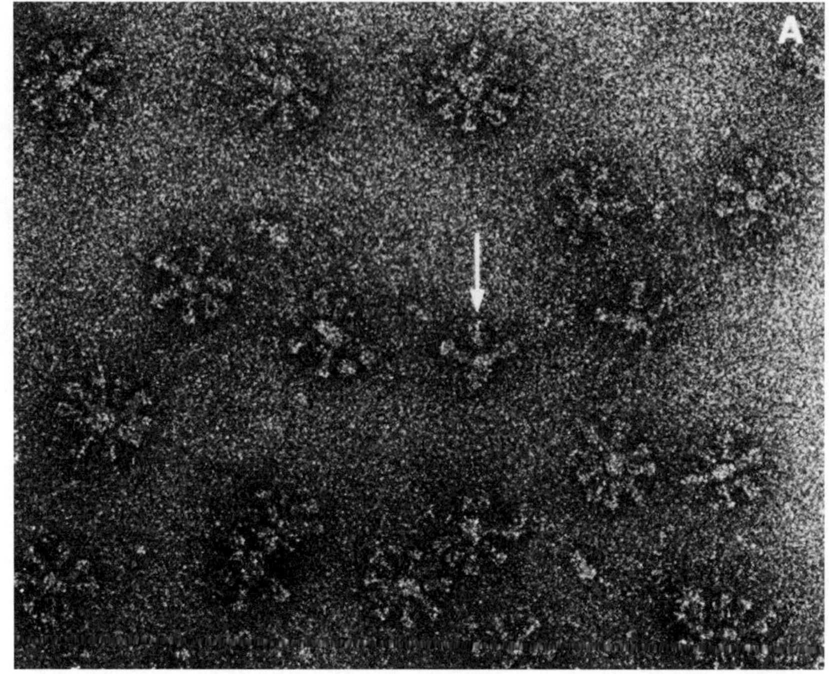

(Figure 31). The experiment proceeds on the basis of the invisible images: first the quality of the images is verified by diffraction (which I will discuss in a moment); then the images are digitized and analyzed using a software package called IMAGIC-5. The software looks for characteristic orientations of individual haemagglutinins within each rosette and groups similar orientations together (Figure 32). The sum of several hundred individual haemagglutinin molecules in different orientations finally yields a three-dimensional reconstruction, which the investigators publish in the typical shiny acid greens, blues, and magentas of their image-analysis software (Color plate 17).[25]

Figure 33 is another example: these are viruses, shown (at top) at the maximum resolution of 13Å, which is made possible by a slight defocus and a huge loss of contrast, and then again (bottom) at a lower resolution of 22Å, which is achieved by putting the microscope out of focus. In this case the lower image was used for data analysis: the choice is always a compromise that depends on the power of the image-analysis software.[26]

It is typical of such work that the raw images were not reproduced. After all, why illustrate a picture that is itself barely visible and that will only

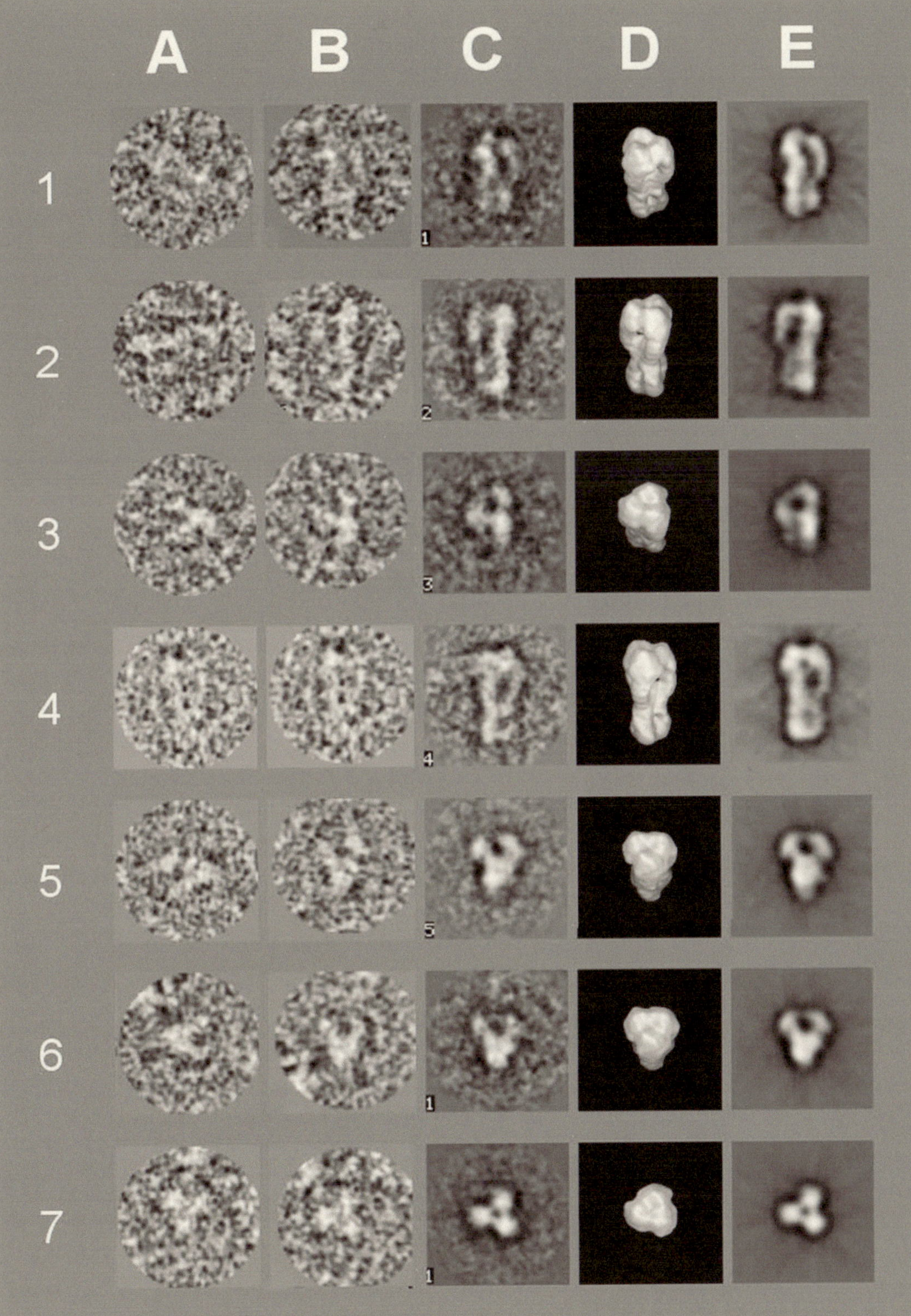

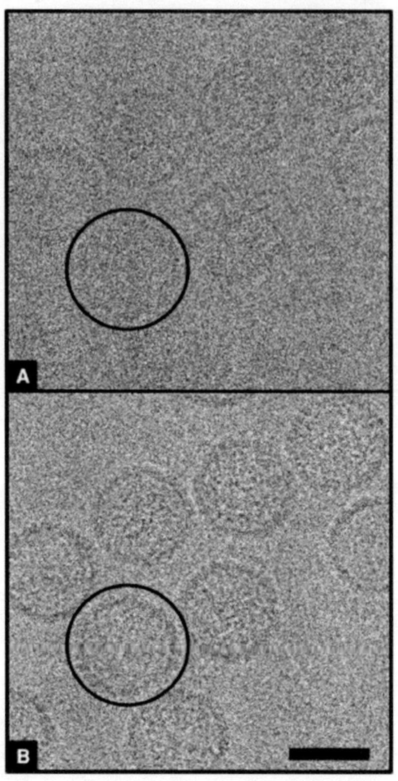

FIGURE 33

Ice-embedded bacteriophase P22 procapsids. Top: close-to-focus image, with defocus approximately 0.95μm. Bottom: more strongly defocused image, at about 2.7μm. Images courtesy Janita Jakana, from P. A. Thuman-Commike and W. Chiu, "Reconstruction Principles of Icosahedral Virus Structure Determination Using Electron Cryomicroscopy," *Micron* 31 (2000): 690, fig. 4. Courtesy Elsevier.

be used as one piece of evidence in a statistical aggregate of similar images, all of them digitized, enhanced, classified, averaged, and reconstructed? The astonishing fact about such work is that it has become routine, for the first time in the history of science, for scientists to begin with images that they cannot see. Ever since telescopes were created, instruments have been providing images the naked eye could not have perceived; and ever since ultraviolet microscopes and X-rays, instruments have been generating images from radiation the eye cannot perceive. But in those cases the images that the instruments delivered were *themselves* visible. Electron microscopy routinely begins with images that are scarcely distinguishable from random noise.

In a sense, reconstructions like Park and Kim's work on the Hubble Deep Field are more radical because they *never* result in forms that can be perceived (see Figure 23). But here, too, a tremendous amount of trust is placed in the algorithms built into software such as IMAGIC-5. In this last example the research team also used a digitization package, Image Science Software GmbH, and they employed other software to perform the optical diffractions needed to verify the image quality: in other words, an entire suite of mathematical operations was needed to turn the featureless "snow" into shiny 3-D models.

31   *Resolution in Electron Microscopy*

As in light microscopy, the theoretical limit of resolution in electron microscopes does not usually match the actual resolution. Aberration aside, the limit of resolution is given by the formula

$$d = 0.6\,\frac{\lambda}{\sin\Theta}$$

where λ is again the wavelength and Θ is the angle of incoming light—the analogue to *N.A.* in light microscopy.[27] In practice the resolution is limited to about 1Å. (The easiest way to imagine the lengths in electron microscopy is to picture the meter divided into 1,000mm, and each millimeter divided into 1,000μ, and each micron divided into 1,000nm. A nanometer is 10Å.) A resolution of 1Å would be approximately 25 times the wavelength of the illuminating electron beam.

That's on paper: in actual experiments the concept of resolution is more involved. The lenses in electron microscopes are magnets, which bend the electron beam. Like glass lenses, magnetic lenses have chromatic aberration (so that different wavelengths are focused at different distances from the lens) and astigmatism (so that rays from different directons are focused at different distances from the lens). Electron microscope lenses are especially affected by spherical aberration, which is different with magnetic lenses than it is with glass lenses. In theory the resolution of an uncorrected electron microscope is

$$d = 0.6(C_3\lambda^3)^{\frac{1}{4}}$$

where C_3 is the objective's third-order spherical aberration and λ is the electron wavelength.[28] To increase resolution it is necessary either to decrease the wavelength (use higher voltage and more energetic electrons) or to decrease the spherical aberration. Unfortunately, higher-energy electrons damage most specimens. Decreasing the spherical aberration is also difficult because electron lenses cannot be adjusted to decrease or eliminate spherical aberration. The history of electron microscopy has been largely a history of increasingly elaborate and sensitive patches to these two limits.

The first images that approached atomic resolution in electron microscopy seemed to follow the laws set down in optical microscopy. In 1975 Victor Phillips succeeded in producing images of dislocations in germanium crystals.[29] His images did not quite have atomic resolution, but they showed rows of atoms, and it seemed for a while as if the ultimate test images for high-resolution optical and electron microscopy looked the same—both were rows of fine lines. By 1983 it was routine to analyze crystals, even in unusual settings such as the "magnetosomes"— tiny magnetite crystals—in magnetotactic bacteria (Figure 34).[30] Even today there are absolute resolution records, which are usually performed on extremely thin crystal lattices.[31]

Unfortunately, resolution in electron microscopy is not just a matter of adjusting the instruments to meet the theoretical limits. Resolution is

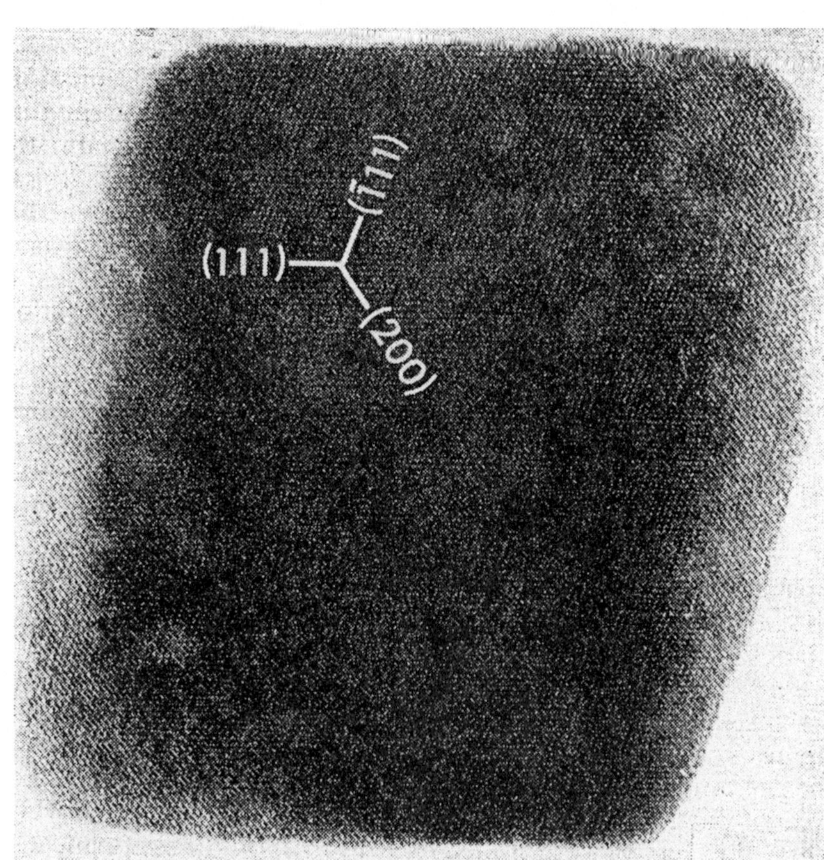

FIGURE 34

Magnetosome (magnetite microcrystal in bacterium). From Tsuyoshi Matsuda et al., "Morphology and Structure of Biogenic Magnetite Particles," *Nature* 302 (May 31, 1983): 412, fig. 3. Courtesy Macmillan Publishers Ltd.

limited by many things: sample thickness, signal and noise curves, source power, the peizo and laser setup and the current voltage converter (which limit time resolution), probe tip configuration, source coherence, contrast, aberration, high temperatures, and even the time it takes to make an image.[32] Atoms can shift, and the illuminating radiation itself damages molecules. Looking, in effect, is also destroying. For some this is a "tiresome" problem that needs to be overcome, usually by cooling the specimen; but it is also a fundamental truth of high-resolution electron microscopy, an issue that can never be entirely solved.[33] Among the many factors that come into play, contrast is paramount, as it is in light microscopy.

32   *The Contrast Transfer Function*

In ordinary light microscopy (without phase contrast or other additions), the dark parts of the object are places where the light has been absorbed. In electron microscopy, few electrons are absorbed: instead, they are scattered. As the specimens get thinner and the magnification gets higher—as the images approach their limits of resolution—fewer electrons are scattered. What matters then is the phase contrast between electrons.

Electron phase contrast is generated in two ways: by *defocusing* the image, and by generating interference between rays that are misfocused by spherically aberrant lenses.[34] Defocusing is a way of forcing scattered rays to follow a longer path to the image than unscattered rays, enhancing the phase difference between them. As in optical phase contrast, transmission electron microscope (TEM) images show fringes wherever the specimen becomes suddenly more or less thick or massive. Fringes create bright lines around underfocused edges and dark lines around overfocused edges. When the object is in perfect focus, the fringes are minimized, and the effect is lost. In its place is a generalized granular background, which gives high-resolution TEM images their typical look.[35]

Focusing an electron microscope is not a simple matter. Often it makes sense to try a series of "through-focus" images at different values of defocus (Figure 35).[36] The object that appears in Figure 35 is an amorphous alloy film, used as a test object to see if the microscope—a high-resolution darkfield scanning transmission microscope, HAADF-STEM—can pick out titanium atoms. There are several optimal points in a through-focus

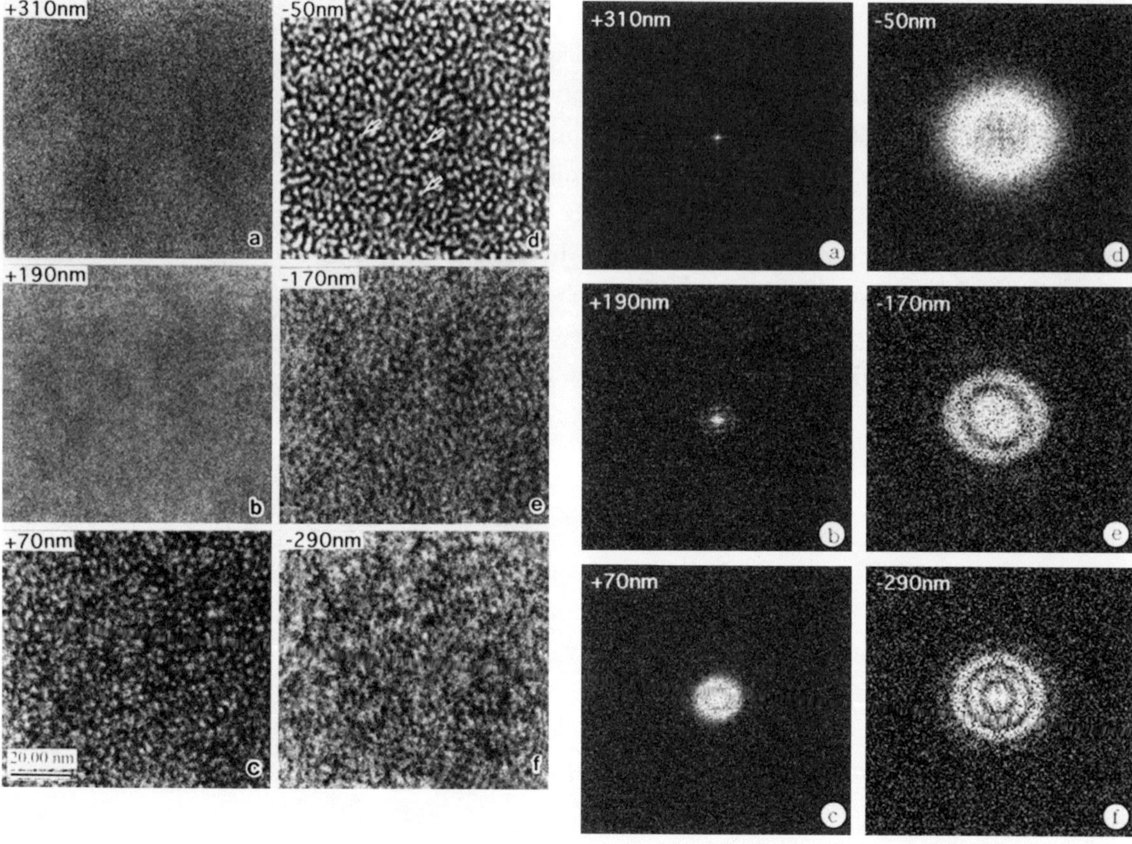

FIGURE 35

Through-focus series of high-angle annular detector darkfield scanning transmission electron microscope (HAADF-STEM) images of an amorphous $Ri_{16}Si_{84}$ alloy film. From N. Tanaka et al., "An 'On-Line' Correction Method of Defocus and Astigmatism in HAADF-STEM," *Ultramicroscopy* 78 (2000): 108, fig. 6. Courtesy Elsevier.

series. The *Lichte focus*, for example, is the point where there is minimal damping of high-frequency information. The *Scherzer focus*, which is approximately one-third of the distance from the optimal focus for resolution and the Lichte focus, is defined in terms of the coefficient of spherical aberration, C_S:

$$Scherzer focus = \sqrt{C_S \lambda}$$

In this example the Scherzer focus is a deliberate underfocus of -50nm, and it allows the experimenters to resolve individual titanium atoms (arrows in the top right image).[37] The overfocused images are not the same as the underfocused ones, which is verified by comparing all six images to their Fourier transformations (Figure 36).[38] The Fourier transform pattern can be thought of as a topographic map: the lighter regions are places where a function called the *contrast transfer function* has higher values, and the distance from the center measures a reciprocal vector

(1/nm). The image at the top right, the optimal one, shows a large area in which the contrast transfer function (CTF) has a high value (at least 0.05). The Fourier transforms in the left column show nearly uniformly low values for the CTF, indicating a lack of contrast; and the last two images, *e* and *f*, show an irregular distribution of the CTF, signifying a lack of resolution.[39]

The contrast transfer function can be graphed as an oscillating curve that repeatedly crosses the *x*-axis. The zeros (places where the *y* value is zero) are spatial frequencies—repeating forms on the image with a certain distance between them—that are imaged with no contrast at all. The CTF is generally a matter of compromise: it is possible to adjust the microscope so that a range of spatial frequencies—forms at varying distances from one another—will be imaged with good contrast, but there will often be spatial frequencies that are imaged without contrast, and others that are imaged in negative contrast. A typical TEM image can be very confusing in this respect: some large patterns might be clearly visible, while others might be "negative," and still others might be invisible or nearly invisible. As the microscope is focused and defocused, the CTF changes: the positions of the zeros shift, and different forms become clear. In Figure 32, the experimenters had to merge micrographs made with different defocus values to produce an optimal image.[40] (It would be as if a camera could only capture objects at certain lateral distances from one another, so photographs had to be taken in different foci and superimposed in a computer.)

Fourier transforms and measurements of the CTF can be made very quickly, so they can guide the process of focusing the microscope. A slower method, which works best when the microscope is first set up, is to make a series of Ronchigrams: essentially shadow-pictures of thin periodic specimens under various degrees of defocus.[41] Figure 37 compares a graphitized carbon film with and without aberration correction.[42] On the left, without correction, the lattice planes of the carbon could only be imaged when the microscope was set at a large underfocus. On the right, the lattice planes—which are 3.4Å apart from one another—show up clearly at a smaller underfocus. In both pictures, Ronchigrams (large images) are compared to Fourier transforms (inset images). The Fourier transform at the right reveals that a range of spatial frequencies is being imaged, while the one on the left shows only a single spatial frequency.

Fourier transform patterns for the images in Figure 35. From N. Tanaka et al., "An 'On-Line' Correction Method of Defocus and Astigmatism in HAADF-STEM," *Ultramicroscopy* 78 (2000): 109, fig. 7. Courtesy Elsevier.

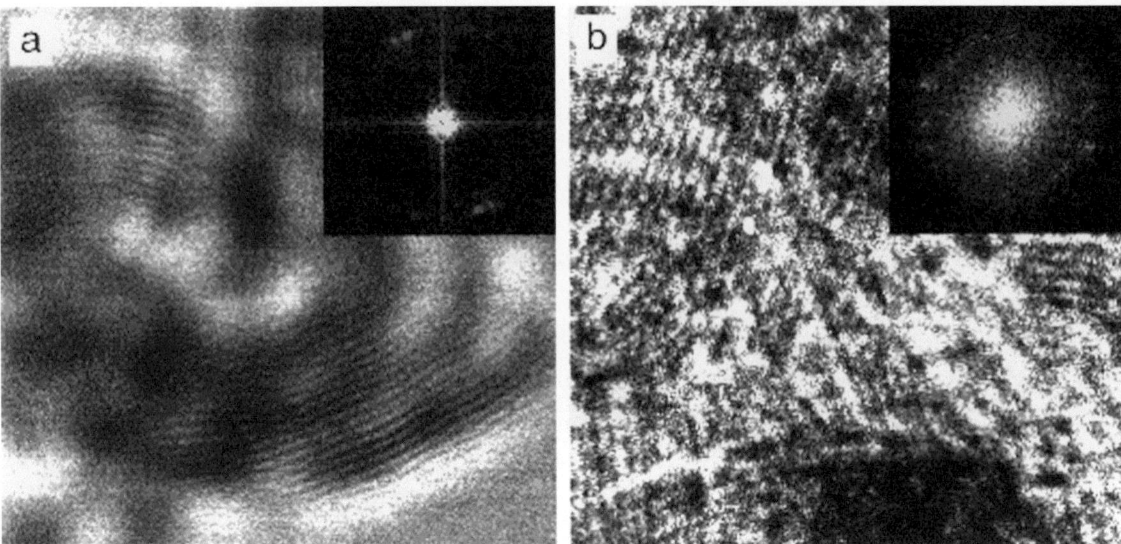

FIGURE 37

Ronchigrams and Fourier transforms (insets) of STEM images of graphitized carbon (002) lattice planes. Left: without aberration corrector. Right: with corrector. From O. L. Krivanek, N. Dellby, and A. R. Lupini, "Towards Sub-Å Electron Beams," *Ultramicroscopy* 78 (1999): 6, fig. 4.

This is just the beginning of the study of the CTF, which is partly a matter of engineering, partly of physics, and partly knack and experience. Defocus is very different from optical microscopy's relatively simple focus, underfocus, and overfocus: the CTF measures a delicate relation between, on the one hand, the average distances between parts of the object and, on the other hand, the contrast and resolution with which they appear: a situation entirely alien to everyday notions about blur and focus.

33   Exit-Wave Reconstruction

Surface chemists who work with high-resolution electron microscopes distinguish between the microscope's point resolution ρ_S, and its information limit ρ_I. The former is the "finest detail that can be interpreted directly in terms of the structure," and the latter is the "finest detail that can be resolved by the instrument, regardless of a possible interpretation."[43] In some situations, such as looking through relatively thick sections of crystals, the focus changes in bewildering ways, scrambling parts of the signal and degrading the information limit. Figure 38 shows a NiO crystal, seen with various values of focus. Among the unexpected phenomena, the top layers of nickel atoms seem to diffuse upward with increasing underfocus—as shown by the compressed image at the right. The best strategy to use in such a situation is *exit wave*

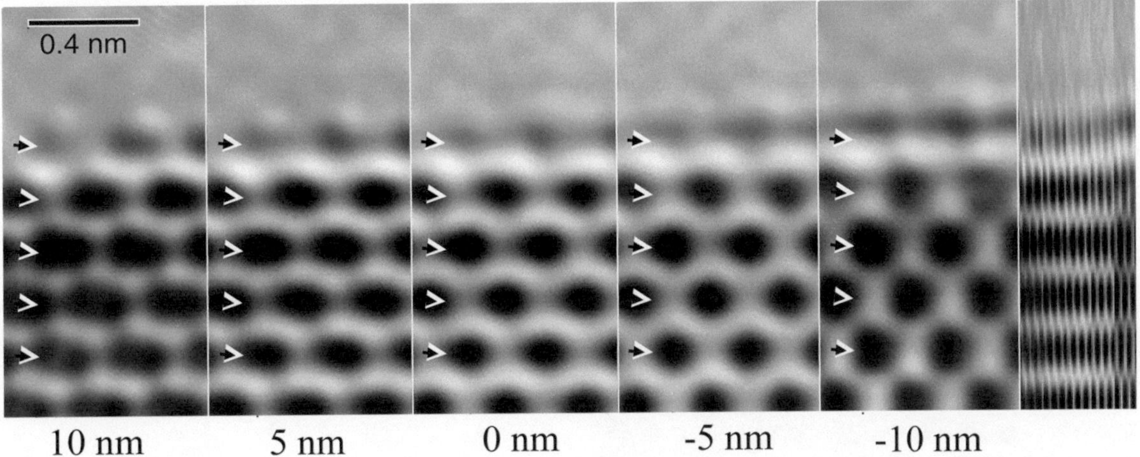

FIGURE 38

Experimental exit wave images (showing amplitude) of the (001) surface of a NiO crystal, shown at five focus values. From H. W. Zandbergen and D. van Dyck, "Exit Wave Reconstructions of Surfaces and Interfaces Using Through Focus Series of HREM Images," *Solid State Ionics* 131 (2000): 43, fig. 6. Courtesy Henny Zandbergen.

reconstruction: a computer analyzes the through-focus series and produces an image of the electron's exit wave, which is clearer than any of the individual images. In that fashion the point resolution is made to approach the information limit. (In this example, the information limit ρ_I is 0.14nm.) Figure 39 is a comparison of one of the frames of the through-focus series (upper left) with the results of the computer-generated exit wave reconstruction (upper right and bottom left). The reconstruction retrieves information that would have been lost, distinguishing the amplitude of the exit wave (upper right) from the phase (bottom left). In the picture of amplitude, the nickel atoms appear as finely resolved black dots; in the picture of the phase, the atoms are black dots in the thinner portion of the crystal (toward the top) and white dots in the thicker portion (toward the bottom). The authors of this study provide sample rectangles to help compare: all three rectangles look the same in the picture on the right—each rectangle is centered on a hexagon of black Ni atoms, with a seventh at its center. In the phase image at the bottom, the same three rectangles enclose different patterns. The rectangle at the left looks like a simple negative of the rectangles in the amplitude image, but in the lowest rectangle the white spots and dark areas have changed places: evidence of a phase shift caused by the thickness of the crystal.

In this kind of work it is common to see images qualified as "experimental," meaning they are made the old-fashioned way—all at once, of one sample, at one focus. The opposite of "experimental" is "calculated." In Figure 39 the two images at the right and bottom could not have been

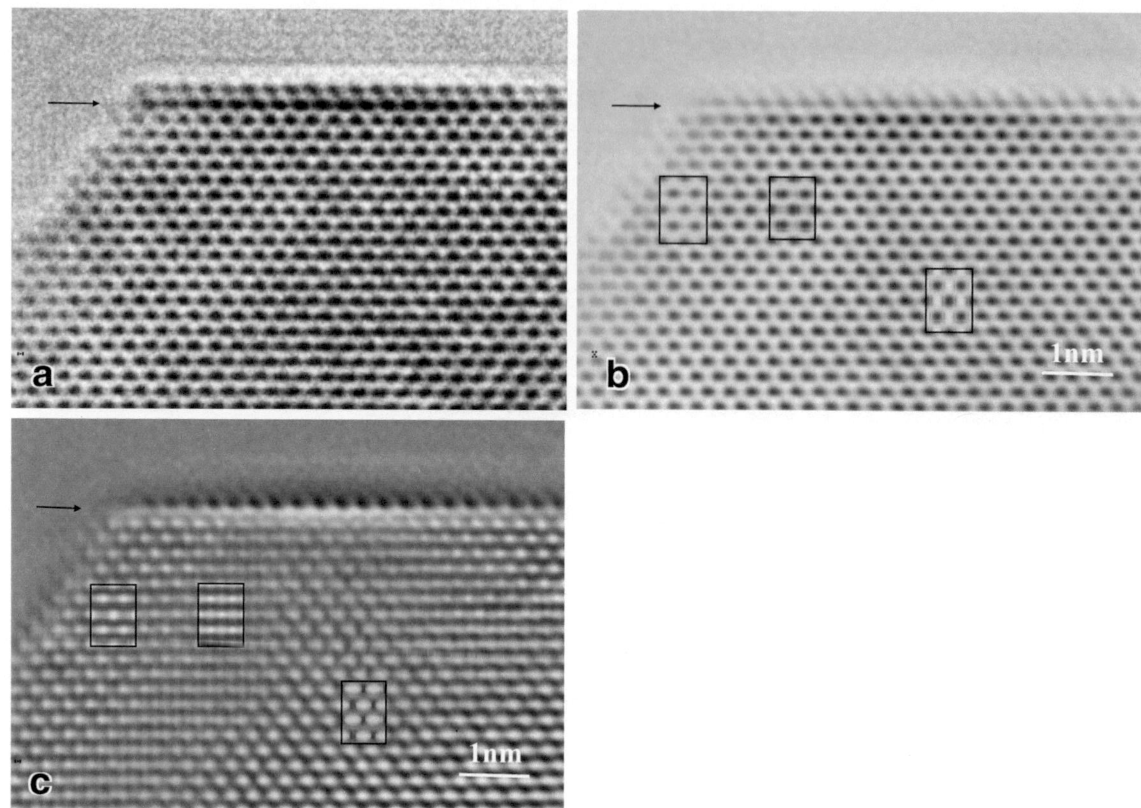

FIGURE 39

Profile images of the (001) surface of a NiO crystal. (a): an experimental image made at −100 nm focus; (b) amplitude of the exit wave, calculated from a series of fifteen through-focus images; (c) phase of the exit wave. From H. W. Zandbergen and D. van Dyck, "Exit Wave Reconstructions of Surfaces and Interfaces Using Through Focus Series of HREM Images," *Solid State Ionics* 131 (2000): 42, fig. 5. Courtesy Henny Zandbergen.

generated by a single exposure: they are products of intensive calculations, performed on fifteen different images—a statistical aggregate of digitally modified images.

Exit wave reconstruction only works when the object is a perfect crystal. In other cases it is necessary to work with the exit wave as it is, or even to look at the electrons reflected back from the object. Some metal catalysts are made of tiny particles of platinum embedded in carbon; to study them it is necessary to make images of the "secondary electrons" that pass through the carbon, as well as the "backscattered electrons" that bounce off the platinum particles and return to the surface.[44] Both kinds of images are shown in Figure 40: those on the left are secondary-electron images, and those on the right are backscattered electron images. (A lower-energy beam was used to make the top two images.) The individual platinum particles are labeled *A*, *B*, *C*, *D*; those close to the surface, such as *A* and *D*, shine brightly no matter how

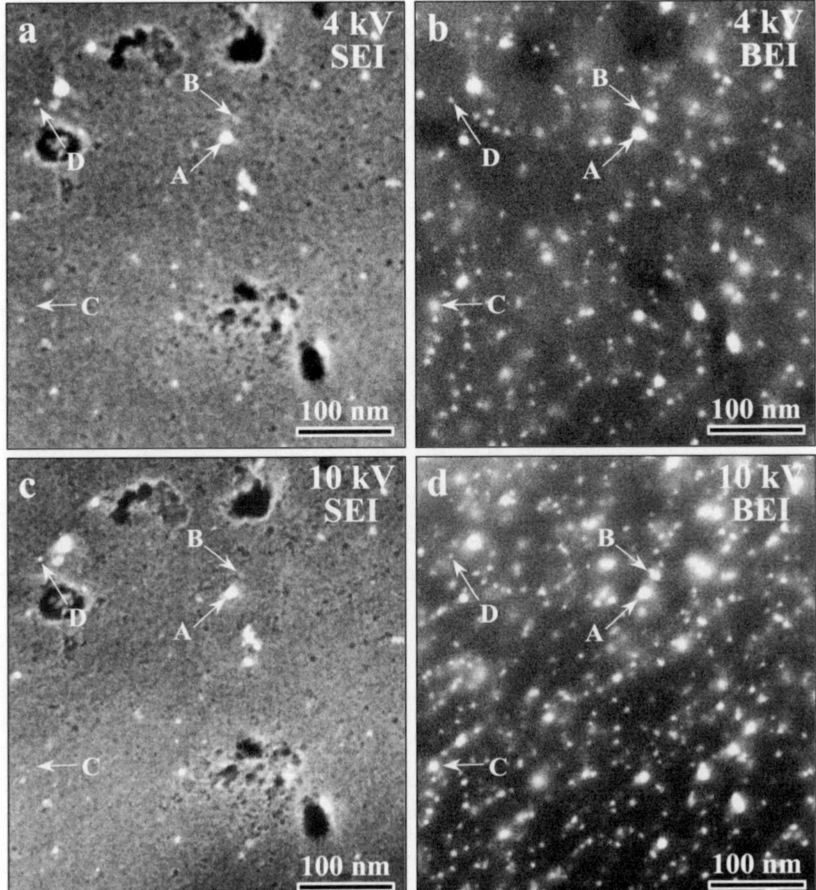

FIGURE 40

Carbon with embedded platinum particles. Top: 4 keV; bottom: 10 keV. Left: secondary-electron image; right: backscattered electron image. From Jingyue Liu, "Contrast of Highly Dispersed Metal Nanoparticles in High-Resolution Secondary Electron and Backscattered Electron Images of Supported Metal Catalysts," *Microscopy and Microanalysis* 6 (2000): 396, fig. 6. Courtesy Cambridge University Press.

much energy is used to illuminate them. Particles deeply embedded in the carbon matrix are not visible in secondary-electron images, but they show up well in high-energy backscattered images (for example, particle *C*, which is bright in the lower-right image). There is no correlation between the brightness of particles in the secondary-electron and backscattering methods, and there is no simple formula to predict the brightness of a particle. The experimenter, Jingyue Liu, concludes that if the energy is high enough (> 10 keV) and the particles are small enough (< 100 nm), the intensity of the images will vary in accord with the particle's size. With less energy, no such relation can be found. In the end, despite Liu's statistical simulations, the problem cannot be quantified, and the objects flicker in and out of visibility like faint stars on successive astronomical plates.

34 *Scanning Probe Microscopes*

The contrast transfer function and exit wave reconstruction apply to TEMs. Beyond them is a bewildering variety of electron microscopes. After TEMs, the next to be developed were the scanning electron microscopes (SEMs). In the last fifteen years of the twentieth century there were also SPMs (scanning probe microscopes), including STMs (scanning tunneling microscopes), AC-STMs, AFMs (atomic force microscopes), CFMs (chemical force microscopes), and—at the very end of the century—NSOMs (nearfield scanning optical microscopes).[45] Those basic kinds subdivide into an astonishing number of evanescent technologies: inelastic tunneling spectroscopy, ballistic electron emission microscopy, scanning spin-precession microscopes, scanning thermal microscopes, and a dozen others just between 1981 and 1995.[46] Most recently there are scanning capacitance microscopes, magnetic resonance force microscopes, and atomic-resolution acoustic microscopes.[47]

Scanning probe microscopes are the most interesting development, because they do away with lenses altogether, substituting a tiny, pencil-shaped tip that hovers just over the atoms in the sample. By various means, the tip registers the atoms' presence, typically by wavering back and forth in response to the surface, and that vibration generates the picture. (It used to be that microscopic objects trembled: now the apparatus also oscillates. Scanning probe microscopes are sometimes suspended in elastic hammocks to dampen their inherent shaking so that the probe is the only thing that moves.)

In theory, the probe tip does away with Abbé's limits on the diffraction-limited resolution of microscopes, because it means that resolution is no longer limited by wavelength. Instead it is limited, in general, by the size of the probe itself. An ultramicroscopic probe can resolve ultramicroscopic detail: moving the probe tip upward by as little as 1Å (roughly the width of an atom) causes the tunneling current to decrease by a factor of ten.[48]

STM pictures are frequently manipulated to look like garishly colored, bumpy moonscapes.[49] Each lump in the topography is said to represent an atom, but more precisely speaking, STM images do not resolve atoms at all: they measure a probability function, which depends on the likelihood of electrons "tunneling" between the atoms of the sample and a tiny probe that hovers overhead.[50] What seem to be atoms in the computer printouts—atoms as solid as little hills—are really mathematical functions of properties of the atoms. (They are not electron orbitals, as is sometimes implied.[51]) In one configuration the tunneling voltage is kept constant, so the three-dimensional surface is a graph of the heights above the surface that produce equal tunneling current—not a map of the contours of the atoms themselves. As one of the inventors of the technology puts it, although an STM image "can be related to topography, in general such an image does not represent pure topography," because many properties of the image can influence the tunneling current aside from topography. Anything that affects tunneling current affects how high or low the probe goes, and that in turn generates the "map." When there is more than one kind of atom present, for example, the STM image will not be a good indicator of topography because each kind of atom has its own tunnel probability. "It is often not clear what property of the specimen causes the contrast," another team writes. "Often the contrast will not depend linearly on that property of the specimen, and the transfer function relating object and image will vary from location to location on the surface." The tip might be responding to the attractive van der Waals force, the meniscus force, the electrostatic force (if it's an aqueous medium), or the hydration force (the work required to dehydrate surface groups when forcing two materials together).[52] A panoply of forces and limitations conspires to prevent STM images from simply being pictures of atoms; this puts some pressure on the word *topography*, which has to stand for whatever *seems* to be there.

The atomic force microscope (AFM) does resolve atoms, by measuring how the probe is repulsed by their electric charges. Even so, the pictures that AFMs produce are not images in the ordinary sense. The microscopic probe hovers over the surface, springing back when it passes over an atom; the probe's vibration is detected by shining a laser onto it. The laser light reflects up, off the probe, and hits a target. The target is calibrated into four regions, and what gets measured and finally imaged as individual atoms is really a mathematical analysis of the laser's motion across the detector.

The most recent addition to the stable of ultramicroscopes is the NSOM, near-field scanning optical microscope. There are two kinds; the "apertureless" one, with the higher resolution, was invented in 1989 by H. Kumar Wickramasinghe.[53] In this method, a tiny light source shines on a microscopic metal tip that is suspended just above the object being studied. Minute quantities of light reflect off the sharp tip, and a detector records how much light passes from the tip to the object. The light varies as the tip goes down to the surface or retracts up and away from it. NSOM is extremely sensitive—it has already achieved resolutions of 1nm—and it can, in theory, image a single molecule or even a reaction in process. If the light source is tuned to a wavelength that is absorbed by one of the molecules in the object, those molecules will show up distinctly against a lighter background. In Color plate 18, the image at the upper left is an AFM picture of individual molecules of the stain called coomassie blue. The upper-center image is a detail, also made with an AFM. At the upper right is an NSOM image of the same detail, showing how the molecules absorb the light reflected from the tip. (The bottom row of images in Color plate 18 is the same, but with molecules of rhodamine. The wavelength was changed to match the molecules' peak absorption, as shown at the far right in each row.) The molecules appear in those two right-hand images as black patches. It is strange to reflect that pictures B and C seem so similar, yet B is a map of electric repulsion, and C is an image of absorption. The images are akin to everyday pictures, because they show dark where light is absorbed (as it is by masses of coomassie blue, rhodamine, or any other colored substance). The NSOM images, for all their unusual and intricate technology, are "pictures" in the ordinary sense. Like Marco Breuer's photographs (see Color plate 6 and Figure 14), the images in Color plate 18 are made without lenses or apertures of any kind, and AFM images even work without light. (As in Breuer's pictures, they record the force of encounters with actual objects.) Yet the NSOM images can almost be understood as pictures of familiar-looking objects that are light and dark.

I say "almost" because NSOM pictures are not images of static objects, captured the way a flash illuminates a scene in ordinary photography. Instead the probe tip and the object are shaken back and forth at high frequency (typically 80MHz). In one configuration, the probe goes up and down while the sample vibrates back and forth. The two movements are at slightly different frequencies, and the signal that is detected is the difference between the two. The idea is to mask out the light scattered from other parts of the sample, and detect only light modulated by both dither motions together. Background "noise" decreases as the sixth power of the distance between the source of the scattered light and the tip of the probe.[54] So an NSOM image, which seems most like an everyday photograph, is really an accumulation of data from intense vibrations at a microscopic scale: hardly a picture in the ordinary sense.

35 *Acoustic Microscopes*

One step further away from familiar vision are atomic-resolution scanning acoustic tunneling microscopes (SATMs). The idea is to make the sample vibrate—using ordinary sound waves of very high frequencies—and to watch how individual atoms move. Much of the work on SATMs has been done in a laboratory headed by Eduard Chilla, at the Paul-Drude-Institute for Solid State Physics in Berlin.[55] Initially, the problem with the process was that the materials tended to vibrate far more quickly than the scanning probe tip could manage, so the images of atoms would be blurred (Figure 41 [b]). The key was to add an alternating current to the curcuit with a slightly different frequency; then the two frequences (the material's own "surface acoustic wave" [SAW] and the added AC frequency) mix, producing a frequency that is much slower and can be detected.

The result is pictures of individual atoms vibrating. Figure 41 (c) and (d) shows the phase and amplitude of the atoms' vibrations, respectively. In such a close view, all the atoms are in phase with one another, so ideally the two pictures are uniform when the sample is perfect and flat. When the phase and amplitude pictures are enlarged, they show some fine structure (Figure 42, left). The images are messy because the experiment was done in ambient air at room temperature: it could have been improved if the gold crystal had been imaged in an ultrahigh vacuum at low temperature.

The two details in Figure 42 (a, c) were then compared with computer predictions (Figure 42 [b, d]). When atoms in a crystal are subjected to a SAW, they move in elliptical orbits—they stay in place, but they vibrate in little ellipses that are perpendicular to the crystal surface. The elliptical shapes in the computer reconstruction of the phase image (Figure 42 [b]) are a reflection of those elliptical orbits. The hexagons in the amplitude image (Figure 42 [d]) are harder to visualize, but they are consequences of the particular eccentricity of the atoms in this experiment. In both cases, the computer predictions are matched—just barely—by the actual images. It is possible to see a hexagon in the bottom right image, and bright ellipses in the top right image.

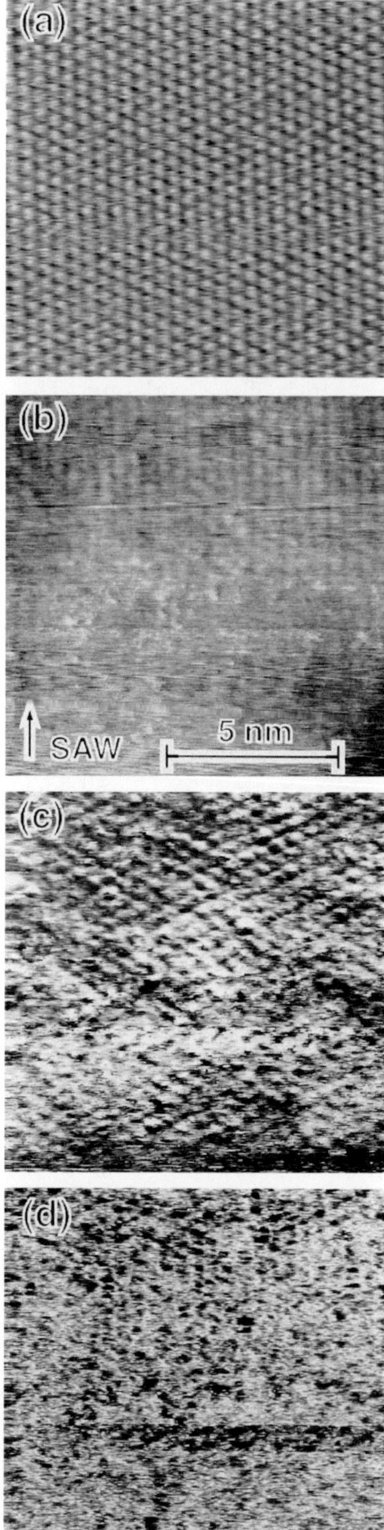

FIGURE 41

Gold crystal (111) surface. (a) STM image; (b) STM image, with surface oscillating; (c) SATM image of the phase of oscillating atoms; (d) SATM image of the amplitude of os‐cillating atoms. From T. Hes‐jedal, E. Chilla, and H‐J. Fröhlich, "Direct Visualization of the Oscillation of Au (111) Surface Atoms," *Applied Physics Letters* 69, no. 3 (July 15, 1996): 355, fig. 1. Courtesy *Applied Physics Letters*.

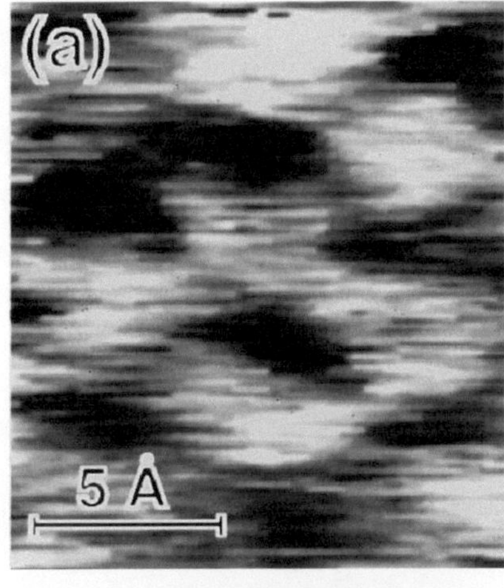

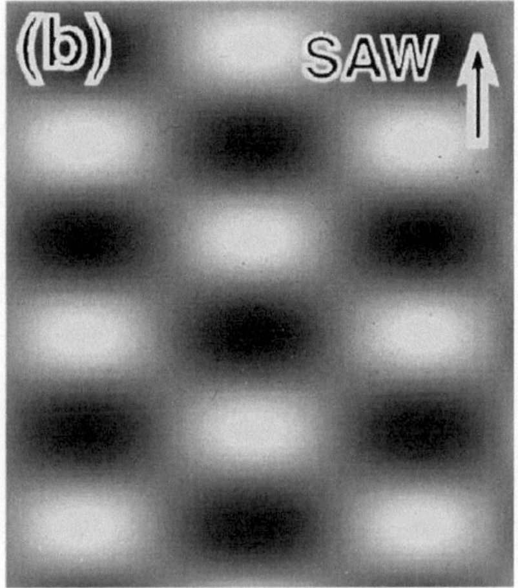

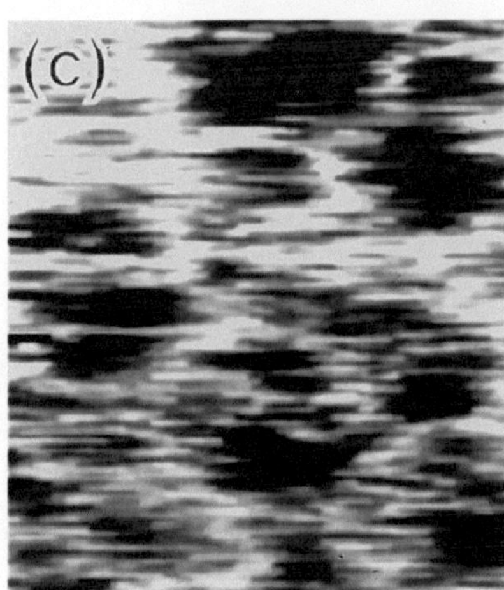

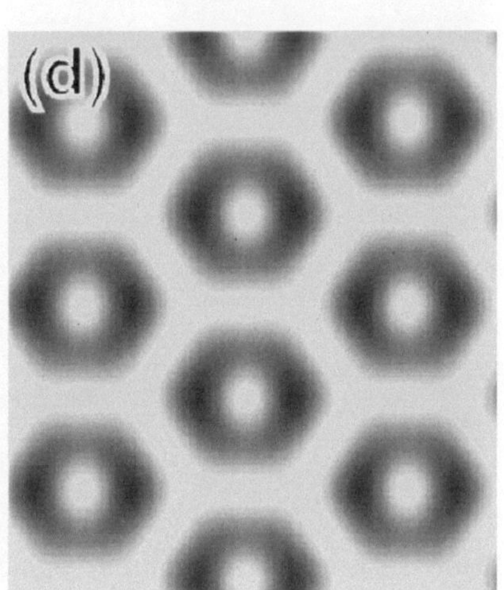

FIGURE 42

Gold crystal (111) surface. (a) detail of Figure 41 (c); (b) computer model of (a); (c) detail of Figure 41 (d); (d) computer model of (c). From T. Hesjedal, E. Chilla, and H-J. Fröhlich, "Direct Visualization of the Oscillation of Au (111) Surface Atoms," *Applied Physics Letters* 69, no. 3 (July 15, 1996): 355, fig. 2. Courtesy *Applied Physics Letters*.

Consider what is being made visible in these images. As the probe tip scans the surface of the gold crystal, a tunneling current passes between the gold atoms and the tip. That current always goes straight from the tip to the "topography"; sometimes the current is vertical, and other times it is slanted. The little ellipses stand up vertically in the surface, and each atom is at a particular place in its elliptical path when the probe approaches. (All the atoms in the sample are in virtually identical places because the image is "stroboscopically" rapid in relation to the size of the SAW.) When the tunneling current is colinear with the atom's position (its displacement vector), then the phase image registers a maximum. When it is non-colinear, some intensity is subtracted. Hence the phase image is a picture of added and subtracted *vectors*, not topography in the ordinary sense. In addition, the bright spots (they should not be called atoms) are elliptical because the atoms are like little spheres half sunk in water: as they vibrate elliptically, they trace an ellipsoidal surface, which is what the probe tip encounters.[56]

What kind of an image is this? First, it isn't a picture of atoms, because it records tunneling current. Second, it isn't a static image of a static object but the mathematical difference of two frequencies—the object and the scanning probe tip are both vibrating in the range of 50,000 times a second, and even the slower frequency that is recorded is in the range of 10,000 vibrations a second. And third, it isn't a picture of heights and depths but of the orientation of vectors. We are very far from light here, and even far from sound.

36   Invisibles: Electron Bubbles, Fluxons

The sensitivities of camera films correspond well enough to the sensitivities of the eye so that snapshots record phenomena that can be seen by the eye. An X-ray film, a gamma ray record made by an astronomer, and an infrared picture made by a surveillance camera have to be understood by analogy because they do not record light that is visible to the eye. The patterns of light and dark, the shadows and highlights, are *analogous* to those our unaided vision might record. By their nature, electron microscopes of various kinds are even farther removed from ordinary vision. Unlike X-rays, gamma rays, and infrared, electrons are not even part of the electromagnetic spectrum. Still, it seems important to retain the remnants of naturalistic representation. Electron microscopy is full of techniques for making the unfamiliar seem analogous to the familiar. For example, hard surfaces are "shadowed," sprayed with metals at an oblique angle so that the resulting images look as if they were surfaces struck by the sun. Sometimes surfaces are observed at an angle, again making them reminiscent of landscapes. (In one book a reflection electron microscopy [REM] image is compared with a sunset view of Mont Saint-Michel in order to show that the alien microscopical object is "really" just a landscape.[57]) In such cases electrons are made to behave *as if* they were photons: they strike or pass through surfaces and are absorbed or scattered. But there are other kinds of microscopy in which electrons record things that no photon ever could. Scanning probe microscopy is an example, because it measures contours of electron density. Such images are still seen in terms of naturalistic representation, even though the objects they record could never be made visible by electromagnetic radiation. There is a wide range of such subjects: images of electric surface charges, elasticity, conductance, magnetization, and friction. The images are often made to look like solid objects illuminated from an ordinary light source.

One recent example involves the visualization of electrostatic potential inside a supercooled fluid known as a quantum Hall liquid. The "liquid" is confined to two dimensions between two semiconducting strips, so that electrons are compelled to move on a flat plane and not up or down. If the tip of a scanning probe microscope is held over a thin sample, a "bubble" of electrons forms underneath the tip and follows the tip as it scans the surface. If the tip and bubble pass an area of high electrostatic potential, an electron is expelled from the bubble; each time that happens, a signal is recorded. The results look exactly like topographic maps, with continuous contours around "peaks" of electrostatic potential. They are beautiful pictures, and entirely unrelated to the maps they seem to resemble. As in other fields from meteorology to gene mapping, the thought of an undulating landscape marked in isobars is a familiar analogy; but here, especially, it is an incomplete analogy: the electron liquid is utterly flat—as flat as any surface can be—and its "isobar lines" are not the abstractions of topographic maps, but discrete areas where electron transfer is prohibited by "Coulomb blockade."[58]

Another example of electron microscopy that reveals things that electromagnetic radiation could not is Lorentz microscopy, which detects magnetic fields. The field was first explored in the 1960s, though it had been theorized earlier.[59] Magnetic flux lines can be resolved in electron microscopes, provided that a magnetic field H is placed perpendicular to the optical axis, causing the electron beam to divert to one side. Remarkably, the resolution of magnetic flux does not depend on the wavelength λ of the illuminating electrons. Instead it matters that the flux quantity $\Delta\Phi$ is large compared to the flux unit hc/e, where h is Planck's constant, c is the speed of light, and e is the charge of the electron. (The constant $hc/e = 4 \times 10^{-7}$ gauss cm^2, and the flux quantity

$$\Delta\Phi = D\delta H$$

where D is the thickness of the volume of the added magnetic field measured along the optical axis, δ is the diameter of the field region illuminated by the electron beam, and H is the magnetic field.[60])

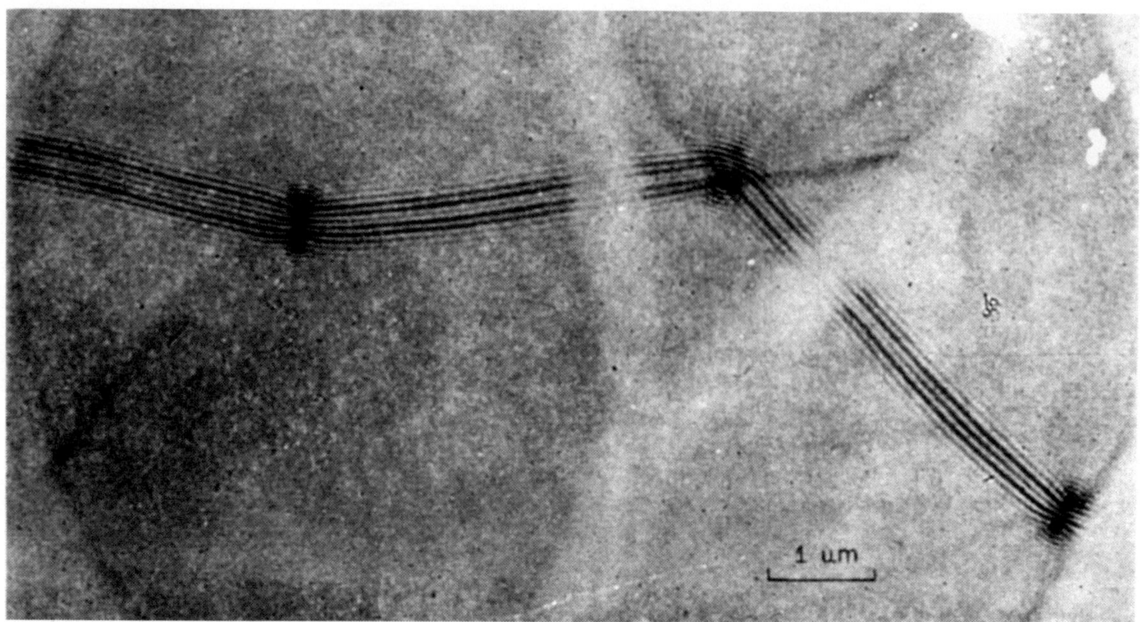

FIGURE 43

Interference fringes in magnetic domain walls, printed in negative contrast. From R. H. Wade, "Lorentz Microscopy or Electron Phase Microscopy of Magnetic Objects," in *Advances in Optical and Electron Microscopy*, vol. 5, ed. R. Barber and V. E. Cosslett (London: Academic Press, 1973), 255, fig. 8. Courtesy Elsevier.

FIGURE 44

Fresnel image of a converging magnetic domain wall region in a permalloy film, printed in negative contrast. From R. H. Wade, "Lorentz Microscopy or Electron Phase Microscopy of Magnetic Objects," in *Advances in Optical and Electron Microscopy*, vol. 5, ed. R. Barber and V. E. Cosslett (London: Academic Press, 1973), 282, fig. 24. Courtesy Elsevier.

The early images of interference fringes between magnetic domain walls are also lovely, but they have low resolution (Figures 43, 44).[61] More recently Lorentz microscopy has been used to investigate the properties of superconductors, and in 1989 a team succeeded in resolving an individual fluxon—a quantum of magnetic energy (Figure 45, top).[62] By increasing the resolution the same team succeeded in dividing the fluxon into quantum parts (Figure 45, bottom). The fluxon is nothing that could ever be seen; even aside from the fact that it is microscopic, and aside

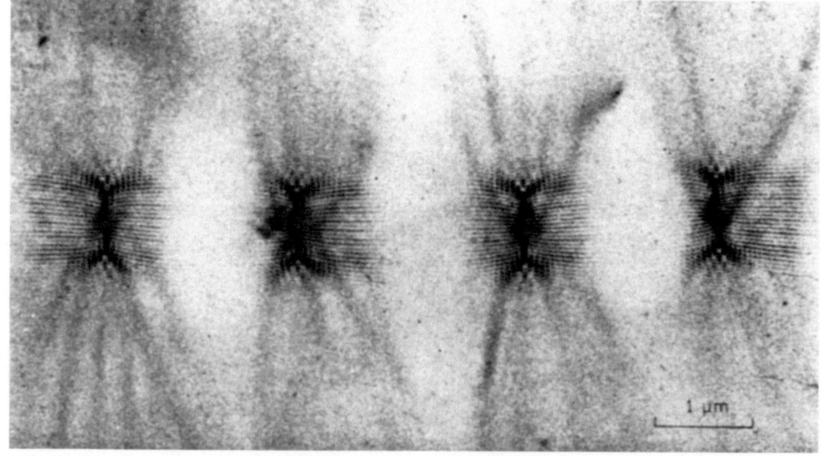

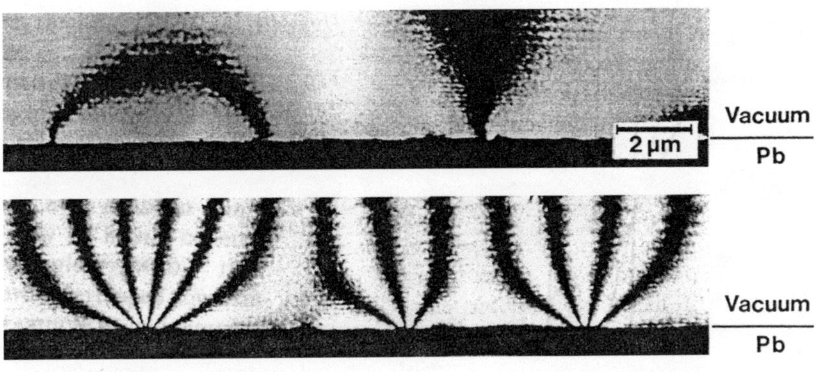

FIGURE 45

Image of a single magnetic flux (fluxon) penetrating a super-conducting Pb film. Top: film thickness 0.2 μm; bottom: thickness 1.0 μm. From Tsuyoshi Matsuda et al., "Magnetic Field Observation of a Single Flux Quantum by Electron-Holographic Interferometry," *Physical Review Letters* 62, no. 21 (May 22, 1989): 2519–22, fig. 1, p. 2519. Copyright 1989 by the American Physical Society.

from the fact that the illuminating "light" is electrons, the "object" itself is a property of the magnetic field: permanently invisible, no matter how much it resembles a tornado.

37 *At the End of Resolution*

At the limits of resolution, chemists and physicists end up peering at grainy plates—exactly as astrophysicists do, except without the image-analysis software that helps astrophysicists lift information from background noise. In 1999 two Japanese physicists devised a way to pull a single atom off its substrate and project it onto a counting device. They made digital field ion microscope (FIM) images of the substrate—a tungsten point—before and after the atom was removed, producing two pictures that look for all the world like details of an astronomical image (Figure 46). The arrow indicates the atom that was evaporated. At first glance, it seems that some other atoms have also disappeared or shifted positions—the atoms along the right-hand margin, for example, seem to have changed. In the experiment, the atom was evaporated onto a detector, which measured the atom's mass (14 amu, corresponding to a single doubly charged Si ion) and its position (at the upper-left of the field). Hence no other atoms could have left the surface except from the position indicated by the arrow. How, then, to account for the difference in the other atoms' positions? The authors suggest that the absence of one atom reconfigured the electric field distribution on the tungsten tip. Where the field becomes weaker, more atoms appear, and where it becomes slightly stronger, fewer are visible. At this scale, the atoms influence one another, and—unlike stars!—they blink in and out of the threshhold of visibility. (Stars blink on account of the earth's atmosphere;

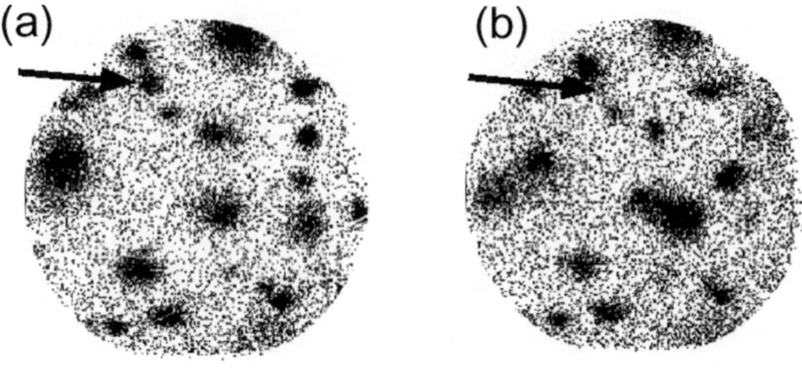

that blinking can be corrected, but this cannot.) The authors have no
method of extending or quantifying their analysis, and their paper ends
with the general observation that the only way to be sure of what happens
in this situation is to know the evaporated atom's mass and position: everything else is necessarily hazy.[63]

Another kind of haze envelops images made by the Swedish chemist
Jan-Olov Bovin (Figures 47, 48). The images show the surfaces of gold
crystals, with the atoms lined up in the familiar rows and columns. From
a distance, so to speak, the crystals look like pyramids myseriously building themselves. At higher magnification it is possible to watch the surfaces of the crystals continuously rearranging themselves: atoms detach
themselves, hover above the surface, and reattach in other places. Bovin
made a series of exposures in rapid succession, revealing how "clouds" of
gold atoms mill about in the vacuum above the crystal. In the top image
in Figure 47, the dark spots in the "air" are gold atoms, moving above the
surface. In the bottom image those atoms have moved (and one has van-
ished). The crystal lattice itself is darker, indicating some atoms from the
cloud have found places in the crystal.

Bovin analyzes sequences of such images, watching to see where the
atoms go—but it is an inexact science. The images are not instantaneous;
Figures 47 and 48 are exposures of 0.13 second each, which smears the
atoms' positions somewhat. Each picture is "some sort of time and spatial
average" of atom positions.[64] Also, the activity is influenced to some degree by the incident beam of electrons in the microscope, as evidenced by
the fact that defocusing the beam reduces the atoms' motions. Like the
images of the single Si atom, these pictures cannot be analyzed any further. Bovin looks very closely, and he writes evocatively of what he sees.

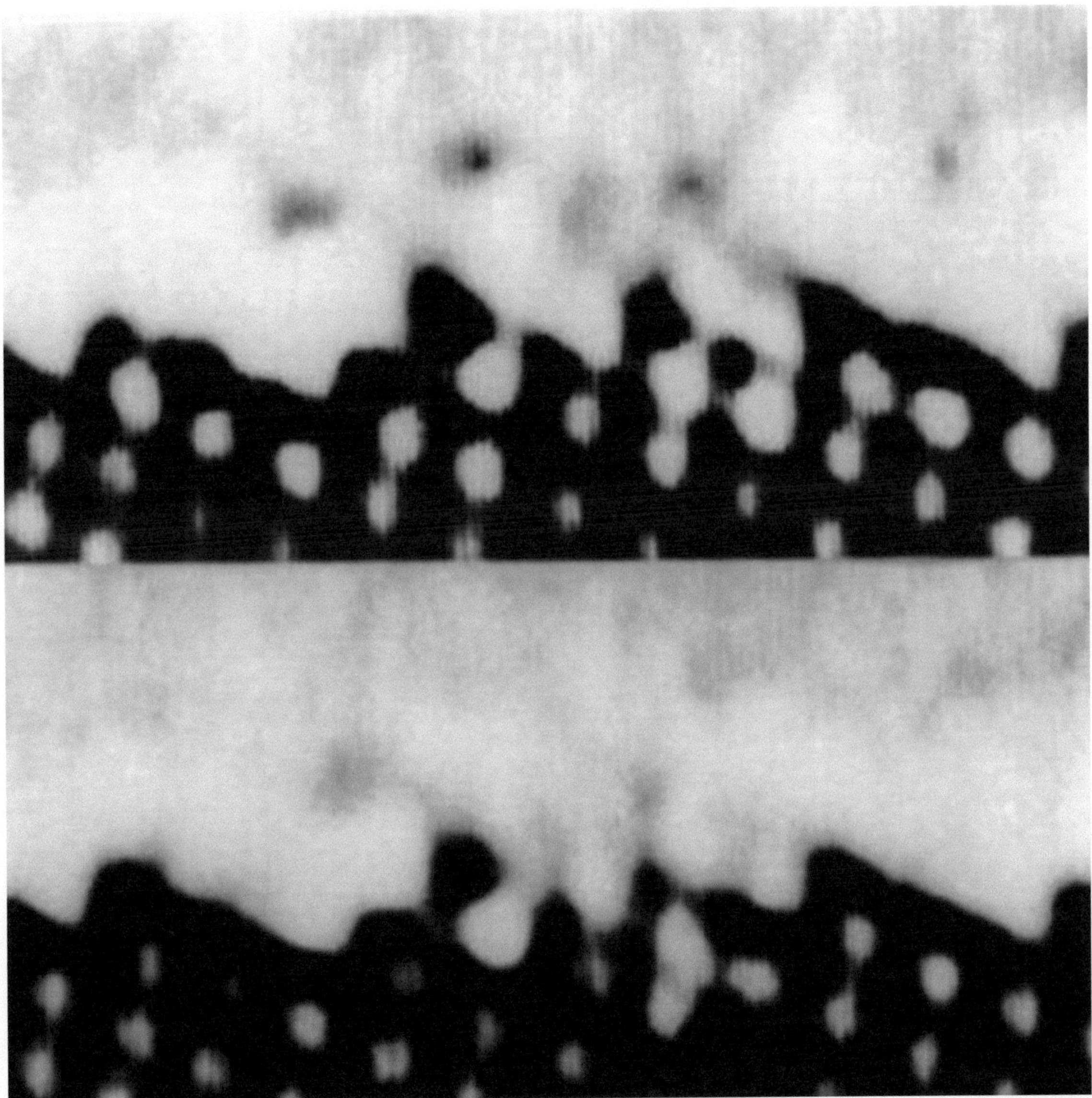

FIGURE 47

Two images showing surface profiles of a gold crystal, viewed on the (331) surface. (The [111] plane runs nearly vertical, up and to the right.) Photo courtesy Jan-Olov Bovin.

A few seconds before these pictures were acquired, he observed parts of the cloud "looking like tornadoes," funneling atoms down into columns of the crystal. In another paper, Bovin describes watching as atoms precipitate from the cloud, lining up in rows and building the crystal outward. The first sign that the crystal is growing, he says, is "the appearance of atoms" near the point of growth, and wherever the crystal grows, "the growth front seems to be preceded by the cloud."[65]

Not much more can be done with the pictures in Figures 47 and 48: the

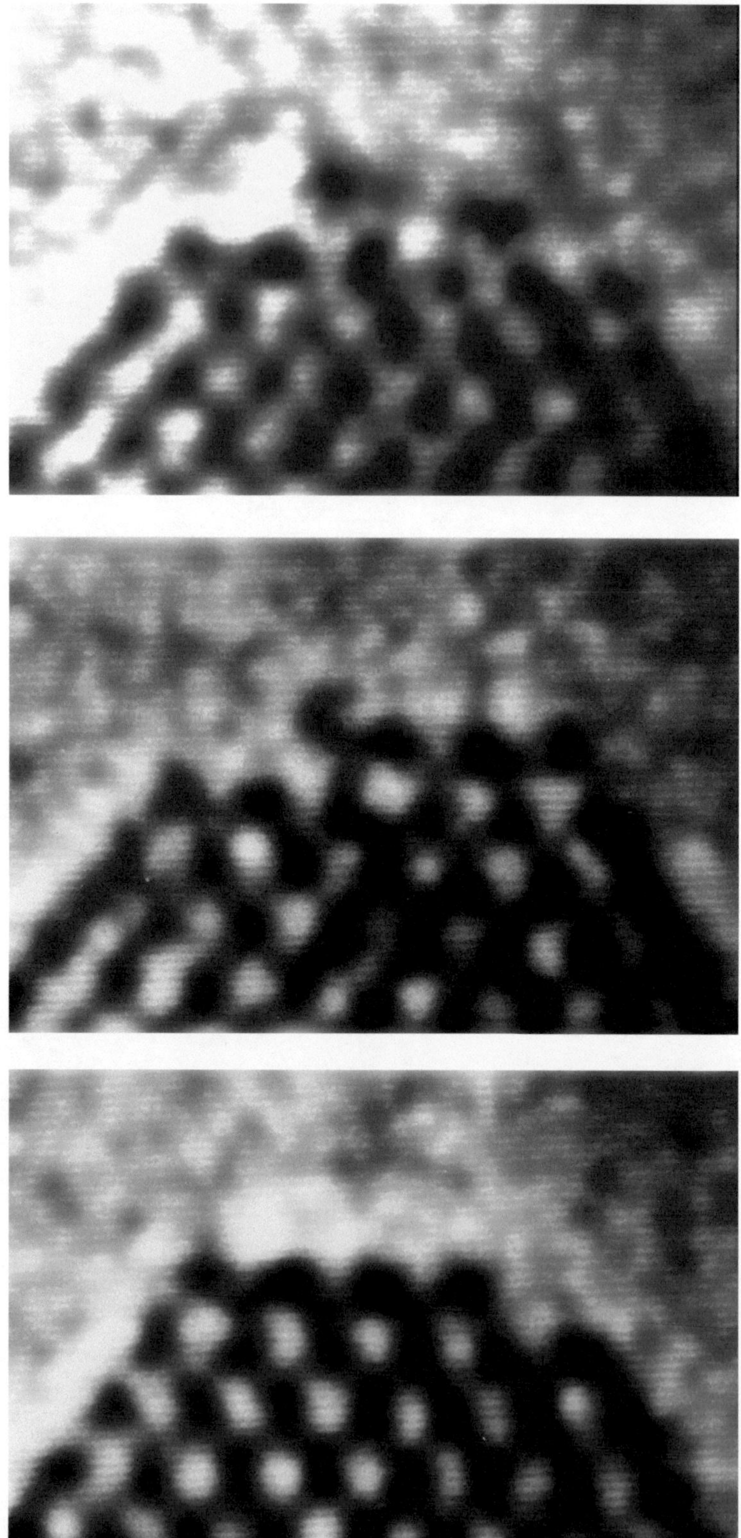

Atoms accumulating on a gold crystal. Photo courtesy Jan-Olov Bovin.

spatial resolution is on the order of 2.2Å, and the two spots on the left of the top pictures are only 5.8Å apart from one another. This is another limit of resolution, where time-averaged exposures add their own blur to the blur of the microscope itself, suggesting movement as much as revealing motion.[66]

38   A New Vocabulary for the End of Images

I hope I have said enough to convey a sense of just how intricate and unprecedented the problems of resolution and image formation have become in both light and electron microscopy. The list of issues is quite diverse: the production of contrast, the diffraction limit, the wavelength of the illuminating beam, the depth of field, exit wave reconstruction, classical lens aberrations, the contrast transfer function, the uses of Fourier transforms, the degradation of the object when electrons become too energetic, assessments of absorption and diffraction, limits set by the length of exposure, and the possibilities raised by video capture and digital image analysis software.

There is a problem winding among these issues that may sound simple: the challenge of relearning how to interpret light and shade. Nearly every image in this chapter appears to be a picture of something fairly solid, lit from a certain direction, casting shadows and sometimes even reflecting light. In nearly every case that appearance is misleading. I started with several examples of optical technologies (phase contrast, Hoffman contrast, and Nomarski contrast) that can produce images where light and dark may no longer have the meanings they have in virtually every other sphere of human experience. I mentioned three examples of technologies (atomic force microscopy, scanning tunneling microscopy, and near-field apertureless microscopy) that do not result in images in the ordinary sense of that word, because they are not pictures of objects that have thickness, opacity, or mass in the usual senses. Three further examples (acoustic microscopy, the quantum Hall liquid, and Lorentz microscopy) result in images that do not record objects in any ordinary sense of that word. These are, I think, increasingly wonderful pictures: the further they get from familiar photography, the more they depend on the limits imposed by everything from contrast to exit wave reconstruction—the vocabulary of the end of images in microscopy.

f i v e Particle Physics

From this point forward, my subject is the submicroscopic realm of elementary particles. Like the atoms of the previous chapter, elementary particles are not objects in anything like the ordinary sense of that word. They do not have bulk and weight in the usual sense, so they can't cast shadows or reflect highlights. Particles and atoms aren't even "things" in Heidegger's sense, or "forms" in Aristotle's, or "substances" in Spinoza's: so far, their only conceptual armature comes from the mathematics itself, aided by a battalion of wobbly metaphors.

Objects become increasingly elusive in this chapter and in Chapter 6. At least electron microscopes can make images of forms that can be associated with atoms, and electron microscopes can be thought of as extensions of optical microscopes, which can be understood as extensions of cameras. Subatomic particles have been imaged by ordinary cameras, but what the photographs show is not the particles themselves, but bubbles and other traces the particles leave behind.

In this chapter and the next, the objects themselves (whatever *objects* might mean in different contexts) do not appear in the pictures. Only their traces are recorded, and those traces provoke a formidable machinery of equations. The thicket of measurements and calculations is the subject of this chapter: it amounts to a move away from the images and into mathematics.

39   Galison's Theory of Particle Physics Images

I will start with a book by the historian of science Peter Galison, called
Image and Logic: Material Culture in Microphysics, because it is the most
concerted attempt to write a history of twentieth-century images in particle physics.[1] Much of Galison's account turns on two traditions of
physics imaging, which he calls "image" and "logic." Image-tradition
pictures are what art historians call naturalistic or illusionistic, and what
literary critics call mimetic: in Galison's words, they "preserve the form
of things as they occur in the world." Often, they have "fine-grained, . . .
persuasive detail" so that "a single picture can serve as evidence for a new
entity or effect." Like photographs of historical events, they can be taken
as proof that an event took place.[2] The logic tradition in twentieth-
century physics produces images that abandon the "sharp focus" of the
image tradition in favor of statistical and digital data. Logic-tradition
images often rely on "aggregate masses of data" instead of single pictures, and they entail "counting" instead of naturalistic picturing.[3] Image-
tradition pictures resemble what they depict, whereas logic-tradition
images may not, and they tend to be statistical or pixellated. The difference
between logic-tradition and image-tradition images, as Galison defines it,
is common in several branches of science; we have seen it for example in
exit wave reconstruction in electron microscopy (see Figures 38 and 39).

Galison's initial story is that as particle physics developed, image-
tradition pictures seemed less and less adequate. It didn't matter if the
pictures were detailed, naturalistic, or convincing. The properties that
counted were exact numbers and cumulative aggregates of statistics,
rather than singular events. Early in the twentieth century, physicists were
content with the visual—with whatever could be captured in illusionistic or naturalistic pictures. As physicists found ways of interpreting such
pictures (and generating more of them), they began to lose interest in the
images as images. Image-tradition pictures seemed limited on account of
their inability to record masses of data from successive experiments, and
because of their dependence on single "golden events." Eventually, in the
logic tradition, pictures became surfaces on which quantitative information was deposited.

There are various ways to reconfigure Galison's schema.[4] Some logic-
tradition images are unlike image-tradition pictures because they are made
of discrete signs such a pixels or dots on a graph. Other logic-tradition

pictures differ from those produced in the image tradition because they record the iterated results of long sequences of individual events, such as collisions in an accelerator. All images of the latter type (which I will call *statistical*) are examples of images of the former type (which I will call *digital*), but the opposite is not the case. Image tradition pictures also differ widely in how quantifiable they are: some can be measured with rulers and grids, some can be quantified with the help of machines, and others can only be quantified in limited or haphazard ways.

What interests me about Galison's dichotomy is the way it points to an ongoing difficulty in twentieth-century particle physics: how to capture, *in images*, the maximum amount of useful information about objects that cannot be naturalistically represented at all. As his story develops, he shows various interdependencies between the two traditions. Here it is useful to note the overall movement away from the image tradition, and the possibility of following an ongoing division within the logic tradition, between statistical and digital strategies.

40   Magnitude, Revisited

In contrast to astronomical objects, subatomic particles live in a kind of reciprocal infinity. It seems they are as infinitesimally tiny as galaxies and galaxy clusters are infinitely vast. But those are only infinities and infinitesimals in the poetic appreciations of nonscientists. In fact, subatomic particles constitute a highly structured field where lengths are relevant, even if they are tiny: 10^{-15} m is significantly different, for instance, from 10^{-30} m. (They point to different objects and distinct specialties within high-energy physics.)

As in the case of astronomical images, intuitively comprehensible sizes (with *intuitive*, as in the previous chapter, meaning relative to contemporaneous and comparable image-making protocols) give way to incomprehensible mathematical magnitudes; and again the photographs themselves are revealing—this time huge plates for tiny objects, instead of vice versa. In the early days of cosmic-ray studies, people used to send up film emulsions in balloons or cart them to the tops of high mountains in order to capture the tracks of particles that are normally absorbed by the atmosphere. In published form, the pictures look very much like other particle images. But the tracks themselves were microscopic, and even the emulsions were small. A typical experiment, done in 1956, involved stacks of emulsions, sandwiched between glass plates; the plates were four inches by ten inches on a side. After the balloon was retrieved, the "strip stacks" were inspected using a jeweler's loupe. When the physicists spotted tracks, they circled them with a grease pencil. Then the stacks were put under a biologist's dissecting microscope, and the emulsion strips were carefully separated and cut for inspection in a conventional optical microscope.[5] A given set of tracks might be two or three inches long, but at the beginning of the event, where the cosmic ray struck anatom in the glass or the emulsion, the event could be concentrated to a micron or less. Figure 49 shows the inception of the most energetic cosmic ray event known in 1956; the scale is fifty microns, or one-twentieth of a millimeter. Four millimeters farther along, the jet of particles starts to diverge (Figure 50). About an inch farther, and the core of the jet is visible, identified as part of a boron nucleus (Figure 51). To analyze this one event, scientists patiently measured 110 charged particles at high magnification under a microscope.[6] The photographs do not preserve the careful work of tracing, cutting, mounting, and photographing the tiny pieces of emulsion, and they end up looking like average-size photos when they are enlarged and published. The only hint of the scale is the out-of-focus smudges typical of light microscopy. Figure 52, a tiny disintegration star recorded at Brookhaven National Laboratory in 1947, was observed through a microscope at 750X. In an optical microscope, the same magnification can spot bacilli, which look very much like the tiny grains in this photograph.

Bubble chambers are very different. They are often large: the chambers themselves can be on the order of a yard or two wide, and they are encased in mechanical equipment, photographic apparatus, and magnets, so they are effectively room- or even house-size objects. Inside the chamber, the bubbles—which string together to form the particle tracks—can be only one-tenth or two-tenths of a millimeter in diameter. The chambers are carefully lit to record the delicate tracks against the much larger void of superheated liquid. Figure 53 shows the interior of a bubble chamber at Argonne National Laboratory. This is known as the twelve-foot bubble chamber, but the inside dimensions are actually only forty inches by twenty inches. The chamber is a cylinder, ringed by glass columns that protect flash tubes. (The image is off-kilter because it is taken from one of several cameras that looks into the chamber from different angles.) Tiny fiduciary crosses are engraved in the glass top of the

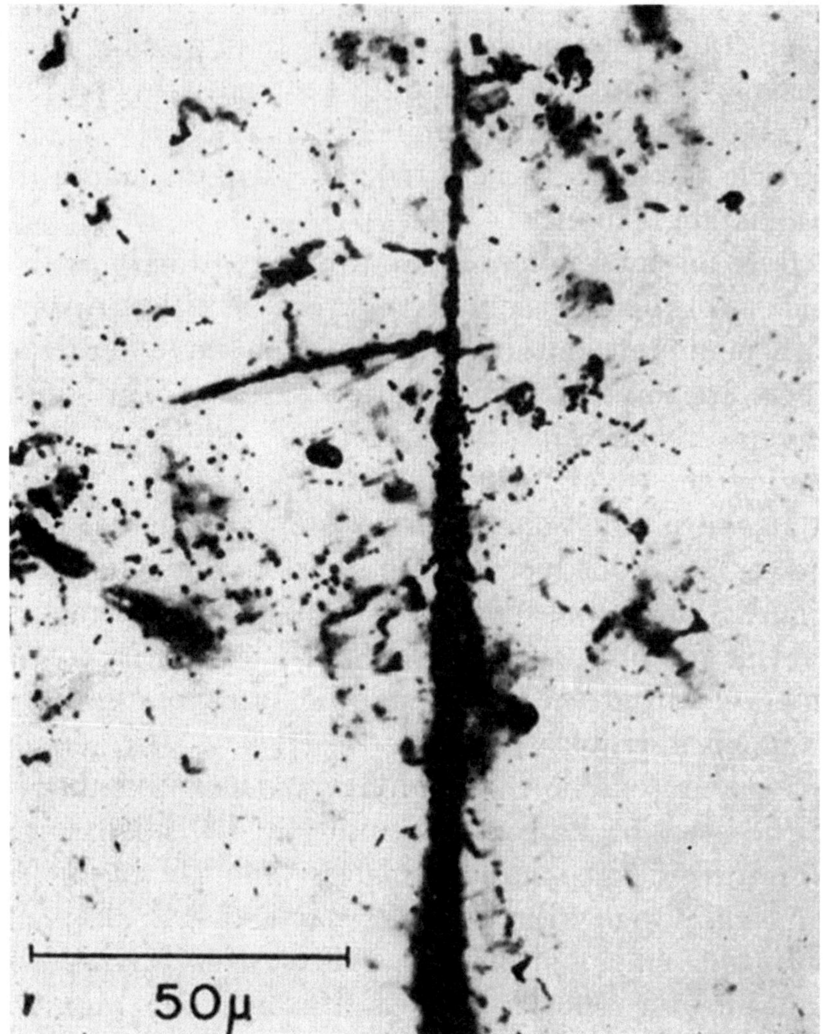

chamber, in contact with the chamber liquid; they help calibrate the stereo photographs that are used to reconstruct the tracks in three dimensions. (The crosses are analogous to the reference dots in Figure 10.) In this experiment, anti-protons are entering the chamber from below, and they are being bent in the 45 kilo-Gauss magnetic field. Their thin tracks are barely visible.

Most bubble-chamber negatives were made on ordinary 35mm film, and tens or hundreds of thousands would be taken in the course of a typical experiment.[7] Interesting events had to be carefully measured, and in the years before computer analyses, a typical working print or projection

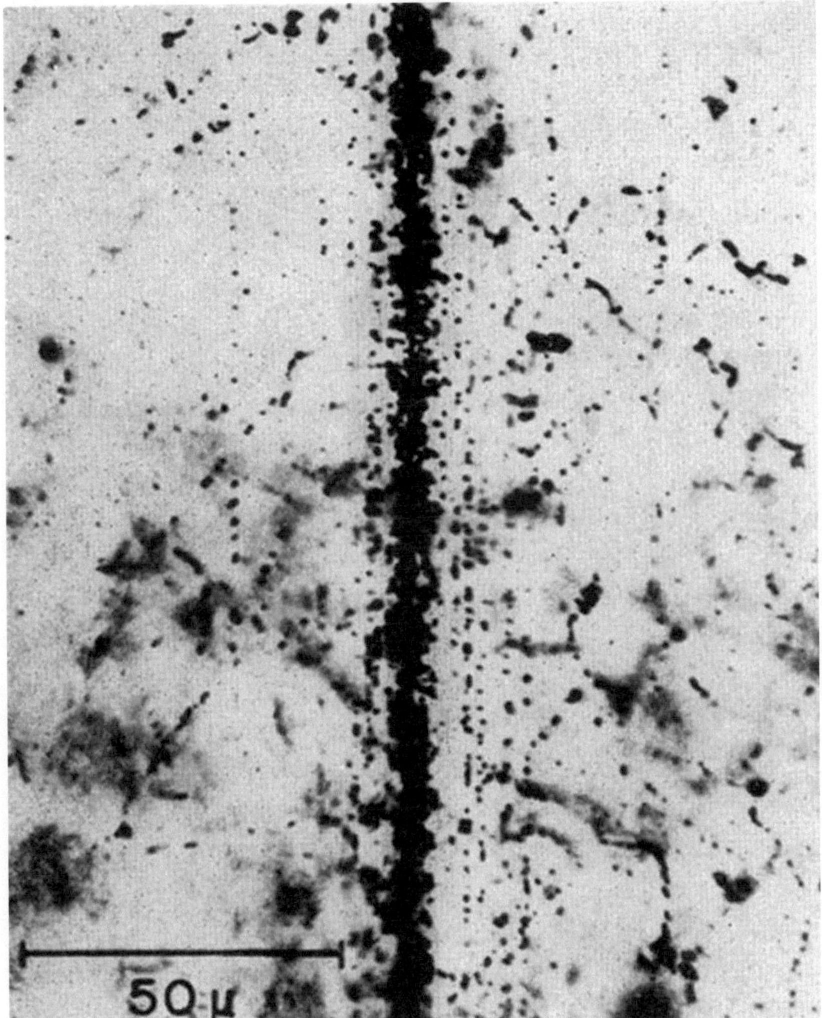

FIGURE 50

Cosmic ray shower in a nuclear emulsion, 4 mm from the point of interaction. From J. Naugle and P. Freier, "High-Energy Phenomena in Nuclear Emulsions," *Physical Review* 104, no. 3 (November 1, 1956): 804–11, fig. 2. Copyright 1989 by the American Physical Society.

was a few feet across, roughly the size of an ordinary poster. (Occasionally, for conferences or celebrations, prints were made two or three yards wide, so they were roughly the size of the actual bubble chambers.) In a poster-size print every bubble can be counted, which can be necessary to determine the particles' ionization and velocity. The optics of the cameras were designed to image individual bubbles as point sources, exactly as if they were stars in an astronomical image. The bubbles are typically packed one per millimeter, so they remain tiny even on large prints, but the tracks themselves have a more human scale—they start to look like ice-skating trails or traces of plant stems.

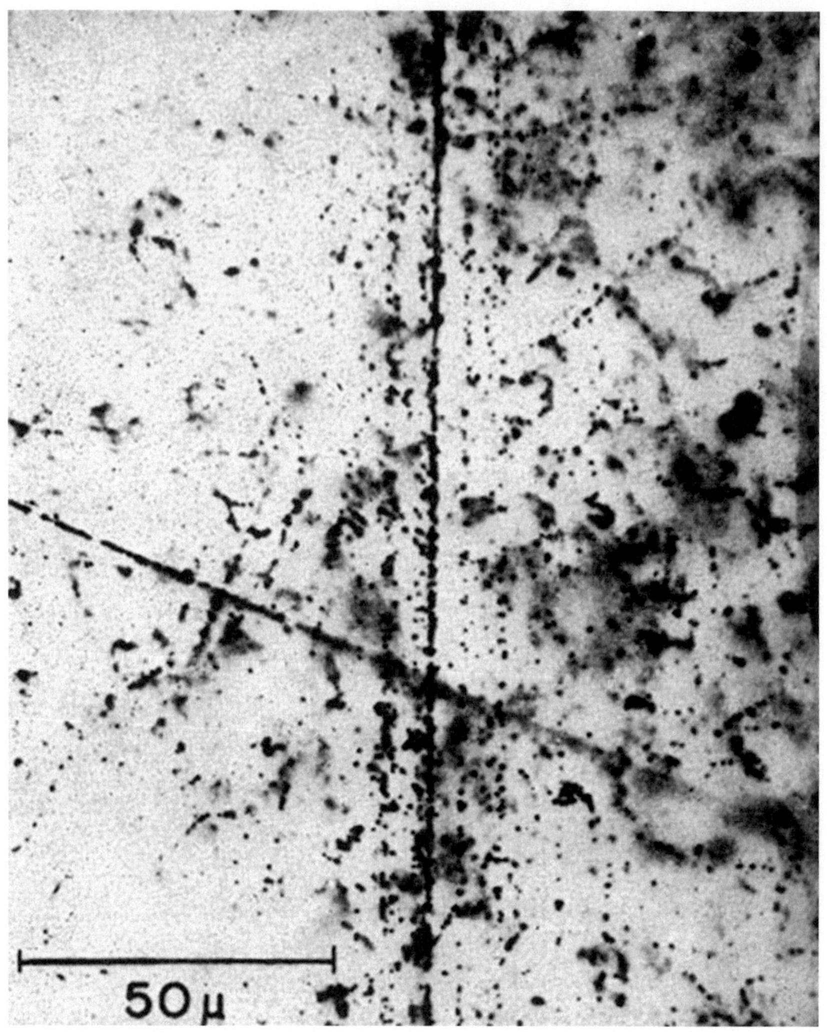

The seventy-two-inch bubble chamber at the Lawrence Berkeley Laboratory produced long, thin images (Figure 54). A small section of the
picture, printed here approximately life-size, shows the scale differences
between modern bubble chambers and the older nuclear emulsions. By
comparison, all three sets of tracks shown in Figures 49–51 would fit inside the period at the end of this sentence. It seems there is a world of
difference between the microscopic tracks in nuclear emulsions and the
poster-size prints associated with bubble chambers. Yet these differences,
which amount to almost three orders of magnitude, are small compared
to the differences between any such image and the particles that produce

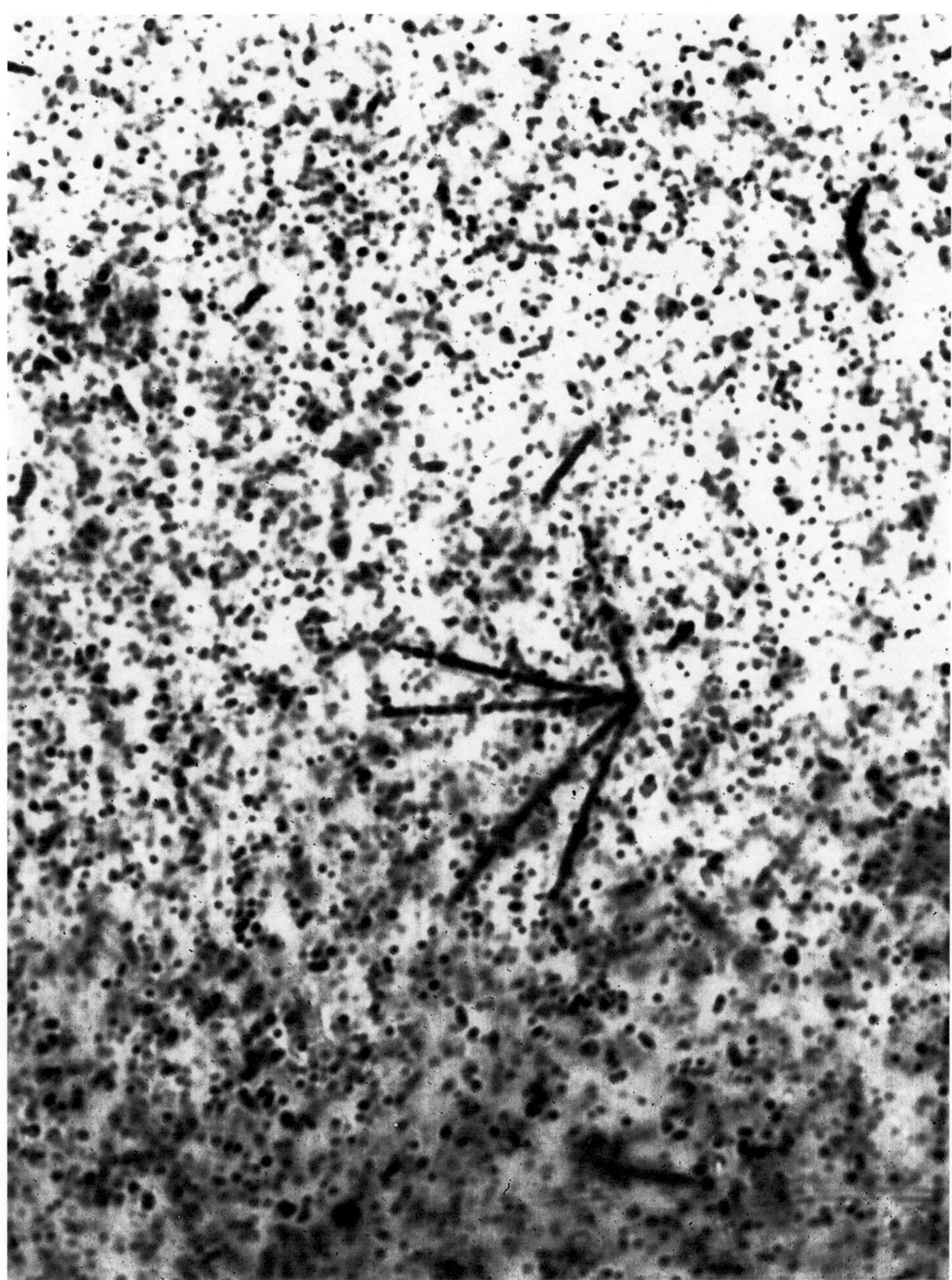

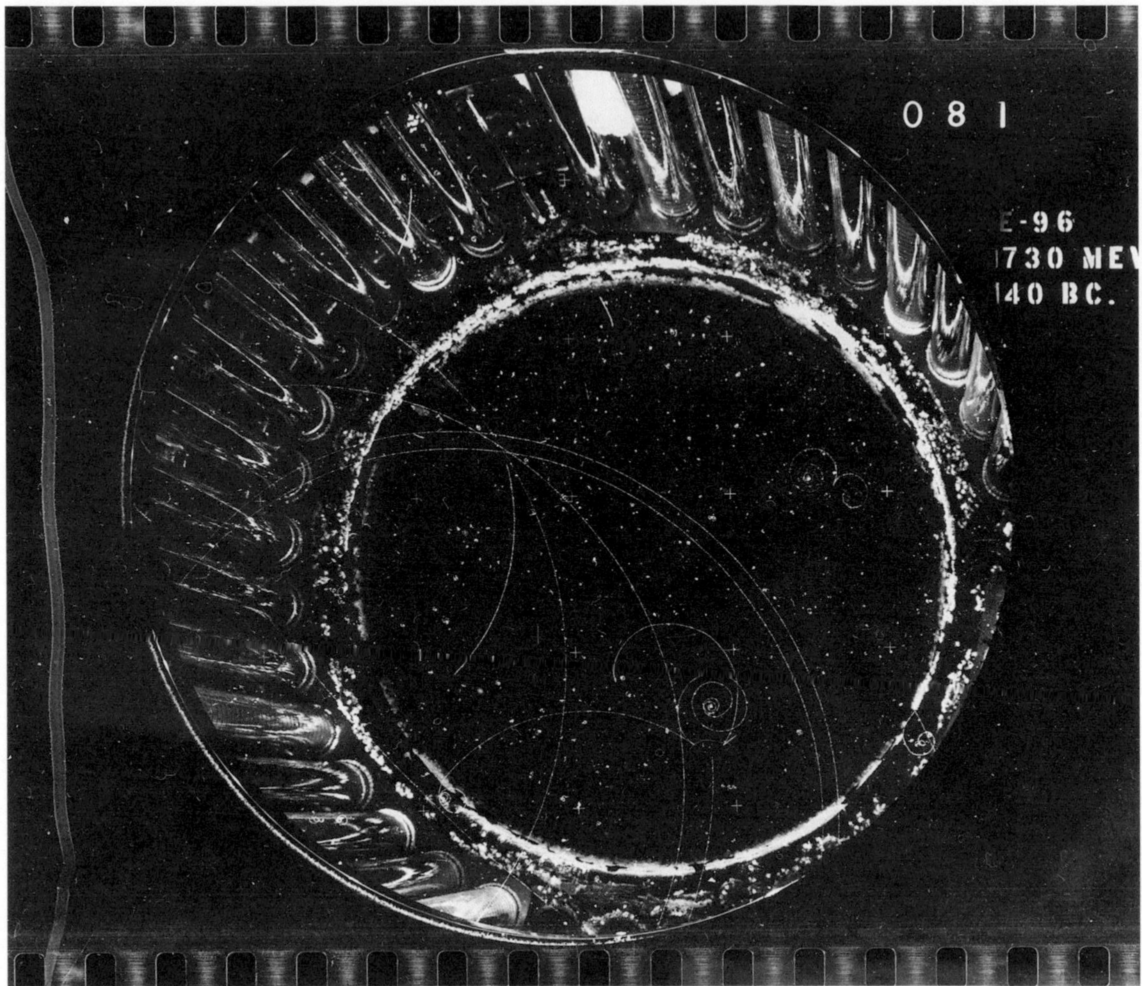

it. The smallest bubble is still unimaginably larger than the particle that precipitated it, even if the liquid, the magnetic field, and the energy of the incoming particles are adjusted to produce razor-sharp dotted lines comprised of nearly microscopic bubbles. The differences between the tracks and the particles take us immediately out of the realm of intuition and into the domain of pure calculation. An electron, for example, is regarded as a pointlike source—in field theory, it is not given any radius at all. For some low-energy calculations, the electron is given a "classical radius" on the order of 10^{-15} m (a figure more adapted to calculation than any putative reality); but pointlike or not, it produces bubbles 10^{-4} to 10^{-3} m, and those bubbles are recorded in chambers whose total dimensions can be around 5m on a side—ten thousand times larger than the tracks, which are over a billion times larger even than the particles' "classical radius."

I mention these cases to exemplify how photographs of particle tracks severely limit what is taken as intuitive understanding. I can come to terms with the scales of the photographs themselves, but never with the disparity between the images and what they represent. It is significant that scales vary so widely even in image-tradition pictures like these. Contemporary logic-tradition detectors are far larger than even the largest bubble chambers, but the images they produce are uniformly the size of computer monitors. Because such pictures are digital, there is no need to enlarge them, and there is little need to even represent them onscreen. Early image-tradition pictures were often reproduced with scales, as in Figures 49–51. Logic-tradition images effectively have no scale: the lengths of the particle tracks still matter, but it is impossible to understand particle phenomena by analogy with human-scaled events, so there is no call for a conceptual bridge—no matter how flimsy— between the scale of the image at hand and the conventional size of the object it records.

Intuition—in the vernacular sense, not the strict Kantian meaning— fails differently for each viewer, and the abandonment of intuition has its history, woven throughout modern physics. For each viewer and for each community of viewers, the limit of intuition is like a thread that winds through the fabric of the images, dividing the familiar from the unaccountable, and the sensible from the unrepresentable.

41   Nostalgia for Normal Pictures

Even before I have really begun to look at images of elementary particles, my own intuitive understanding—which I bring to these images as a student of an undefinable range of Western and non-Western images outside the subatomic realm—has been constricted so far that it is effectively banished. The only way forward leads to the realm of calculation. The details of the visible image begin not to matter, and instead the image becomes an opportunity for calculation. Even in the face of that hard truth, the many centuries of human-scaled images condition the ways viewers want to approach these pictures. The Western tradition of images incorporates a deeply ingrained expectation of resemblance that remains even in images that have abjured it.

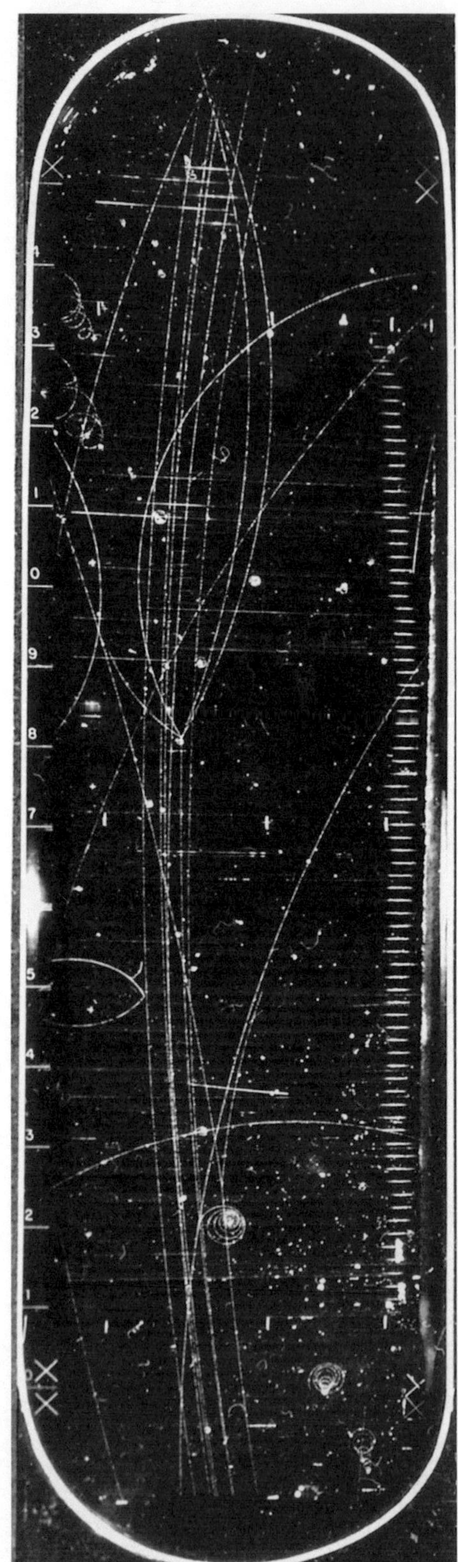

FIGURE 54

Particle tracks in the seventy-
two-inch bubble chamber,
Lawrence Berkeley Laboratory.
Left: entire. Right: detail. Courtesy Lawrence Berkeley Laboratory Photographic Services, no.
X1309607-03037.

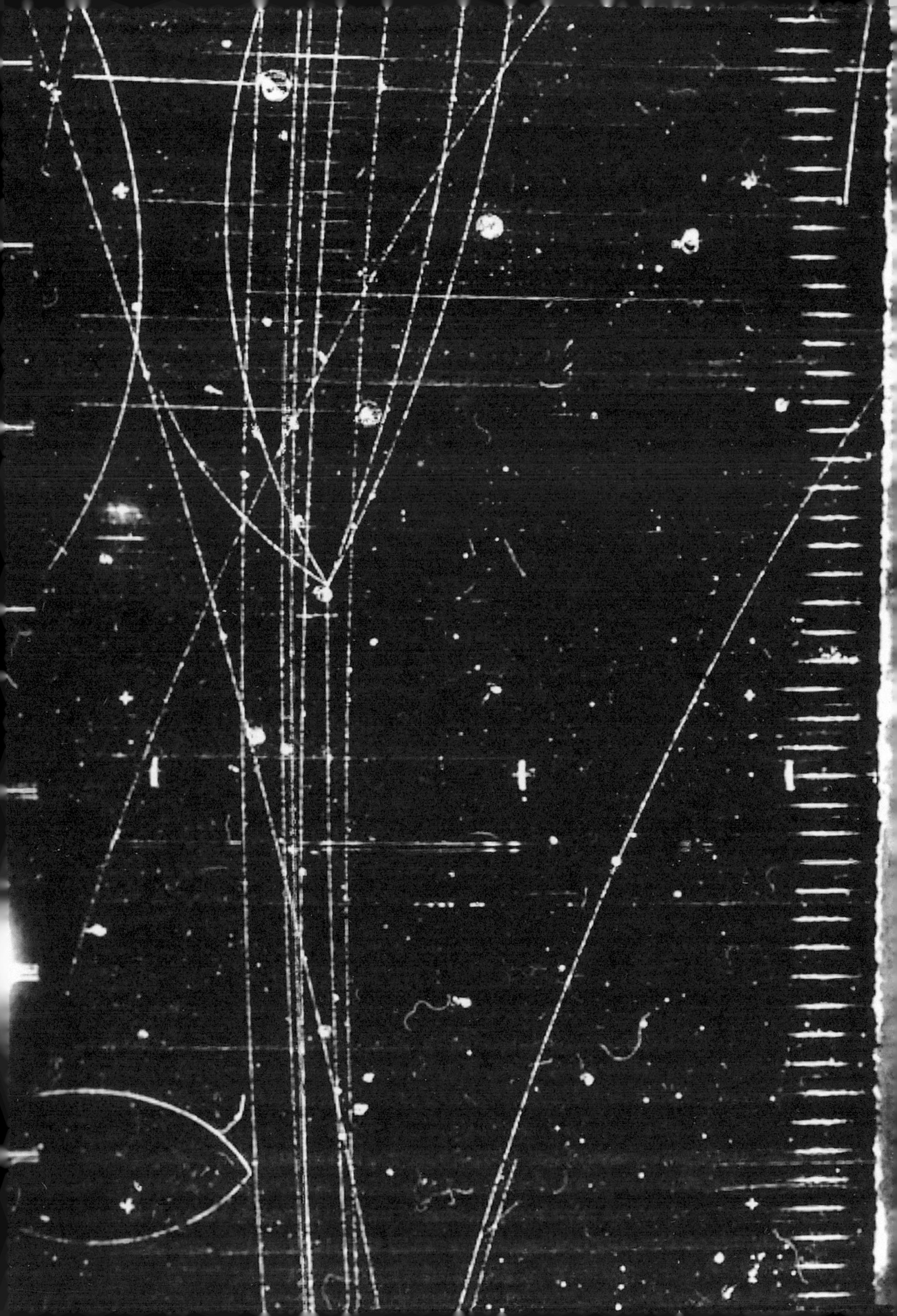

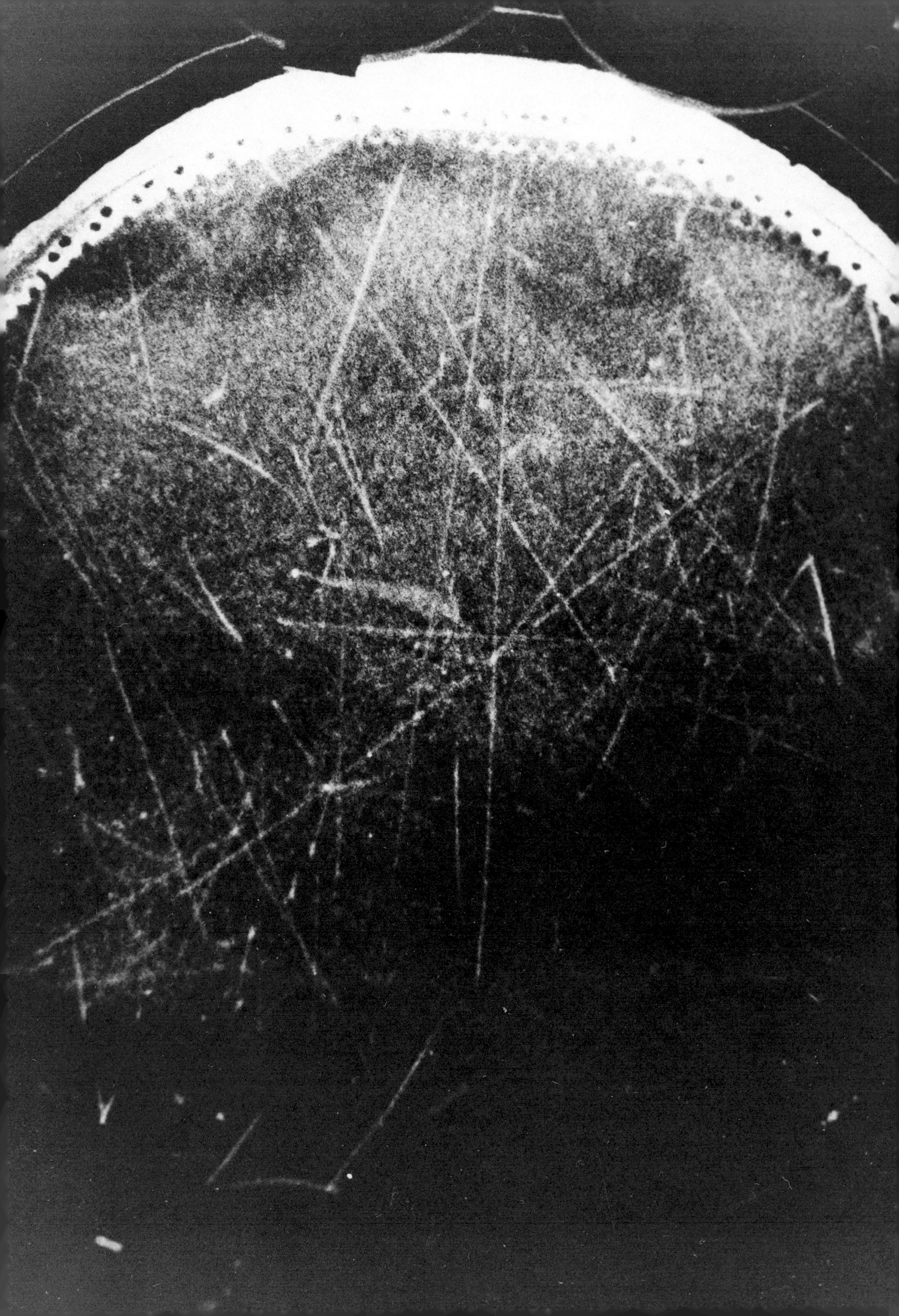

The desire to equate pictures with realistic representation takes many forms. My interest in the sizes of images is one example, and others have to do with the look and "feel" of objects that are—or seem to be—represented in pictures. Cloud chambers, the immediate precursors of bubble chambers, are an example. They were especially hard to calibrate and maintain, giving rise to a range of pictorial phenomena that seem related to paintings and photographs. Sometimes particles would flood the chamber with collisions, like a sudden downpour (Figure 55). Other times the momentum would be wrong, producing collisions on a very small scale, useless for measurement (Figure 56). In Figure 56, "soft," unenergetic gamma rays make small "blobs" of ionization. In Figure 55, the bombarding neutrons are too numerous and too energetic to leave decipherable tracks. (Chain reactions are common in cloud chambers and bubble chambers, and the back of this photograph is labeled "may be of interest under 'neutron bomb.'")

The meteorological look of these photographs is not coincidental. Galison and Alexi Assmus have demonstrated how Charles Wilson, the inventor of the cloud chamber, began as a meteorologist, taking photographs of cloud formations.[8] Wilson's interests, Galison says, "oscillated between thunderstorms and atoms," and he had a long-standing fascination with clouds, vapors, condensation, and related phenomena.[9] Pictures like Figures 55 and 56 belong in a long line of images of clouds, from the /clouds/ studied by the art historian Hubert Damisch (discussed in Chapter 1) to studies by Cozens, Constable, Turner, Valenciennes, and Bierstadt. It is no accident that *improperly set* cloud chambers resemble early photographs of clouds, but the interest in such mistakes quickly evaporated when the physics got under way. (The two plates shown here are among the hundreds preserved in archives that record failed experiments: such errors produce the only images that really resemble meteorological phenomena.)

A deeper link to the history of Western images is the fact that the hazy marks in these cloud-chamber images represent lines so fine they are effectively pure geometry. Somewhere among the coarse blobs of silver in Figures 55 and 56 are the unimaginably tiny and wholly invisible paths of the particles themselves. Bubble chambers and cloud chambers are therefore lovely examples of the ancient distinction between the mundane artist's mark and the impossibly thin Ideal Platonic line. It is a difference that surfaces repeatedly in art history.[10] Baroque physiognomists, for ex-

FIGURE 55

Cloud chamber, showing a flux of fast neutrons. Courtesy Argonne National Laboratory, negative no. 109-7719.

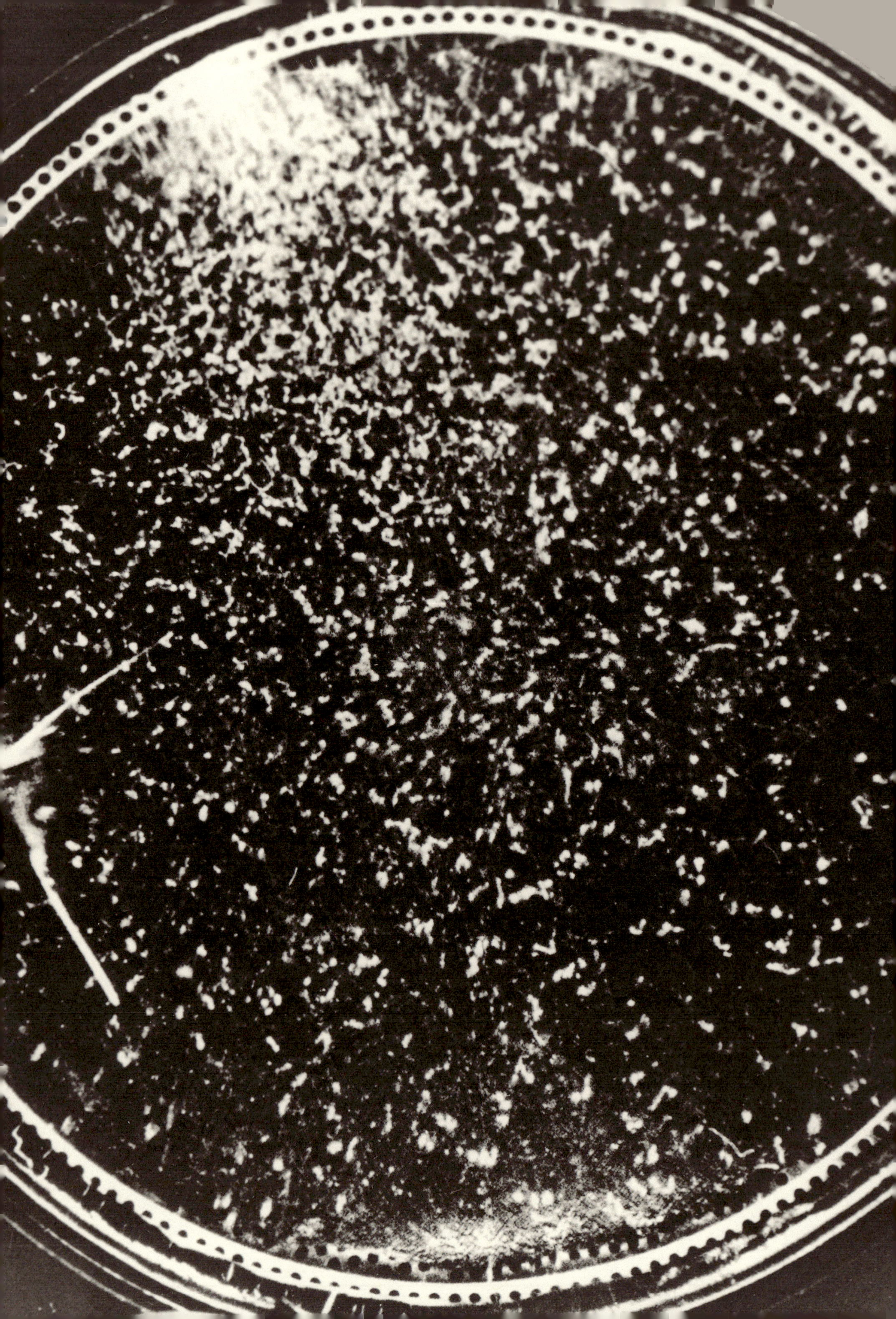

ample, reasoned that behind every actual wrinkled face there is an ideal arrangement of lines that gives it meaning. It was the artist's function to find such schemata and put flesh on them, producing nonmathematical paintings grounded in ideal geometry.[11] Damisch has argued that the transition from the drawn mark (*trait*) to the ethereal Platonic line (*ligne*) is a "dream of the limit" (*rêverie de la limite*).[12] He says that the baroque theorists had it wrong, and that expressions depend on the force of the mark, mingled with the impossible hope that there might be such a thing as a pure line. The issue has no clear-cut answer, because the mark and the line are entangled in the history of writing on pictures.[13] Yet Damisch is certainly right that Western artists and theorists have been fascinated by the contrast between the mazy confusions of real-world marks and the faint intuition of the impossibly fine geometry that seems to lie beneath them. The sloppier the cloud-chamber image, the more it feels like an image from Western art history, made poignant by the spectacular mismatch between the clouds of the representable world and the wholly unrepresentable particles.

These physics images, which seem so remote from Western image making, are related to earlier naturalistic pictures in many ways. Some parallels, like Wilson's cloud chambers, are historically demonstrable; others, like the opposition of mark and line, depend on more general assumptions about images. If the images aren't read or understood, a viewer can be lulled into thinking that these *are* pictures from the Western academic tradition that valued realism, painterliness, atmosphere, chiaroscuro, composition, and gestural expression: in other words, that they are art images in disguise. Cloud- and bubble-chamber images can seem to represent common human experiences: water spraying from a garden hose, or snow or rain. Some particles look like they are gently falling through the air, spiraling down like autumn leaves (Figure 57). Of course all of that is illusory, if only because the particles cross the chamber in less than one-millionth of a second. The bursts, sprays, and jets, and even the "lazy" curlicues are so close to instantaneous, so relativistic, that they effectively occur all at once. For that reason, it doesn't matter how fast the shutter is in a bubble-chamber camera, provided that it catches the tracks before the bubbles grow too large or the fluids in the chamber distort the tracks. Ordinarily a bubble chamber was photographed, "erased," and reset for the next injection of particles in less than forty-thousandths of a second. In Figure 57 the gentle "downward"

FIGURE 56

Cloud chamber, showing soft gamma rays, from K capture in niobium, producing small blobs of ionization. Courtesy Argonne National Laboratory, negative no. 109-7720.

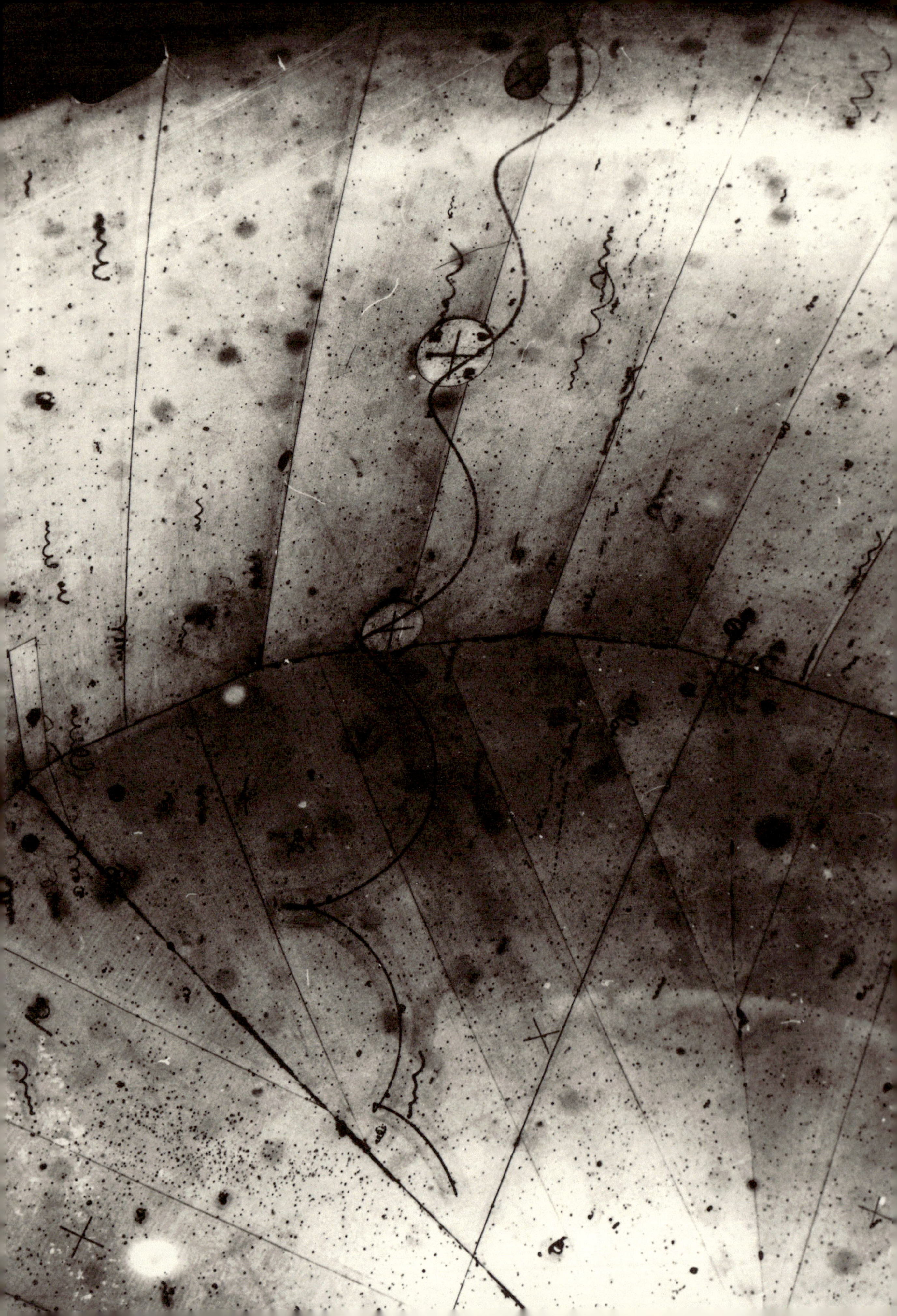

spiral is an unidentified, long-lived charged particle, turning in the magnetic field; and the smaller pieces of "confetti" are electrons, which lose energy more quickly and are absorbed by atoms in the fluid. In some astrophysical images, it makes sense to apply analogies: for example, "jets" and expanding envelopes of gas can be understood in part by studying fluid dynamics. Nothing in this image falls or drifts in the ordinary senses of those words, and there is no room for nostalgia for naturalistic painting and photography. Once again intuition is silenced. The power and rapidity of that silencing, I think, is a defining trait of images in high-energy physics.

Yet nostalgia persists—if that is the right word for an often unnoticed set of expectations about images. Bubble-chamber and cloud-chamber pictures do much more than just weaken "links between that which can be felt and that which can be understood," as the philosopher Jean-François Lyotard says. These images break those links, strictly consigning all unquantified understanding to the realms of error or aesthetics.[14] This is the kind of distinction often missed in the impoverished debates between scientists and historians or philosophers of science: the images can be *read* without nostalgically mistaking what they do for earlier realism; but they cannot be *produced* without drawing on the longer history of mimetic art. For that reason, they also cannot be fully interpreted without tracing the lingering effects of modern and premodern pictures—as Galison has done for Wilson's cloud chambers, and as I have suggested in more general terms. It's important as well that physics images do not have to look like clouds or falling leaves to have these misleading connections to earlier practices. In the next chapter I'll present even less naturalistic images that have been used as if they were "by-products of a romantic nostalgia" in Lyotard's phrase.[15]

Western naturalistic image making has a long history, and these pictures of particles are newborn by comparison: that is one reason they are produced in conditions familiar in the history of Western art. The other reason is that it is still hard to believe that these images are attempts to represent objects so far outside any possible experience that they effectively prohibit all analogic understanding—they belong, in other words, outside the pictorial tradition that still provides the conventions that generate them. These images are harsh, and their phantom realism is only part of their apparent coldness and inhumanity.

FIGURE 57

Bubble chamber. Courtesy Argonne National Laboratory, negative no. 137-77-264.

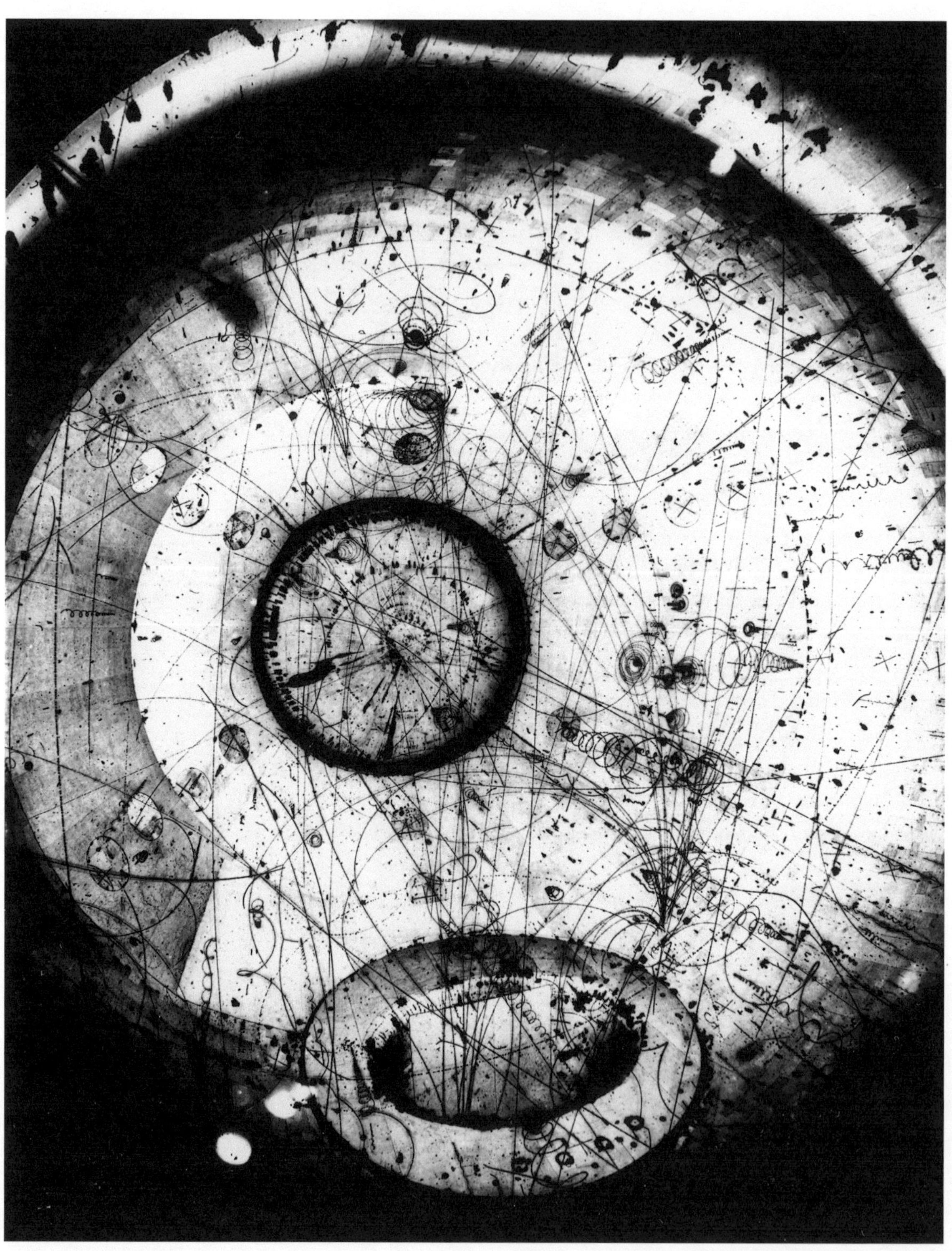

42 *The Concept of Complexity*

Complexity is part of what shuts down what I am calling a viewer's intuitive understanding of images of particles: complexity is also the thicket of representation in which the naive eye gets lost. Yet unexpectedly, the scientific content of a picture that looks overwhelmingly complex, such as Figure 58, may turn out to be relatively simple. This image has all the characteristics of a painting or fine-art photograph: lovely soft stains, atmospheric depth, mysterious dark clouds, intricate calligraphic traces, hidden lights, and a landscape-and-cloud composition, all playing against a distorted and dysfunctional grid. The image could easily be correlated with mid-twentieth-century painting, from Cy Twombly's graffiti to Ferdinand Wols's scratches. It is even possible—as it is with Twombly or Wols—to zoom in on parts of the image and find new levels of seductive detail (Figure 59). The image seems prefabricated for modernist art criticism. It is no surprise, then, that it was chosen from the Fermilab archives by an artist, Jeannie Therrian-Gottschalk, and printed by a professional archivist at the Art Institute of Chicago. (It is one of a few purely physics images that have found their way into the photographic collection of the Art Institute, where they live a kind of shadow life. They aren't usable as physics, because they have become detached from the experimental contexts that would allow them to be read; and they aren't exactly art, because they were only chosen by an artist, and not manipulated or reshown as art.)

Particle physics images can easily be taken as art, provided they are interpreted wholly in the light of nonscientific art-world criteria. Bubble-chamber tracks are occasionally presented this way, and from an artist's point of view there is nothing illegitimate about it: the images are simply reused as art, just as any image might be. But when the appropriation and recontextualization involves the claim that the images are still being presented as physics images, then the operation is open to serious criticism. It is too easy to look at the curling path of an electron as if it were a calligraphic mark by a painter, and too complacent to hope that any of the relevant meanings of the physics image are captured by art criticism. In art-historical terms, taking a physics image as an art image requires an impoverished sense of the meanings and contexts of works by artists such as Twombly, Wols, or LeWitt. In philosophic terms, such an appropriation does exactly what Jean-Luc Nancy, Jeffrey Librett, and others want to

FIGURE 58

Fifteen-foot bubble-chamber event. Fermilab, Batavia, Illinois. Negative chosen by Jeannie Therrian-Gottschalk, printed by Kristen Merrill. Courtesy Art Institute of Chicago, print 1998.69, E37852. Froom Fermilab roll 1012B, negative 08345. Silver gelatin print, printed 1997.

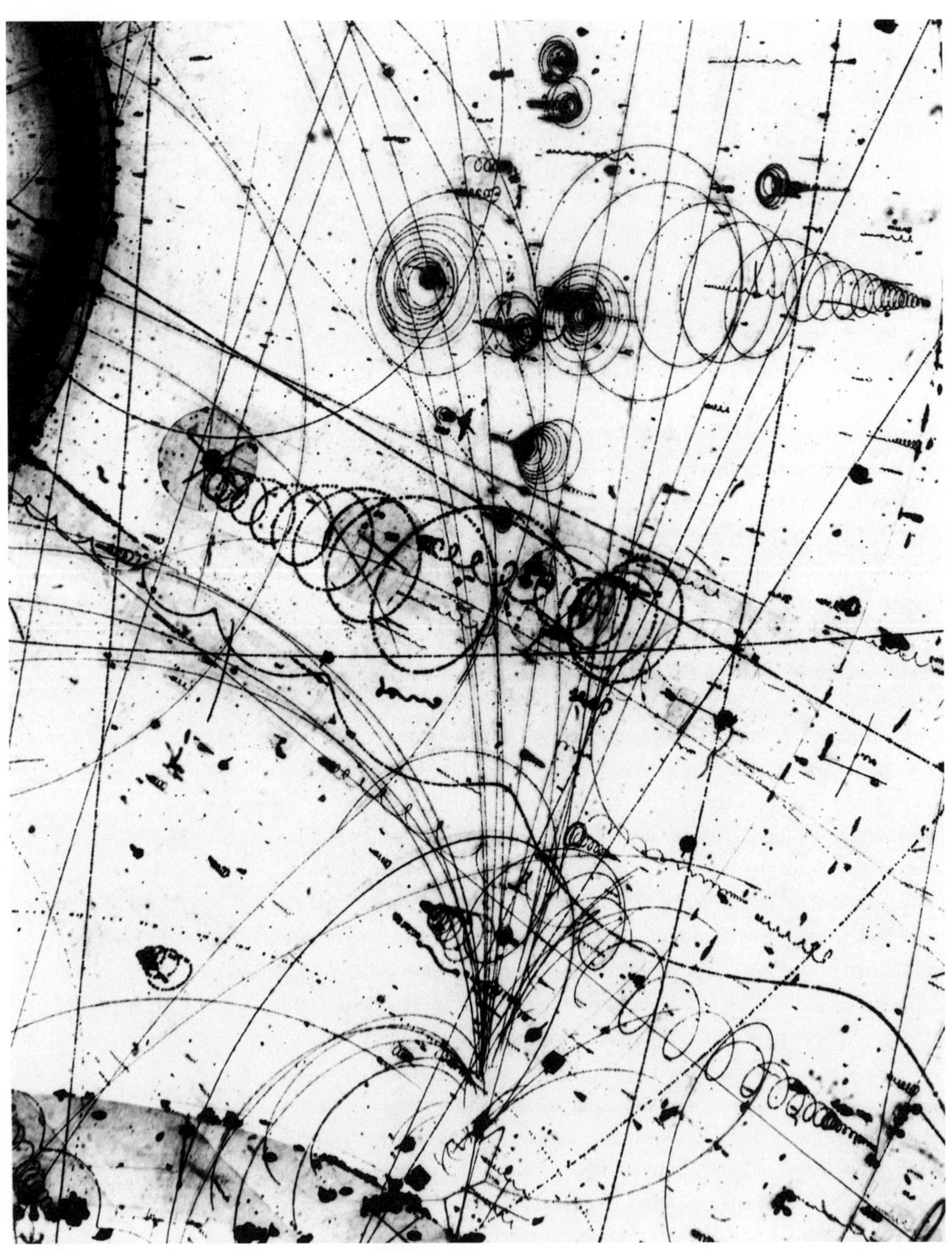

avoid: it aestheticizes the sublime, making it into a game of taste and re-
finement rather than a foundational account of the limits of representation.[16] And from a physicist's perspective, such a reading misses the point
of the image itself by failing to *read* it.

Lyotard's exhibition, called *Les Immatériaux*, did something of the
same by including bubble-chamber images along with paintings in various postmodern and modern styles, including abstract expressionism,
post-painterly abstraction, French support-surface abstraction, and geometric color field painting.[17] The physics images were not read or interpreted in any way, so they could only look like physicists' versions of
postmodern painting. Lyotard meant to show that many kinds of images
participate in a postmodern rejection of the values of naturalistic, painterly, romantic, atmospheric, or otherwise nostalgic art-historical images.
He chose particle physics images that look flat and linear and have a minimum of space or texture. *Les Immatériaux* was a nostalgic exhibition
despite itself, because Lyotard couldn't help implying that even physics
images can be recuperated, brought into art-world image making.[18] I
doubt they can. Despite their many entanglements with older traditions
of Western art, bubble-chamber images cannot be adequately interpreted, even using the language of the most radical anti-representational
art criticism.

Nothing is as it appears in bubble- and cloud-chamber images—
even complexity itself. The half-art images, Figures 58 and 59, are actually simple by physics standards. Their bewildering lines are the traces
of just a few kinds of stable particles, and—as in the cloud-chamber
pictures—their lovely atmospheric effects are unwanted artifacts. That
does not vitiate Galison's points about the origins of cloud chambers
and bubble chambers in meteorology, but it means that the atmospheric
and naturalistic effects are errors, and other forms of complexity are
going unread.

Complexity, like intuition, varies with the viewer: for a physicist, the
image in Figure 58 is simple because it records many examples of just a
few kinds of events. The two images shown in Figures 60 and 61 are
from the same bubble chamber at Fermilab (notice the converging seg-
ments of the chamber walls), but they actually record a far more complex
interaction. They were published as a *jeux d'esprit* at a bubble-chamber
conference in Geneva, Switzerland, in 1994.[19] Figures 60 and 61 may seem
like two different events, but a close look reveals that they are pictures of

FIGURE 59

Detail of Figure 58.

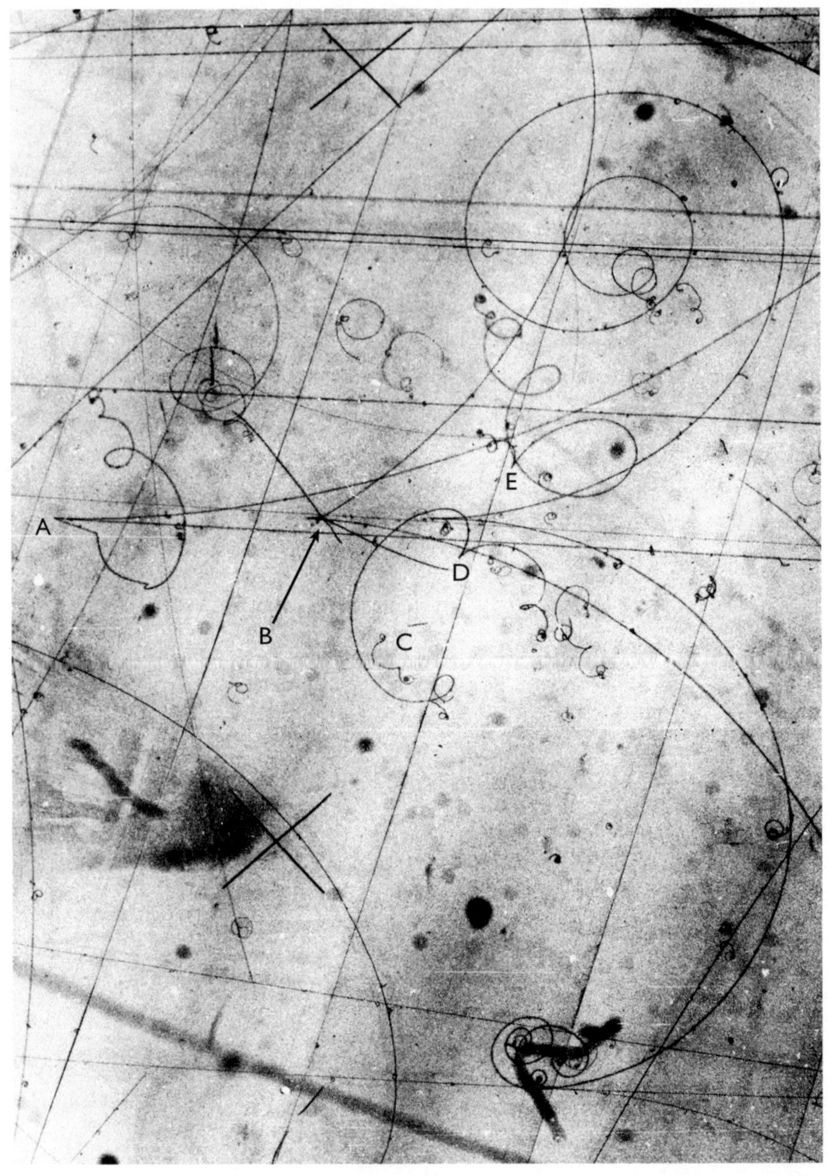

FIGURE 61 *(opposite)*

Figure 60, from another angle. Courtesy Goronwy Tudor Jones.

the same set of tracks, taken from two widely separated perspectives. A spray of particles, called a "vertex," is at the left center of each image. There are three visible particle tracks: one goes straight to the right, a second curls upward, and a third goes straight for a moment, then "kinks" several times and spirals off. In Figure 60, the third track appears to go upward, and in Figure 61, it seems to go to the right, but that is just a trick of perspective.

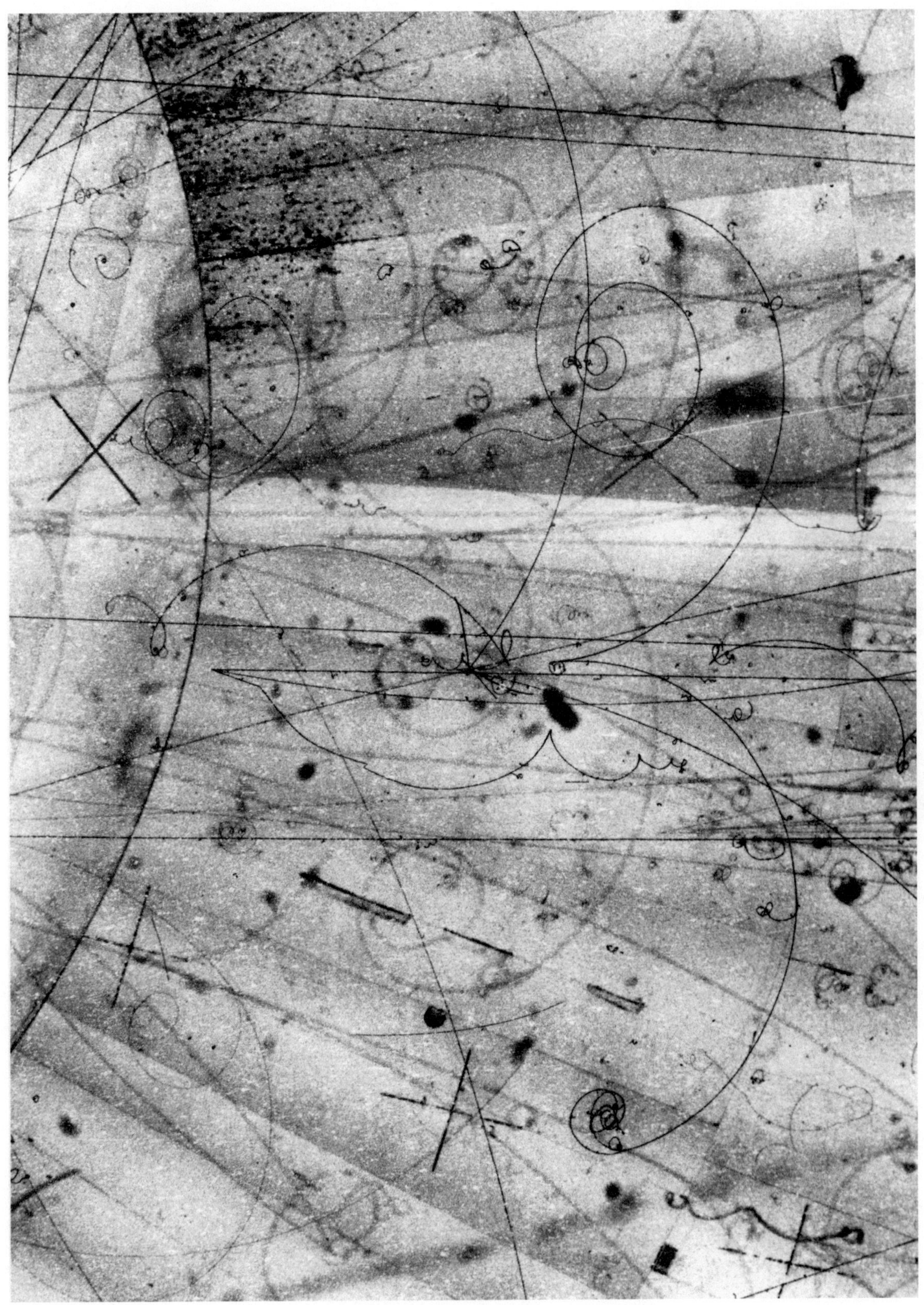

The author, Goronwy Tudor Jones, calls his one-and-a-half page essay "A Remarkably Rich Bubble Chamber Picture (The Competition!)." He says the tracks in these images represent at least fifteen particles: at the initial vertex A, the lower path is a sigma particle, Σ^+, which decays at the first little kink into a positive muon, μ^+, which in turn decays into a positron, e^+. Each decay also creates a neutrino (which leaves no track, because it is not charged). In all, therefore, the lower path from vertex A involves six particles: Σ^+, π^+, μ^+, ν_μ, $\overline{\nu}_\mu$, ν_e. The nearly straight horizontal line leading from vertex A straight across the plate is a negative muon, μ^-, and the line that curves gently upward from the originating vertex is a hadron, probably a pion, π^+. A fourth, invisible particle—a "k-zero," K^0—also originated at vertex A and then decayed at point B into positive and negative pions, π^+ and (π^-. The trickiest part of the interpretation involves noticing that the short, straight track starting at vertex A—the one left by the Σ^+—points to the place labeled C, where there is a very short track left by a stopping proton. The invisible particle leading from the place where the Σ^+ decays is a neutron, n, which collided with the proton, producing a neutral pion, π^0; the neutral pion rapidly decayed into two photons, γ, which decayed at points D and E, bringing the list of particles to:

$$\Sigma^+, \pi^+, \mu^+, \nu_\mu, \overline{\nu}_\mu, \nu_e, p, e^+, \pi^-, \mu^-, e^-, n, K^0, \pi^0, \gamma.$$

In addition to this plethora of particles, Jones points out that the event also involves all four fundamental forces of nature. The strong, weak, and electromagnetic forces are all represented in the decays that begin at vertex A. In addition, Figure 61 demonstrates the force of gravity, because the dark, smudged lines in the upper-left-hand corner are flakes of ice that had sunk to the bottom of the chamber. That last isn't a serious observation (ideally, there is no ice in the chamber, and it isn't relevant to the interactions), but it does make the point that this event is an unusually complete demonstration of many kinds of interaction.

Hence complexity in such pictures is not what it may seem. The bare list I have given is only the entrance to the thicket. Beyond it there are formidable calculations, and such images will often be supplemented by additional diagrams and images to support the analysis. It goes without saying that neither the list of particles nor the calculations are sources of complexity that are normally applied to fine art: they are evidence of realms beyond the merely visible that fine art has scarcely begun to explore.

43   *From Looking to Calculating*

Bubble-chamber images are really intended to be *measured*, not just seen. A viewer doesn't stay with a bubble-chamber image the way she might with a painting or photograph, because it fails to represent what is really of interest. The point is to use the image—effectively, to use it up—in order to learn about processes that could never be captured on film.

The actual protocols for measuring particle tracks have become routine; as Galison has shown, women who were not physicists were once given the task of analyzing the curvatures and lengths of the tracks.[20] A small number of equations govern many of the measurements of particle tracks. Charged particles follow curved paths because a magnetic field is applied to the entire bubble chamber; they move in accord with the Lorentz force law, which can be used in the approximation

$$p = \frac{r}{.0003B}$$

where p is momentum of the particle, B is the magnetic field strength, and r is the track's radius.[21] The larger the radius of the curve, the higher the particle's momentum. In Figures 60 and 61, the gently curved line arcing upward from vertex A is positively charged, and the line leading to the right hardly curves at all, so it is a very high-momentum particle. (The line curves ever so slightly downward, so it is negatively charged; sight along the pages—put your eye close to the surface of the page—to see the curve.) In Jones's calculation, the upward-curving path, a pion, π^+, has momentum 939 MeV/c, and the very slightly curved horizontal track, a muon, μ^-, has momentum 45,272 MeV/c.

Particles lose energy as they travel, mostly by ionizing the atoms they pass. "Delta rays"—electrons knocked out of their orbits by passing particles—form characteristic spirals. As an electron spins in the chamber's magnetic field, the electron loses energy by ionizing the atoms it passes: it slows, its spiral seems to tighten, and finally it is absorbed by an atom in the chamber. Delta rays "fall" like confetti throughout Figure 57. (They can also form little thorns or tendrils on either side of a particle track, and if they have very low energy they appear as closely spaced beads along the track itself.)

A particle's energy can be calculated by measuring its path length. The rate of change in energy per unit path length is given by

$$\frac{dE}{dx} = \frac{a}{v^2}\left[\mathrm{In}\ bv^2\right]$$

where v is the particle's velocity, and a and b are constants that depend on the material inside the bubble chamber.[22] This is a rough approximation. To make it more exact, it is necessary to take account of the fact that the particles will be traveling close to the speed of light. With relativistic particles, the expression becomes:

$$\frac{dE}{dx} = \frac{a}{v^2}\left[\mathrm{In}\left(\frac{bv^2}{1-\frac{v^2}{c^2}}\right) - \frac{v^2}{c^2}\right]$$

where c is the speed of light.[23] (Like all relativistic expressions, this one reduces to the nonrelativistic form when v is significantly less than c.)

Electrons behave somewhat differently from uncharged particles, because they also lose energy by radiating photons whenever they are slowed down by the electric fields of atoms in the chamber. That process, called *bremsstrahlung* ("braking radiation"), has its own equations and has to be factored into calculations of the energies of electrons. (*Bremsstrahlung* is not as important for other charged particles at the energies and speeds common in bubble chambers.) If the photons have enough energy, they convert into electron-positron pairs, and when those pairs decay, they emit more photons, leading to an "electromagnetic shower." A typical electron-positron pair can be seen in Figure 58: the electron and positron start out nearly parallel, and then they curve apart in opposite directions.

The individual bubbles, visible in Figures 53 and 59, were often counted. The usual formula, good for hydrogen bubble chambers, is:

$$n = k\left[\frac{v}{c}\right]^{-1.92}$$

where n is the bubble density, v is the velocity of the particle, c is the speed of light, and k is a constant that had to be continuously recalibrated from photo to photo, by comparison with known particle tracks. When a particle slows down, the bubbles get closer together—a sure sign the particle is losing energy.[24]

Using relations like these, physicists can calculate the particles' momenta, their energies, and sometimes their velocities. From combinations of those three values, they can identify the particles and calculate their masses. Few identifications of particles in bubble chambers are entirely certain; they are usually best guesses consistent with the tracks. A central limitation is that most particles can decay into several different combinations of particles. Jones's analysis of Figures 60 and 61 is a reasonable interpretation based on the most likely kinds of decay; as physicists say, he chooses events with the highest branching ratios. He identifies the short, nearly straight track of the particle that goes down from vertex A as a Σ^+, decaying into a π^+ (the more strongly curved track) and a neutrino, n. The neutrino is invisible and heads to point C. Measurement confirms his guess in this case; in other instances Σ^+ particles decay into different combinations of particles:

$$\Sigma^+ \rightarrow p + \pi^0$$
$$\Sigma^+ \rightarrow p + \gamma$$
$$\Sigma^+ \rightarrow p + \pi^+$$
$$\Sigma^+ \rightarrow n + e^+ + \nu_e$$
$$\Sigma^+ \rightarrow \Lambda + e^+ + \nu_e$$

where Λ is a lambda particle, the rarest of the options. Each of these possibilities (and others I haven't listed) has a branching ratio that is readily available in the standard handbook of particle properties. Figures 60 and 61 probably show exactly what Jones proposes, but there is always the possibility that something rarer has occurred (especially with the pion, π^+, that curves gently upward from vertex A).[25]

That is a taste of the kind of interpretation that actually goes into particle-chamber photographs. It has little to do with naturalistic representation. Bubble-chamber images don't lead to nostalgic tours of atmospheric landscapes but to forests of uncertainty and thickets of calculation.

44   *Objects Too Complex for a Single Picture*

Calculations make these images substantially more complex than fine-art images because they add numerical values to unaided looking. Bubble-chamber images also become complex when they are accompanied by ancillary representations such as traces of the particles' paths, explanatory graphs, and flow charts.

Even without such diagrams, a single bubble-chamber photograph, considered by itself, will often require supplementary views. The images shown in Figures 60 and 61 are typical examples. Early cloud- and bubble-chamber photographs were reproduced as stereo pairs, so readers had to cross their eyes to see them. (Figures 60 and 61 are from vantage points too far apart to be recuperated by human eyes: they are intended for machine analysis.) In the earliest examples, the three-dimensional information was not used for measurement as much as to aid visualization; it was easier to see the directions of paths and the orientation of curves in three dimensions. Stereo pairs in old physics journals produce wonderful effects: the pictures spring off the page, but the effect doesn't add much to the analysis. Today, three-dimensional information is crucial for the analysis of momenta, and single images are virtually unknown. Contemporary digital images in the logic tradition typically come in two views, called beam-view and orthogonal, but because they are digital they can be rotated to any perspective.[26]

The use of stereo pairs and multiple images already takes particle physics images out of the ordinary history of Western image making, where the single image has long been the norm. Contemporary bubble-chamber experiments still use stereo pairs, but not *as* stereo pairs. Instead the 3-D data are reconstructed using very exact perspectival measurements (which are far more exact measurements in all but a handful of Western fine-art images since the Renaissance).

Other kinds of supporting images include the tracing and the flow chart, both used in Figure 62. They are educational aids, used for introductory textbooks and when the event is unusual or especially important. In this case, the rare particle is a "charmed meson" called a D-star (D★). It is too short-lived to show up in the photograph, decaying almost immediately into another short-lived meson, the D^0, which in turn decays into a jet of other long-lived particles. The D★ lasts less than 10^{-19} seconds, and the D^0 lasts less than 10^{-12} seconds, too short a time to leave any measurable track. The flow chart is necessary to show the first instants involving the D★ and D^0; the rest can be followed in the tracing.[27]

Even Figure 62 omits detail, because the tracing does not include the invisible neutral particles. Compare, for example, the spiral at the center

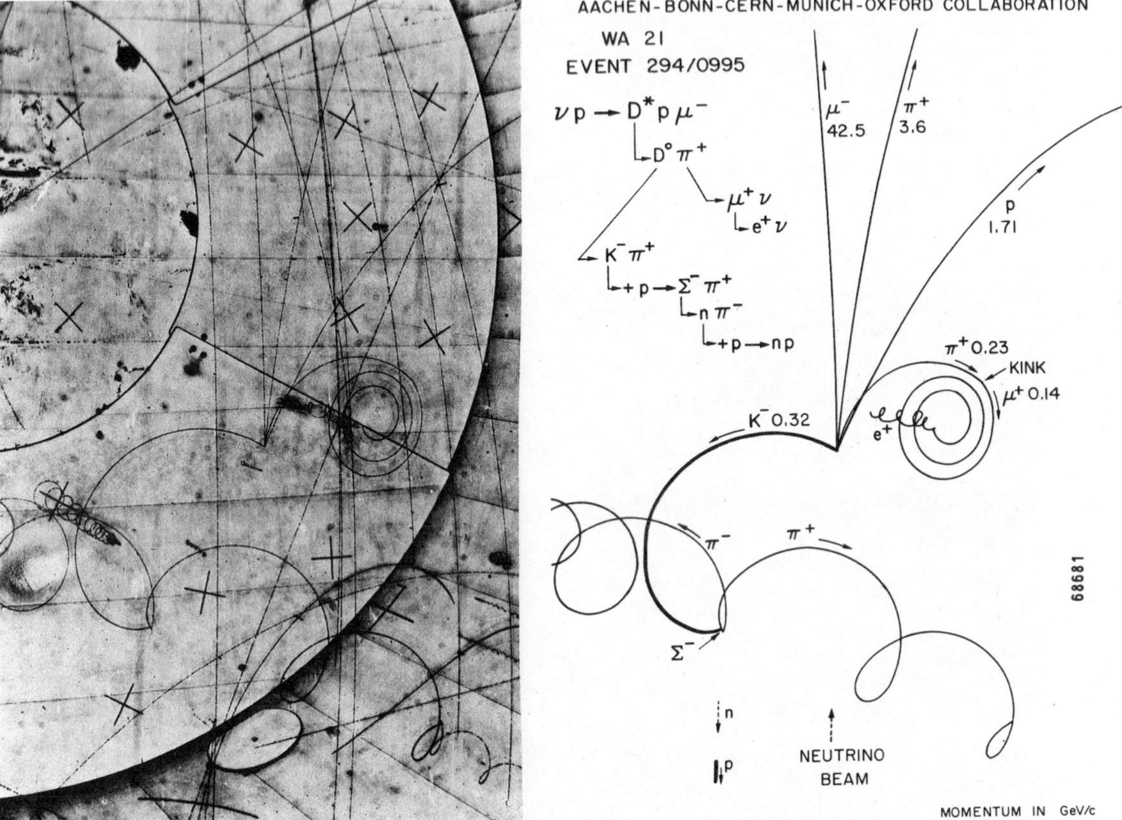

right with its representation in the flow chart (also at the center right). At
the point labeled "KINK," the pion, π^+, decays and produces a muon, μ^+,
and a neutrino, ν. The neutrino leaves no track, so it is omitted from the
tracing. Figures 63 and 64 show the effect of adding neutral particles.
The initial collision of a proton and an antiproton at the vertex labeled A
produces four particles:

$$\bar{p} + p \rightarrow K^0 + K^- + \pi^+ + \pi^0$$
$$\rightarrow \pi^+ + p \rightarrow \pi^+ + p$$
$$\rightarrow K^- + p \rightarrow \Lambda^0 + \pi^0$$
$$\rightarrow \pi^- + p$$
$$\rightarrow K^0 \rightarrow \pi^+ + \pi^-$$
$$\rightarrow \mu^+ + \nu$$
$$\rightarrow e^+ + \nu + \bar{\nu}$$

FIGURE 62

Production of a charmed
meson. A BEBC liquid hydro-
gen event. Aachen-Bonn-
CERN-Munich-Oxford
Collaboration, December 1978.
Courtesy CERN, Geneva.
Photo 68681.

Diagram of the interaction in
Figures 63 and 64

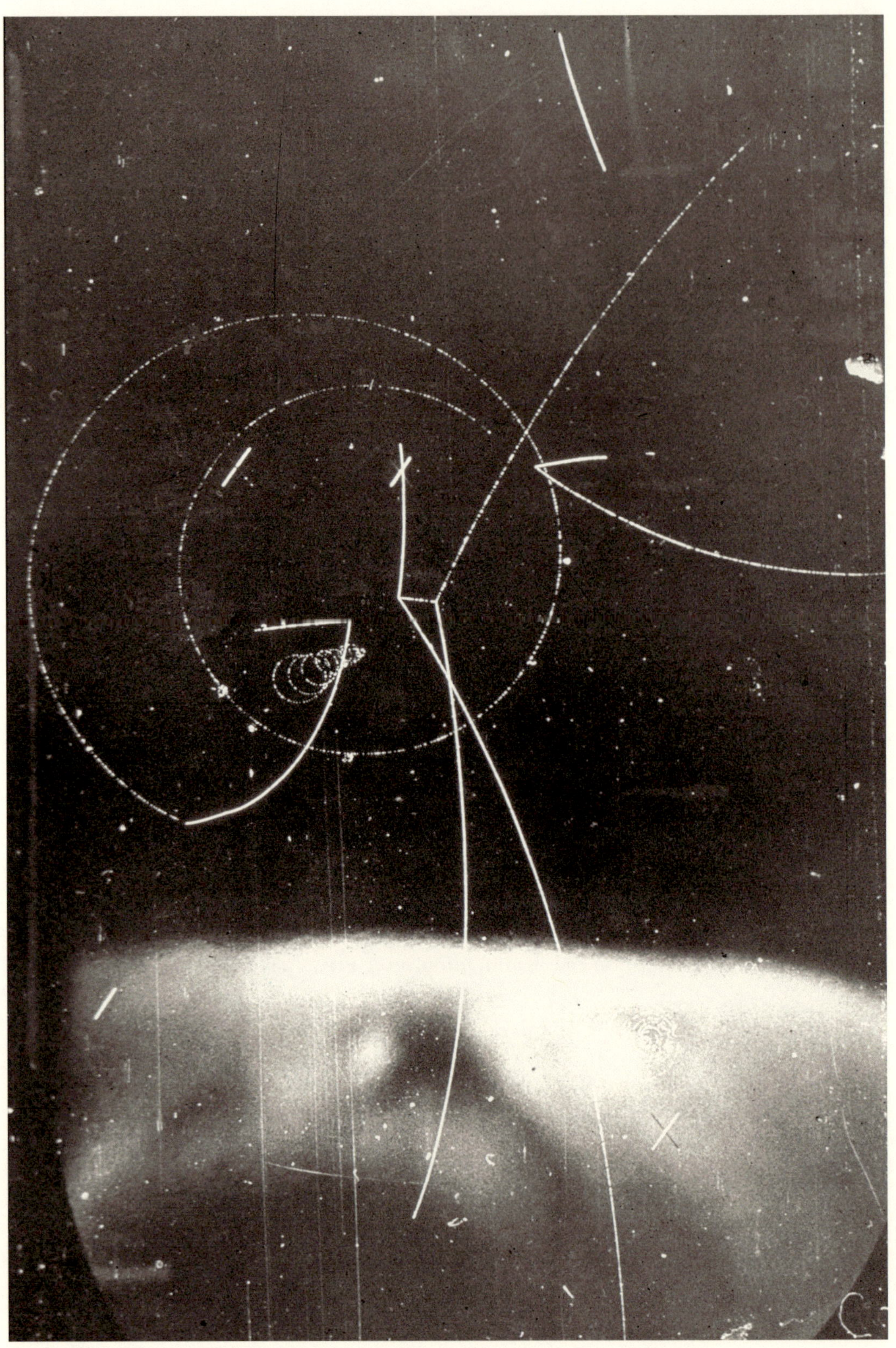

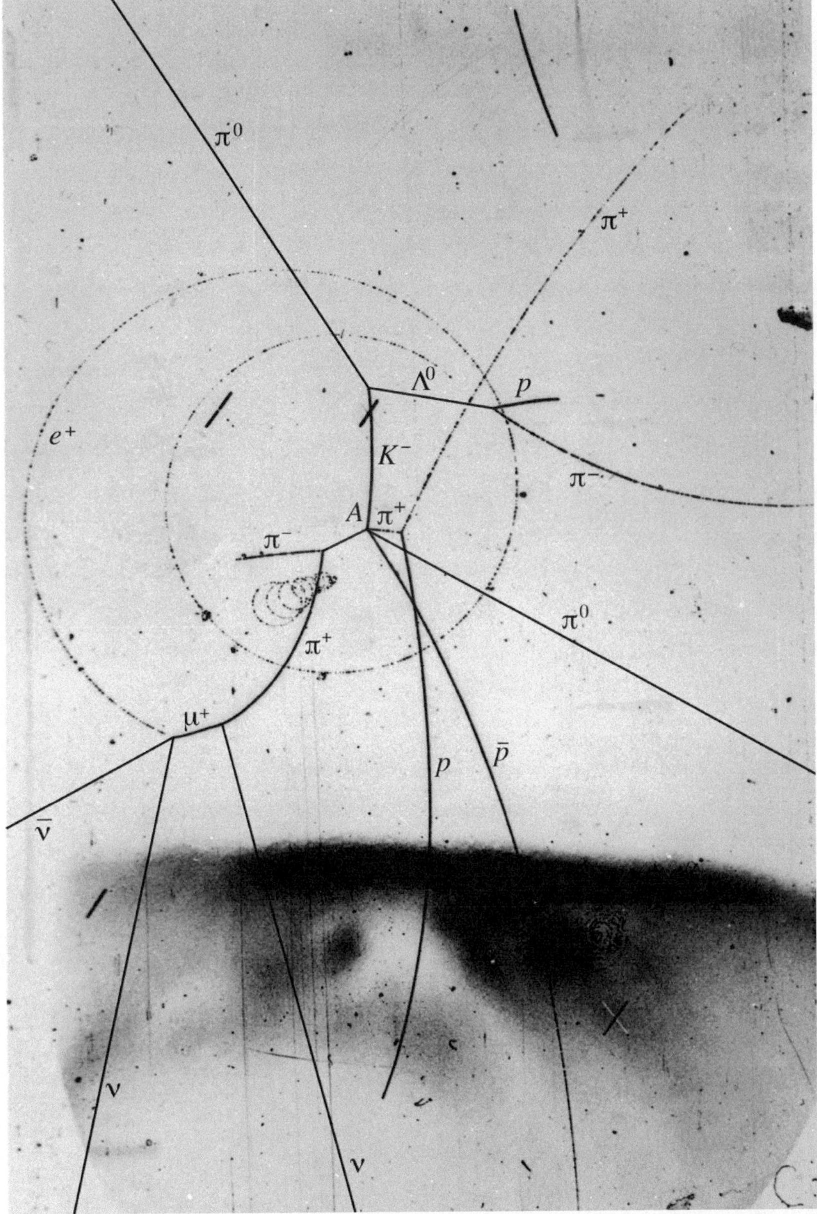

FIGURE 63 (opposite)

Proton-antiproton reaction, producing a K^0. c. 1971. Courtesy CERN, Geneva, photo 13668.

FIGURE 64

Analysis of Figure 63, with neutral particles added. Courtesy CERN, Geneva. Diagram: author.

three of which can be followed on the original photograph, Figure 63. In Figure 64 I have added the track of the neutral particle K^0 and the other neutral particles. They complete the picture, making it more spidery and complex.[28]

Particle physics is not the only science that requires multiple representations. The study of genes involves a half dozen entirely different kinds of images that are often used together. The philosopher and biologist William Wimsatt calls the practice the "thicket of representation," and he notes that some kinds of pictures may actually be incommensurate with others.[29] The objects themselves (genes, DNA and its associated proteins) are not amenable to any one pictorial strategy, and it takes specialized visual translations to move from one kind of image to the other. The recent discovery that DNA is "hypertexted" (read by different proteins when it is in Z-DNA, A-DNA, and B-DNA forms) will only make the challenge of adequate representation more acute, as will the increasingly accurate understanding of real-time chemical processes like folding, unfolding, and duplication.

There are other examples in biochemistry, which seems to lend itself to these problems of representation. An excellent text is Stephen Harrison's lecture "What Do Viruses Look Like?" which sets out nearly a dozen different ways of visualizing viruses, from electron micrographs to ribbon diagrams. The differences between the pictures are sometimes so great that it takes additional schemata to show how they are related.[30]

In fine art, on the other hand, it is still normative to represent an object with a single image, although it could be argued that when Seurat painted fifty-odd canvases to prepare for *La Grande Jatte*, or when Picasso filled notebooks with dozens of drawings for *Les Demoiselles d'Avignon*, they were also thinking of the inadequacy of any single representation. These scientific images are not independent in the sense of the fine-art images they fortuitously resemble: they need help from other images and from calculations. Much less of what is interesting is visible, and much more lies outside two-dimensional visual representation.

45   *Complexity in the Logic Tradition*

In comparison to the silver emulsions of the image tradition, the computer monitors used in the logic tradition seem to display distinctly simpler pictures (Color plate 19). Most are blurry, as all computer monitors are, and they have significantly less detail. Color plate 19 is part of a well-known image, recording the discovery of the W particle. As a whole, the image *looks* like a silver-gelatin picture, but it is digital from beginning to end. The dots do not need to be sharp because their positions are known with precision. They are not random bubbles or droplets that would need to be measured from an enlargement, but discrete signals in the detector that have already been located and mapped.[31]

Galison writes of logic-tradition images in terms of the move away from the notion that a single picture will provide the authentic evidence of a single "golden event." Logic-tradition images are made and analyzed much faster—by computer—and so there can be many more of them. Their cumulative evidence is often what counts—not the single unique image. That is one aspect of the ongoing move, among some scientific imaging, away from an older frame of mind (which includes fine art), in which the picture itself is the repository of all salient meaning. Logic-tradition images are available by the thousands on the Internet, and they can have little value in isolation from other images taken from the same experiments. That makes sense in light of their use in statistical calculations, and also because they come after the fact—they are nothing more than convenient forms in which to display quantitative information that has already been gathered. Virtually nothing is seen in such images that cannot be better understood by looking at the raw data and the ensuing calculations.

Where an image atrophies, calculation thrives. A tremendous amount more calculation goes into the interpretation of a typical contemporary experiment than went into even the most complex bubble-chamber or cloud-chamber photographs. A typical logic-tradition image can be seen in Color plate 20. It is a wire-frame diagram of a warehouse-size detector called L3, at CERN, outside Geneva.[32] As the labels indicate, the image was generated on May 2, 1991, at 11:11 P.M., and it records event number 322 of run 284,602. The detector is composed of a massive series of nested metal cylinders, each one designed to track a different kind of particle. The whole detector is shown naturalistically, in axonometric perspective, from one side. But the particle tracks, shown in green and blue, are not all naturalistic. They are superimposed on the wire frame by an interactive software program called REL3, making a composite image that is neither an X-ray view through the detector nor a graph, but a little of each.

At the center, where electrons collide with positrons, several jets of particles can be seen spraying through the "time expansion chamber" in a fairly naturalistic fashion: those are the particles generated by the collision. Above and below the red cylinder are green and blue boxes stacked in towers: they are three-dimensional bar charts, whose heights are proportional to the particles' energies. At the top and bottom of this image are red wire-frame representations of another part of the L3 detector called the hadronic calorimeter; there the particles are registered by little green and blue squares, distorted a little by the perspective. The area of each square is proportional to the particle's energy—another non-naturalistic convention. The image is a rather intricate concoction of realism and diagrams, of graphs and image-tradition pictures. In the end the whole display is not really necessary: what matters are the quantitative measurements, not the picture itself. Pictures like this are not often published in physics journals, but they can be found on the Internet, in scrapbooks, on lab monitors, on plastic overhead-projector sheets used in classrooms and for lectures, and in public-outreach material. The event in question—number 322, showing the decay of a Z^0 particle—simply takes its place on a table of statistics of similar events.[33]

It is not always clear, even to physicists, how it's best to talk about images like these last two. J. H. Mulvey uses a curious sequence of scare quotes to describe images like Color plate 19, as if he does not quite want to call them pictures. "To complete the story of unification," he writes, "we must add the superb 'pictures' which demonstrated the existence of the W and Z^0 bosons, 'taken' with Rubbia's 'electronic bubble chamber.'"[34] What could that mean, I wonder: a picture that's really only a "picture," not taken but "taken," in an "electronic bubble chamber"? Could he mean the very thing that doesn't exist, that could bridge Galison's two traditions? Mulvey's vacillating terminology is evidence of the lingering image tradition and the aura of truth such images continue to have. Why else would someone want a "superb 'picture'" even when they couldn't get a "superb *picture*"? Mulvey is really only being anecdotal, as a breather between calculations. It is clear he is aware that naturalistic pictures (not in quotation marks) are no longer useful.

The history of twentieth-century images in particle physics is the history of a field whose interests have moved away from what can be shown in pictures: away from resemblance and representation, and into many other kinds of images that are ultimately mathematical. The real complexity of these images is the complexity of *leaving* the image and finding meanings that the image can no longer contain. Never since the inception of Western images has the subject matter of the image been so far beyond the reach of the image itself.

six Quantum Mechanics

Now I come to a final possibility. So far I have been talking about short-falls of representation. In art, that means pictures have incomplete realism because the artist doesn't want to fully represent the object. In the case of scientific images, representation falls short because of a lack of hardware or software, or because the object is physically unrepresentable (for example, when the illumination has too long a wavelength in relation to the object). These obstacles can reach the point where the majority of the interesting properties of the object are not present in the image, making the picture little more than a set of hints at things that are not shown. That was the case in the previous chapter, and it may seem to mark a kind of limit in the abandonment of the naturalistic image.

I have been calling these shortfalls, and the concomitant inadequacy of the images (whether that inadequacy is intended, as it is in art, or not), the *unrepresentable* elements of pictures. The forms in question are generally known to viewers and makers, even if they do not appear in the image in question. A viewer of Martin's or Krane's paintings will know what they decline to represent, and a viewer of a blurry picture of a galaxy will have an idea how the galaxy would look at higher resolution.

There are two other kinds of lack in relation to representation that are even more stringent. I call the first of these the *unpicturable*, meaning whatever is not conceived (by the maker or viewer) as a potentially visualizable or picturable form. In the bubble-chamber and cloud-chamber images, the absent particles are unrepresentable, but they are also unpicturable because there is no pictorial form adequate to them. The final kind of lack is the *inconceivable*, meaning whatever does not present itself to the viewer's imagination at all, either as a picture or as an unpicturable property.[1] (I use the term *inconceivable* rather than *unconceived* or *unconceptualized* in order to stress the passive nature of the absence: these are not forms or properties that the makers or viewers of pictures try to find or represent; they are simply overlooked as pictorial possibilities.) The inconceivable can still leave traces in images, but only in a negative sense, because it can be deduced by considering what the picture does represent or propose as potentially representable. Both the unpicturable and the inconceivable are relevant properties in quantum physics images, which I think are the most conceptually challenging representations that the twentieth century produced. In this final chapter, I will approach these problems of the unpicturable and the inconceivable by considering how certain kinds of images have been taken to work, and in particular what has made them seem useful or useless. That will ultimately lead me back to what might be considered unpicturable or inconceivable about such images.

46   The Last Remnants of "Normal" Pictures

Quantum mechanics texts pelt readers with one impossibility after another, wearing down even the most concerted efforts to match the mathematics with anything that could possibly seem familiar. There is no solid ground for analogies with experience outside quantum mechanics, and teachers are often at pains to cut away any remnants of what is taken as intuitive familiarity that still cling to the weird equations. Students begin with such concepts as black-body radiation, the wave equations of classical optics, and Maxwell's equations; but often enough as soon as those subjects are reviewed, the texts say analogies to thermodynamics and optics have to be abandoned. The objects in question—the subatomic particles and fields—are absolutely, immediately, and hopelessly beyond what students commonly identify as their intuition or imagination. No model works, no analogy fits.

The entire enterprise of particle physics can be described, without much exaggeration, as an elaborate set of labels, symbols, and rules, to describe objects that are permanently inaccessible to the senses. The labels (quarks, gluons, leptons) are well known; the symbols run very quickly through the entire Greek and Roman alphabets and begin again in capitals, boldface, superscript, and subscript. (A working physicist can easily tell you at least one conventional meaning of every letter in the Greek and Roman alphabets, uppercase and lowercase. A particle physicist at Johns Hopkins University once told me he could name at least two properties designated by every letter in both alphabets.) The rules are presented in various formalisms, and they grow so unwieldy that physicists continuously condense long equations into shorter and shorter forms, making very brief expressions stand for very long ones. Thus the Schrödinger equation

$$E\psi = -\frac{\hbar^2}{2m}\frac{\partial^2\psi}{\partial x^2} + V(x)\psi$$

can be written in many forms. Its core is an expression for the sum of kinetic and potential energy, $E_{tot} = T + V$. (The core form is visible in the equation above.) Often it is useful to assume plane-wave solutions, which means the Schrödinger equation operates on a wave function such as $\psi = e^{\frac{i}{\hbar}(px-Et)}$, in which p is momentum, E is energy, t is time, and x is position. In that case, the Schrödinger equation could look like this:

$$i\hbar\frac{\partial}{\partial t}e^{\frac{i}{\hbar}(px-Et)} = \frac{-\hbar^2}{2m}\frac{\partial^2}{\partial x^2}e^{\frac{i}{\hbar}(px-Et)} + V(x,t)e^{\frac{i}{\hbar}(px-Et)}$$

in which form the basic equation for energy, $E_{tot} = T + V$, is expressed in terms of the wave function. In other contexts it would be enough to write $|\psi\rangle$, which is Dirac's abbreviation for the wave function.[2] Yet even in its more prolix forms, the Schrödinger equation opens the way to a practical infinity of calculations. The farther into particle physics a student goes, the more she is surrounded by an outlandish scaffolding that doesn't pertain to anything human. In my experience, the only solution is to go forward as Dirac first counseled, trusting the mathematics and leaving understanding to one side.[3]

(The same interest in reducing intricate equations to simpler formalisms can be observed in physicists' use of graphs. The graphs in the *Physical Review* and elsewhere are often x-y coordinate plots, but they can compress a tremendous amount of information if the axes and functions are chosen carefully. Humanists tend to use unnecessarily complex graphical forms, partly for their expressive potential. When a subject is genuinely complex, a reductive formalism is a practical necessity, and there is no call to spell out the "expressive" details.[4])

Physics' dense formalism promotes a sense of irreality, which physicists convey especially well when they allow themselves to speak informally. David Griffiths's *Introduction to Elementary Particles* (1987) isn't written so much as it is spoken, in a styleless and sometimes corny conversational idiom that was popular in late-twentieth-century science. (The conversational tone gives books of that period a pleasant directness, although it could be argued that it avoids the problem of *writing* altogether.) Griffiths tends to put words like *visualize* and *explain* in quotation marks. He notes that a newly minted phrase like *quark confinement* really "doesn't explain anything," leaving it unsaid how rules and formulas actually do "explain" things. The rule "all naturally occurring particles are colorless" is "clever," he says, and it " 'explains' (if that's the word for it)" why particles are not made of one, two, or four quarks.[5] In fact, Griffiths puts a lot of words in quotation marks in his text: "if we could just get in close enough to see the 'true' strong, electric, and weak charges," he writes; or "conservation of charge and color are more 'fundamental' than the conservation of baryon number and lepton number"; or again, quarks are " 'red,' 'green,' and 'blue' "; or "anything with a lifetime greater than 10^{-17} sec or so [is] a 'stable' particle."[6] Griffith's styleless manner brings out the latent, unwritten question marks that permeate quantum physics. The inadequacies of English-language concepts and mathematical formalisms are well known and widely debated, and they lead to philosophies of nominalism and realism, and toward debates about the nature of description and observation. This book is not a forum for the question of what the quantum theory "really" describes, or how the practice of physics has been influenced by various theories about what should count as an adequate representation of physical reality. Even if I could explore such problems, I wouldn't try: too many books do that already.[7] Here I only want to show how the same questions are being played out, silently, in the pictures that are found in physics textbooks and professional journals.

Images have interesting roles in quantum-mechanical assaults on the intuition. An image in a textbook can be a friendly thing: it can bear the stamp of familiarity, or at the least it can promise that the relevant information can literally be taken in at a glance. The images in the previous chapter appear naturalistic, and they are, even if most of their meaning and interest lies outside the naturalistic qualities of the image. Quantum-mechanics images that appear naturalistic are often far from it, and in fact they wear their realism lightly, the way a ghost wears its sheet. A picture may seem to represent something familiar, but seen more closely, what looked familiar disappears. The images seem easy to interpret, so people try to talk about them as if they were naturalistic images. But the talk can be derailed by the realization that the pictures cannot possibly be representing things the way they are. Philosophically, the direction is the opposite of the uncanny: first these images seem familiar, then they grow increasingly alien. (An uncanny object seems alien, but comes to seem intimately familiar.)

Quantum mechanics is taught by an iterated process in which an intuitive-looking picture is advanced and then retracted, with the warning that intuition won't ultimately help. As soon as the student has grasped what appears to be a process that can be visualized, the author or teacher intervenes, explaining that visualization is just a way of introducing the material. There is an inherent, systemic contradiction in this practice: it says both "quantum phenomena need to be visualized" and "quantum phenomena aren't amenable to visualization." In this chapter I will be tracing several increasingly rapid and radical disavowals of the image, leading to the point where images seem not only hopelessly inadequate and misleading, but deeply misguided—and yet still indispensable. First, however, I want to look one more time at the problem of nostalgia and the ongoing hope that images—even in quantum mechanics—might be "normal," naturalistic, and adequate to their tasks.

The pictures known as Feynman diagrams are an interesting example (Figure 65). They represent interactions of particles, spiderweb-fashion. Time, as physicists say, runs upward in this diagram: earlier events are at the bottom of each diagram, and later ones are on top. Here a muon, μ, and an electron neutrino, v_e, enter from the bottom, exchange a W^- particle, and exit as a muon neutrino and an electron at the top.

The picture looks fairly realistic, but we would be wrong to think of it as a realist picture akin to a perspective drawing or an engineer's diagram of a machine. The particles don't necessarily approach one another at this angle, and they don't stay a certain distance from each other. (Nor does time always run upward: in some diagrams time is omitted altogether and the axes represent momentum and position, with no defined direction for time.) Feynman diagrams are actually instruments that help physicists

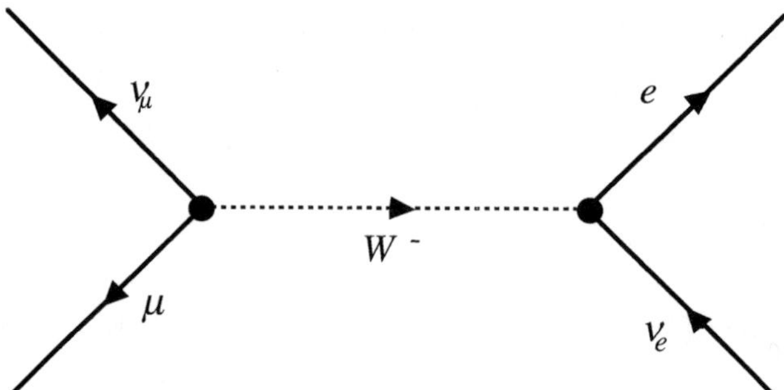

FIGURE 65

Feynman diagram. Diagram: author.

make calculations. Each diagram corresponds to a number, and from the diagram a physicist can calculate things such as the lifetimes of individual particles, the speed of a reaction, and the probability of a reaction taking place. Feynman diagrams can be rotated and twisted to make other diagrams, and in that respect they are more like bits of electrical diagrams (symbols for diodes, resistors, and the like) than engineering drawings. If the right hand portion of the diagram in Figure 65 is turned ninety degrees counterclockwise, it produces a diagram representing the decay of a muon into a muon neutrino, an electron neutrino, and an electron (Figure 66).

There are a number of reasons why Feynman diagrams should not be taken as realistic images. They make it seem that the operators—the equations represented by the vertices—define the creation of quanta at specific points in space and time. But the mathematics entails no such thing: the particles don't come into being all at once, at specific places and times. Perhaps most seriously misleading, the diagrams make it seem as if the virtual particles (dotted lines that begin and end within the diagrams) are on a par with the real particles (solid lines that begin or end outside the diagram). Virtual particles cannot be observed, but that is not because they are smaller than real particles, as the diagrams imply. It's because they do not exist except as part of the wave equation of the entire interaction: they are entailed in the "superposition" of all possible diagrams for a given situation. Virtual particles are not independent in the same sense as real particles, even though the diagrams make it seem that way.

(The tracks of bubble chambers and cloud chambers can also lead to analogous misconceptions. Their crisp linear tracks also make it look as if particles are simply created and annihilated. To put the matter more ac-

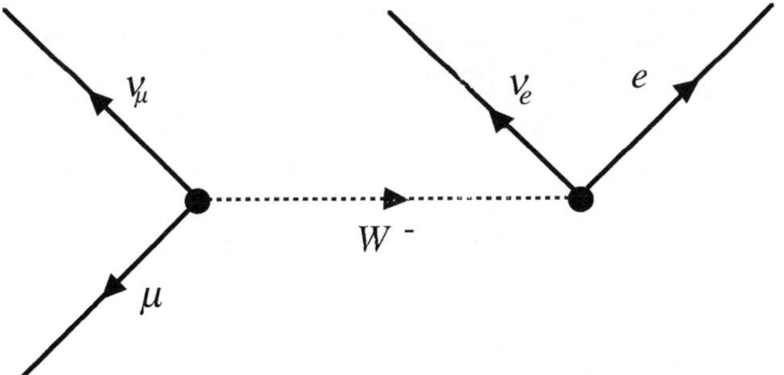

curately, each individual bubble or droplet is an act of measurement. When a particle decays into two others, there is a gap between the final single bubble, just before the Y splits, and the first twin bubbles, one on each branch of the Y. Between those two events, the "state" of the system evolves, as the philosopher Paul Teller puts it, "from a one-quantum system into a superposition involving at least a small amplitude [i.e., probability] for the new two-quantum state."[8] Bubble-chamber and cloud-chamber tracks can't capture that, and in that regard they can be as misleading as Feynman diagrams.)

Feynman diagrams are not realistic pictures, but the nostalgia for "normal" realist pictures remains strong. The historian of science David Kaiser has shown that physicists have bent Feynman diagrams to more naturalistic uses.[9] The "quark flow diagram" is a nonrigorous version of the Feynman diagram that starts to resemble the tracks of actual particles. Such pictures are drawn in several styles, more or less naturalistically to suit the purpose. Figures 67 and 68 both depict the beta decay of a neutron, in slightly different styles (they are adapted from different textbooks). The neutron, made of three quarks, is at the bottom left in each diagram. Two of the quarks are "spectators" and the third (a "down quark," labeled d) decays into an anti-electron neutrino, $\overline{\nu}e$, an electron, e, and an "up quark," u. With one of its quarks changed, the neutron becomes a proton (top left). Both versions of this diagram have the flavor of a realistic picture, or at least a diagram of one: in both, the intermediate W^- boson looks less substantial, and it appears as if the quarks are moving evenly along a little racetrack. It is easy to visualize the proton and the electron going off at an acute angle, as in a bubble-chamber photograph. The images are all the stranger, then, because neither version is a picture of something that could ever look like this, and neither diagram can be used to calculate. Physicists don't bother to draw the comforting

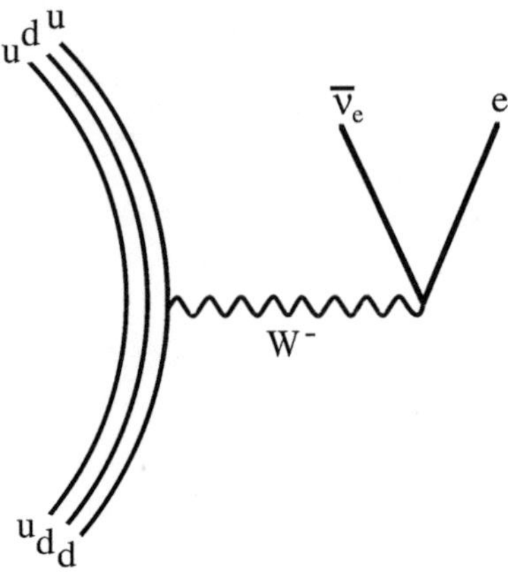

FIGURE 67

Quark flow diagram of beta decay. Diagram: author.

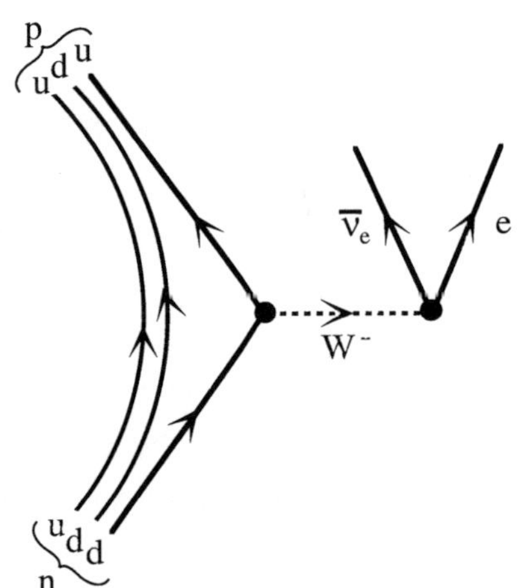

FIGURE 68

Quark flow diagram of beta decay, alternate version. Diagram:
author.

little billiard balls used in popular-science texts, but they are just as useless
and imprecise (Figure 69).

Kaiser has shown how many ways there are to decompress the rigorous calculational intent of the Feynman diagrams so they can work more
intuitively, as if they were realistic pictures. The opposite is also the case:
Feynman diagrams have been pressed into service to describe the theory
of supersymmetry, a proposed extension to the Standard Model of particle physics. Figure 70 is an example: on the left is an interaction between
a muon, a photon, an electron (μ, γ, e—all elements of the standard
model), and three hypothetical particles predicted by the Minimal Standard Supersymmetric Model (MSSM), including a slepton, $\tilde{e}$, and a virtual bino, $\tilde{B}$. The slepton and bino are "superpartners"—particles that are
predicted to correspond with others known to exist. There is a whole
stable of such "sparticles." The right-hand diagram has two quarks (d, s),

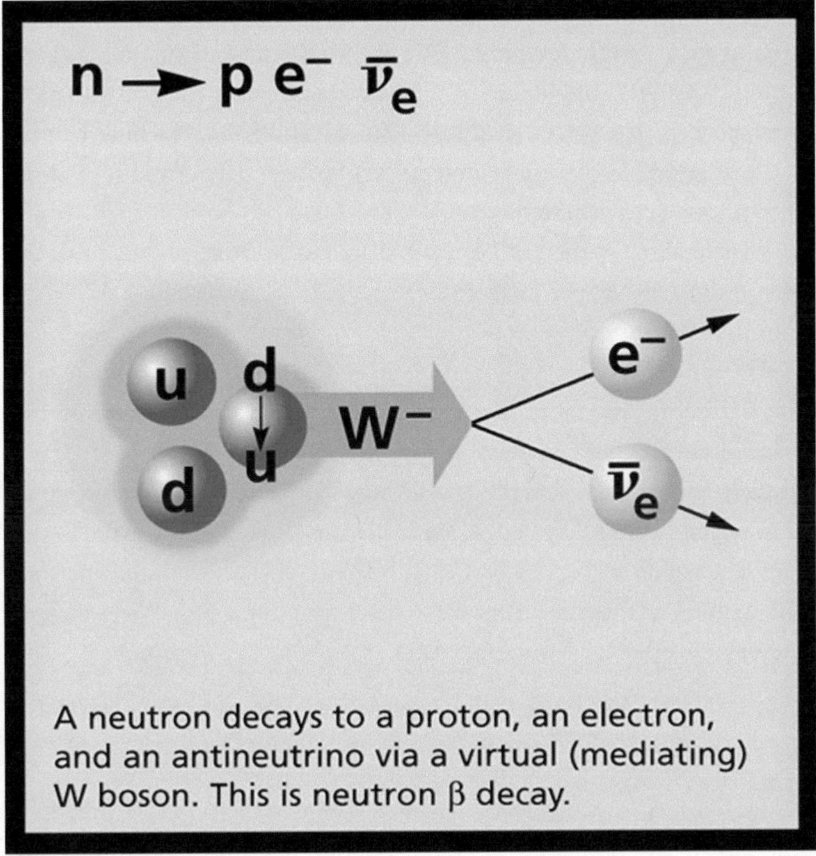

FIGURE 69

Billiard-ball picture of beta decay. From www.cpepweb.org. Used with permission.

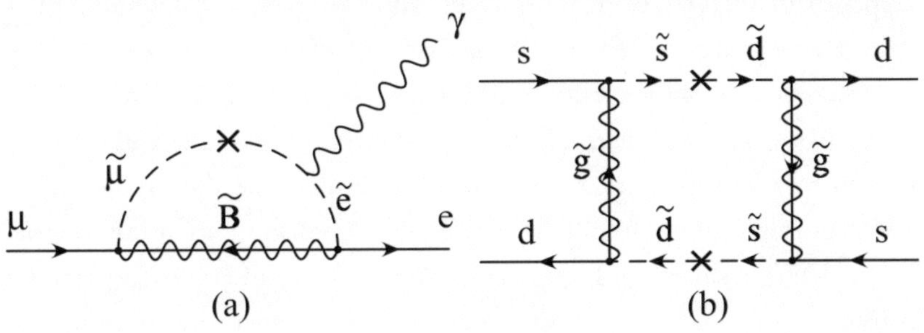

FIGURE 70

Supersymmetric diagrams that cause flavor violation. From Stephen Martin, "A Supersymmetry Primer," arxiv.org/ abs/hep-ph/9709356 (April 7, 1999), fig. 12.

but also several gluinos and scalar quarks (called squarks); the MSSM theory even predicts a particle with the unfortunate name "wino." All this is done with great rigor, and the predictions of the theory largely follow from the nature of the symmetry itself.[10] In this way Feynman diagrams continue to evolve, both as calculational engines and as intuitive pictures.[11]

Even severe critics of Feynman diagrams such as Teller admit that the diagrams have "heuristic" value, helping students to think "in terms of sequences of particle creation and annihilation."[12] Teller says Feynman diagrams are a "useful learning tool"; but it is hard to see, in Teller's own account, why that should be so. Why not begin with the mathematics? Feynman diagrams are a useful calculational shorthand, but how helpful is a "learning tool" that misrepresents its object so deeply and leads people to misuse it so persistently?

The naturalistic misuses of Feynman diagrams are more evidence that the earlier history of Western images permeates current practice. It is also possible to think of these misuses as last-ditch attempts to make what might be called "ordinary human sense" of pictures that have moved beyond such sense. Traditional modes of image making were stretched to their limits in the creation of cloud chambers and bubble chambers. Feynman diagrams and their offspring are something else: they are pictures that no longer set out to resemble what they depict. Yet at the same time (and this is the source of my fascination with these images), physicists, teachers, and students continue to think of them as if they could *also* have realistic properties—as if to say every picture must have some realism simply to be a picture.

47   *Deception*

Feynman diagrams were not designed to resemble their objects; they have sometimes been contaminated by nostalgia for "normal" pictures. It also happens that pictures in quantum mechanics are set up to resemble their objects, even when it becomes clear that they cannot. In those cases the pictures tend to lose their value and give way to nonvisual calculations. The more obvious it is that resemblance won't work, the more the authors have to excuse their own pictures, saying they are inadequate or even deceptive.

An example is the graph of permissible angular momentum vectors for an electron, shown in Figure 71.[13] The arrows, which can spin around in four cones or on the flat disk, represent the vectors (the strengths) of the angular momentum of an electron in a specific quantum state. No other vectors are possible—no other values of the angular momentum are permitted—so the diagram is an example of quantization. The problem with the picture is that it does not resemble its object: electrons do not actually spin, so they can't spin in different directions. "Spin" is a metaphor for an observable property that comes in different orientations and strengths. What looks like a picture of the precession of a spinning top is actually a trope, a metaphor for something else. Students find this

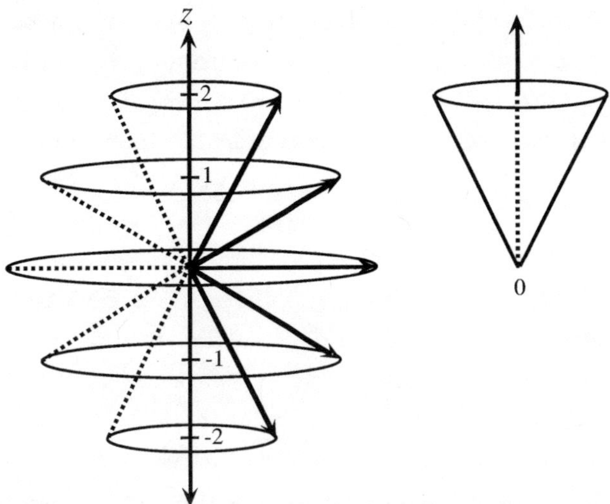

FIGURE 71

Permissible angular momentum
vectors for an electron with
quantum numbers $l=2$ and
$m=\{-2,-1,0,+1,+2\}$. Diagram: author.

metaphor useful because the different values are initially confusing, and because the picture looks a bit familiar from their basic mechanics courses. The picture makes it clear that the x-y direction of the spin vector isn't fixed, because the arrows can point anywhere on one of their cones or on the disk. At the same time, the z direction, the upward tilt of the vectors, is fixed in increments. That is the extent of the picture's truth-value, as well as the extent of its use.

The picture is also used to visualize another abstract property, the electron's "spin angular momentum." Ordinary angular momentum is often introduced by analogy with the earth's orbit around the sun; spin angular momentum is then like the earth's rotation about its own axis. But the angular momentum of the earth's rotation is caused by the sum total of rocks in the earth, rotating around the earth's axis: in other words, it is caused by the earth's structure. The electron, on the other hand, is assumed to be a "point source" that has no internal structure. Hence, when this diagram is used to visualize the spin angular momentum, it is even further from anything three-dimensional, anything spinning, or anything that has angular momentum in the classical sense. For those reasons, physicists only use this diagram to help students get started. Robert Gouiran's well-written primer of particle physics says simply, "this picture is exaggerated for the nature of spin remains mysterious, however much one tries to imagine it."[14] Perhaps "exaggerated" isn't quite the right word, because the picture doesn't emphasize things as much as it fundamentally misrepresents properties that are not representable in this fashion. In that regard the cosmologist Andreas Albrecht may be closer to the truth when he calls the spin angular momentum diagram a "lame attempt" to visualize something that can't be visualized.[15]

In everyday practice, the electron's spin angular momentum is written in this formalized fashion, as "kets": $\left|\frac{1}{2} - \frac{1}{2}\right\rangle, \left|\frac{1}{2}\,\frac{1}{2}\right\rangle$. The angular momentum, which corresponds to the picture I am reproducing here, is written: $|2-2\rangle$, $|2-1\rangle$, $|2\ 0\rangle$, $|2\ 1\rangle$, $|2\ 2\rangle$. (In each case, that's the length of the vector arrow and the height of the cone, for a total of four cones and a central disk.) Ket notation avoids the deceptive picture altogether. But can such a picture ever be entirely left behind?

This sitation is not entirely without its parallels in twentieth-century painting. Modernism also has its misleading pictures that seem to resemble things, but don't. I have mentioned Martin's disavowal of landscape: she named her paintings after landscape features but said the images were purely abstract. In a similar way, Krane's paintings are about natural phenomena like the color of flowers, but they decline to represent flowers. Kandinsky would have been surprised and probably distressed to see his "pure" abstractions decoded by reference to the secret figurative sketches he used to make them. Even so, his early, organic abstractions are irresistibly reminiscent of mountainous landscapes.[16] Although Mondrian named the urban landscape as his inspiration, and even linked it to abstraction, he would not have wanted his viewers to see his works as street maps. Interpretations like these can be thought of as misleading because the paintings were intended to leave their sources and analogies behind; but in art history the secret analogies and inspirations are not discarded: rather, they are woven into the work's meaning. The experience of looking at one of Martin's paintings creates the sense that you're being reminded of something that is closed off. You're not being misled, exactly, but you are reminded of a portion of the work's overall meaning that cannot, in the end, be the work's whole meaning.

By comparison, some physics images are radically misleading, because after a student's initial encounter with a picture, the reference to realism becomes little more than an impediment. Physics teachers and writers introduce pictures like the spin angular momentum graph and then say, as quickly as possible, that they should be forgotten. I have heard this kind of thing said in the classroom: Here is a very useful image, but be careful! It is deeply misleading, it's a "wrong picture," and you should forget it as soon as possible. One lecturer told me that Figure 71 is a "semiclassical picture," by which he meant that it pretends that life on the quantum level mostly works by analogy to life on the human scale. The electron's spin angular momentum is quantized, and classical angular momentum is not. Pretending that things are quantized (discretely arranged) in space, he said, is a "classical way of having a picture." The diagram is seriously misleading, and because it can't be used to calculate spin states, it is both misleading and useless. Why, then, is it so helpful? At the least it's because some abstract properties are easier to memorize if they are first seen in this format; but that explanation, if it's the best one available, should be cause for concern because it implies that even nonpictorial truths are best taken in as images. The question remains whether the specifically quantum-mechanical properties are memorized in a misleading fashion—whether remnants of the "semiclassical picture" will continue to guide a student's sense of electron spin.

48   Uselessness

The electron spin diagram introduces a new class of pictures: those that are not only inadequate (all the images in this book are inadequate to some degree), but also useless once their heuristic or prefatory function has been completed.

Occasionally a modicum of truth inheres in "realistic" quantum-physics images, giving authors an excuse to retain them. An example is the pictures of gluon-gluon bonds, or "color flux lines," in mesons and baryons (Figure 72).[17] Gluons are force carriers, like photons and W bosons; but unlike those particles, gluons can also bond to one another. The force they carry is the "color force," on the analogy of the electro-

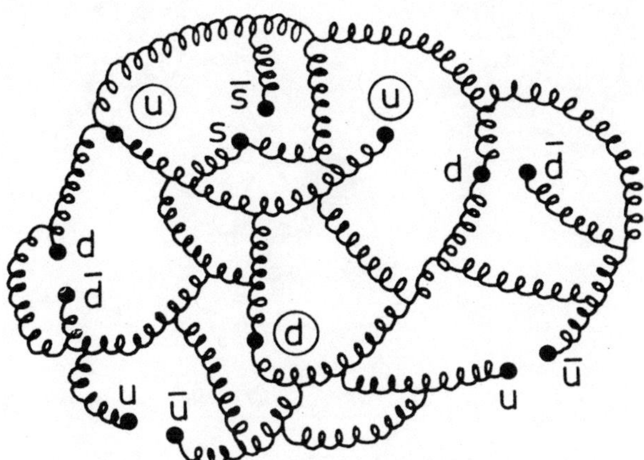

FIGURE 72

A proton with its valence and sea quarks as well as gluons. From Brian G. Duff, *Fundamental Particles: An Introduction to Quarks and Leptons* (London: Taylor and Francis, 1980), 86, fig. 6.13.

magnetic force carried by photons. Physicists sometimes find it useful to picture those bonds in the form of little springs attached to one another. The springs are informal symbols for the forces involved, and they are also a conventional way of representing color flux lines. Color force diagrams are to Feynman diagrams what a scribble by Cy Twombly is to actual calligraphy: they are free versions of a rigorous practice, intended to be suggestive and approximate rather than dependable and exact. Any number of color flux lines, gluon–gluon bonds, can form and disappear inside a baryon, so a picture like Figure 72 need not have any fixed shape. On the other hand, the picture is rule-bound in several specific ways.

Figure 72 represents the interior of a proton, which is itself a baryon, meaning it is made up of three quarks. Two are "up quarks," *u*, and one is a "down quark," *d*. The three are called "valence quarks" because they effectively exist continuously, but there are also "sea quarks," which are transitory and whose number is not fixed.[18] Sea quarks come in and out of existence, and they are inferred by the effects of collisions in accelerators. Figure 73 shows ten sea quarks (the ones that aren't circled), not because there are always ten but because the person who drew the picture wanted to give an impression of some of the possible configurations. Each sea quark is paired with its antiparticle: a "down quark" with an "anti-down quark," and so forth. The gluons that bind the quarks are

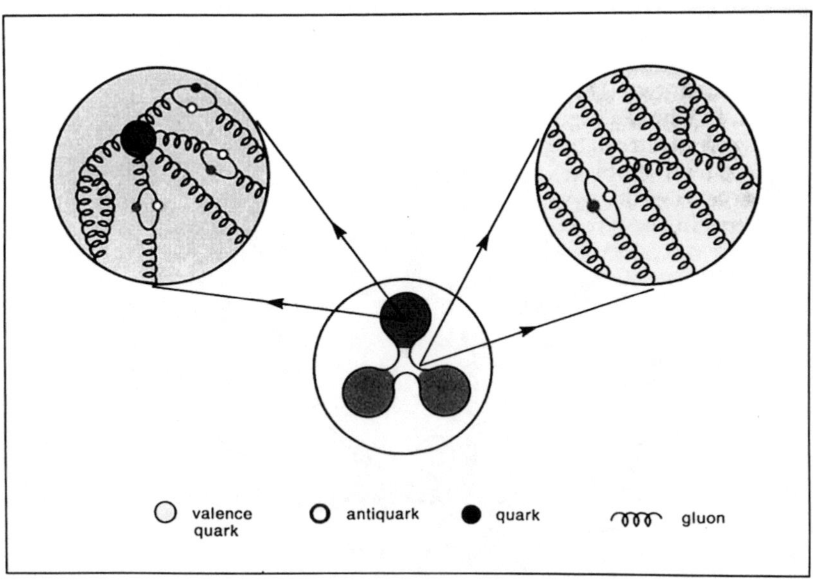

FIGURE 73

Schematic structure of the nucleon. From Bernard Frois and Marek Kaliner, "Where Does the Nucleon Spin Come From?" *Physics World* 7 (July 1994): 45, fig. 1. Courtesy *Physics World*.

also shown, with some attached to other gluons. Laws determine the number of valence quarks, the existence of pairs of sea quarks, and the presence of gluon-gluon bonds. Everything else is up to the person who draws the picture, in this case the physicist Bernard Frois.

Such an image is conventional in its use of springs and balls (leftovers from high school–level billiard-ball pictures and physics demonstrations), and by its implication that the contents of a proton can be inventoried or captured in a snapshot. Yet that's only the beginning of its conventions, because the objects themselves aren't objects disposed in space—they are signs of forces and the experimental effects they have in collision with other objects. Frois's sketch of a nucleon, Figure 73, represents the same object as Figure 72: a baryon made of three "'constituent' quarks interacting via the color force 'tubes.'"[19] Frois magnifies one of the three "constituent" quarks and one "tube," revealing sea quarks and gluon-gluon bonds. In the context of Frois's article, the image shows the "quark sea" from which nucleon spin arises, but it is not strictly related to the calculations he reports. Such a picture does not resemble the objects it depicts, but it also isn't a diagram of logical relations between parts. A physicist can't calculate using pictures like Figures 72 and 73, but then again, in some sense a physicist can't calculate well without having images like these in mind. They are useless, but not entirely false.

In one recent theory, nucleons like the proton include not only the basic three-quark configuration but a higher-energy state that involves five quarks. Thus, instead of being composed of two up quarks and a down quark, a proton could be a combination of two up quarks, a down quark, and a five-quark configuration:

$$p = c_0 |uud\rangle + c_1 |uud c\bar{c}\rangle$$

meaning that the proton is a linear combination of two states, with probability $c_0^2(c_1^2)$ of being in either.[20] Figures 72 and 73 could be adapted to illustrate this theory, but I am not aware of the existence of any such adaptation. Theories about nucleon structure are rehearsed and tested using mathematical notations, and there is no call for an intuitive picture that would not contribute to the theory.[21]

Another image that is occasionally reproduced, but not used to do physics, is the "quark confinement" diagram (Figure 74). It shows two things at once in order to suggest they may be related. On the one hand, it represents a phenomenon in superconductive materials: in a cavity

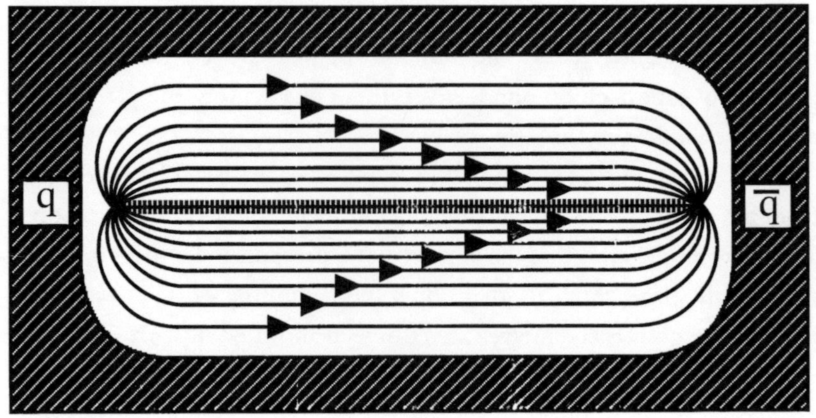

inside a superconducting material (depicted by the heavy dashed lines), a monopole (the confluence at the left) and an antimonopole (on the right) are tied by magnetic flux lines (the curving lines with arrows). Because superconductors exclude flux lines, the lines are confined to an "Abrikosov Gorkov flux tube." It takes energy to pull one monopole from the other, as if the flux lines were rubber bands. Yet the lines can never be snapped like rubber bands, because they only get stronger as they are stretched.

On the other hand, the same diagram is used to visualize how a quark and an antiquark behave: they can't be pulled apart because the force that binds them only increases as they are separated. The curving "color flux lines" (as opposed to magnetic flux lines) are again confined in a "mysterious 'non–Abelian flux tube.'"[22] From the late 1970s onward, when physicists have wondered why it isn't possible to observe a single quark in isolation, they have occasionally turned to this heuristic comparison. Brian Duff, author of a book on fundamental particles, introduces a picture like Figure 74 to help readers think about quark confinement by analogy with a dipole in a superconductor. Then he immediately adds, "these pictures of 'confinement' mechanisms are not meant to be taken too literally"—and that's because they do not prove any similarities between the realms of superconductivity and quark confinement, or even reliably represent the mathematics of either.[23] For Duff, I think, the picture itself is suggestive but almost completely useless.

Yet this case is significantly different from the pictures of quarks and color flux lines, because it turns out the analogy between quark confinement and superconductivity is a good one. It has led to work on the parallels between two fields that had been thought to be unrelated: quantum chromodynamics (the study of the interactions of quarks) and superconductivity. There are also several ways to redraw the pictures so they are more quantitative.[24] Beginning from some "phenomenological guesses" about "interquark static potential, inspired from a naive flux tube picture," physicists have been able to find deep similarities between the two states.[25] Apparently the vacuum can behave as a superconductor, so quark confinement can be understood as a mixture of contributions, including a monopole effect. Color plate 21 shows the contribution of the monopole effect to the energy distribution of a quark-antiquark pair: it's a quantified version of Figure 74. As the physicist Gunnar Bali says, "first there was the intuitive graphical representation of a flux tube which is depicted in many textbooks," then there was a theory, and finally there was a demonstrable connection between quark confinement and superconductivity.[26]

The "naive flux tube picture" has had an afterlife as artwork: the book *Quark Confinement and the Hadron Spectrum II* has a cover illustrated with an abstract painting, showing pooled colors against a thin, desiccated field (Figure 75). The painting, made by a friend of one of the authors, is entirely acceptable by international art standards: it is not unlike some of Jean Fautrier's canvases. It expresses confinement by depicting intertwined blue and yellow regions, isolated from a thin background. Certainly the pools of paint suggest the complexity of the physics (and complexities beyond the physics), and they provide a refreshing contrast to the rigorous mathematics. But that's about it: the discourses of painted meaning and physical laws do not speak to one another. On the dust jacket, the book's title and the ISBN barcode have a jarring effect.[27] To me, Figure 75 is an especially unpleasant image, with its juxtaposition of gestural painting, commercial design, and technical subject matter. It's an apt emblem for the difficulty of speaking about painting and physics together; it is a rum image, ruined by the conflation of two very different discourses.

Quark confinement and superconductivity are not the only cases in which an image suggested properties that later proved true.[28] But most of the time, images that their makers know are inadequate remain that way. The picture of the electron's permissible spin angular momentum vectors, as well as the pictures of sea quarks and gluon-gluon bonds, are matters of misguided intuition. A picture presents itself to the mind, and then it becomes irresistible. For a painter, that can be a fruitful state to be in, and in science, too, it can be productive to be guided by a picture. But

Massimo Brambilla, *Colour Confinement II*, as it appears on the cover of *Quark Confinement and the Hadron Spectrum II*, ed. Nora Brambilla and G. M. Prosperi (Singapore: World Scientific, 1997). Courtesy Nora Brambilla.

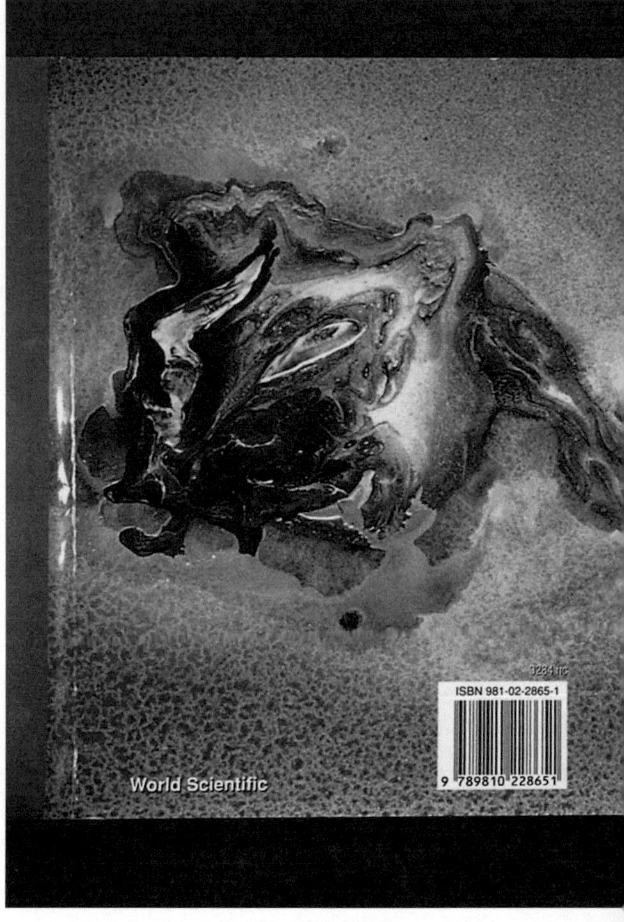

in quantum physics, chances are good that the intuitive image will be so far from the mathematics that it will end up being useless, "lame," and even misleading.

49   Rhetoric

Uselessness is a tricky category, and so far I have been using it in a strict sense: a picture is useless if it cannot be used to calculate, predict, or model a phenomenon. The examples I have adduced so far show that uselessness, like complexity and intuition, is a historically variable quantity: it changes depending on who is doing the reading. The diagram of electron spin (Figure 71) is useful, but in the context of introductory textbooks it can be hard to say how, because students need to be shepherded as expeditiously as possible

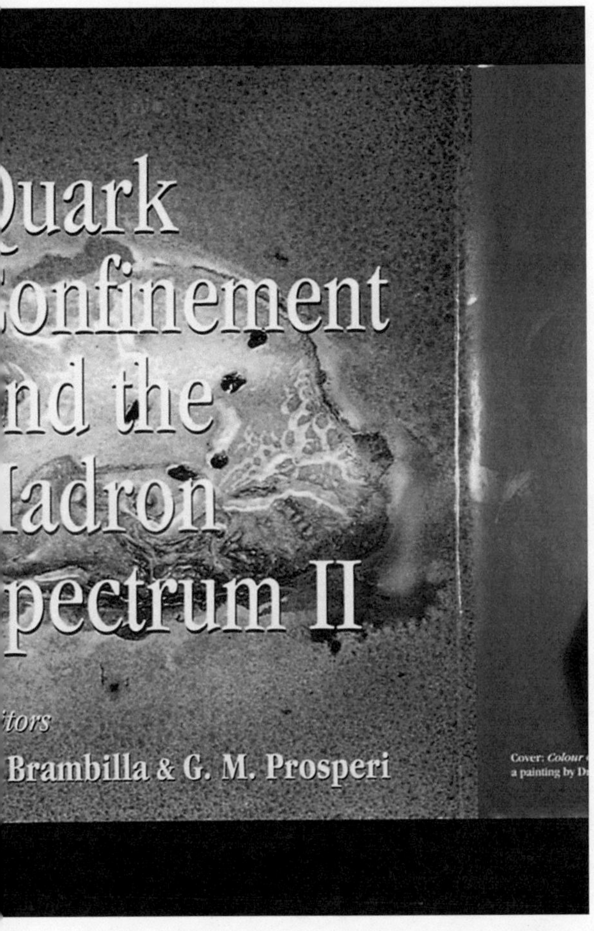

to the point where they can solve numerical problems about quantum numbers. The pictures of color flux lines (Figures 72 and 73) are useful, but the language for explaining that sense of "use" is not often available in advanced physics textbooks—it depends on day-to-day estimations of what might aid inspiration as well as calculation. As the mathematician Sha Xin Wei points out, in this wider sense of "use," a picture need not be isomorphic "to an algebraic procedure or even a logical deductive process in order to prove 'useful.'"[29] It is possible, for example, that a picture might function as an aid to memorization, a stepping-stone to an intuition that cannot be fully quantified, or a prompt for thoughts that cannot be immediately crystallized into formulae. In that wider sphere, pure uselessness may not exist; but there are images whose uses are so hard to describe that they can only be presented, not justified. Even a picture that is useless for all known and practical purposes, and is unarguably deceptive or untruthful, can draw a reader's eye and offer wordless support for a quantified argument. When the actual object of inquiry is far enough beyond the picturable, this may be the most sensible option.

The most interesting cases of nearly pure uselessness in physics are the scenarios for the inflationary universe that have been developed in the last decade by Andrei Linde.[30] Linde calls some of his scenarios "Pollock universes" and others "Kandinsky universes" (Figure 76).[31] He has also produced graphics depicting the "self-reproducing universe," composed of "exponentially large domains, each of which has different laws of physics" (Figure 77). Whenever he can, Linde reproduces brightly colored versions of his pictures: in this case the colors denote boundaries between domains of the inflationary universe; the colors fluctuate at the peaks of the graph, indicating that the "laws of physics there are not yet settled."[32] Linde also maintains a Web site with brightly colored versions of the "Kandinsky" and "Pollock" universes.[33]

Sha Xin Wei, who assisted Linde with some of these images, suggests that they have a purely rhetorical function: they help Linde win converts

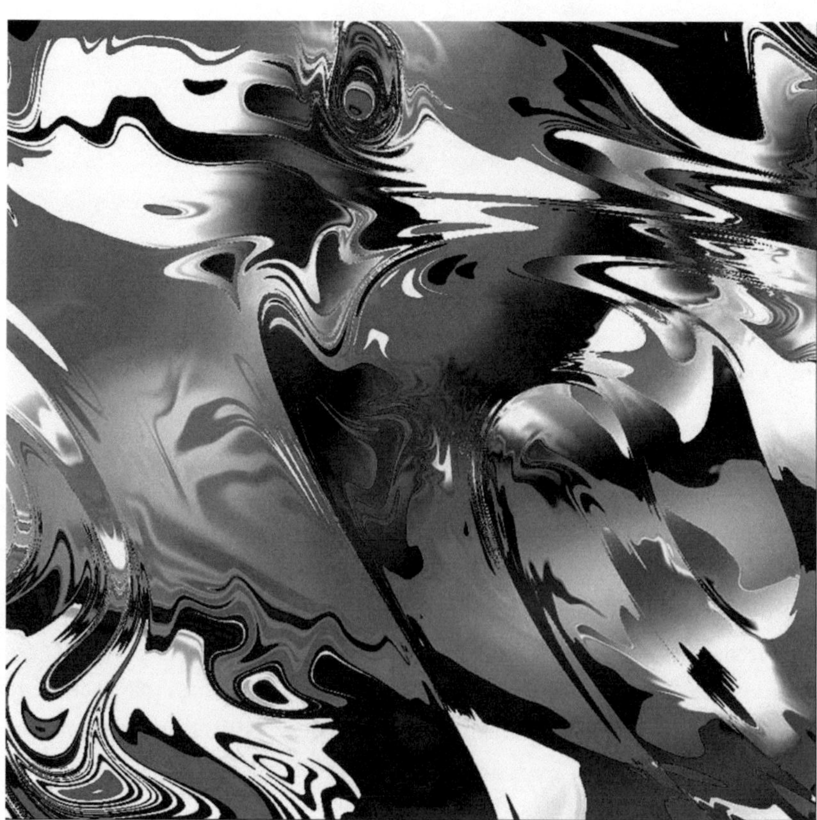

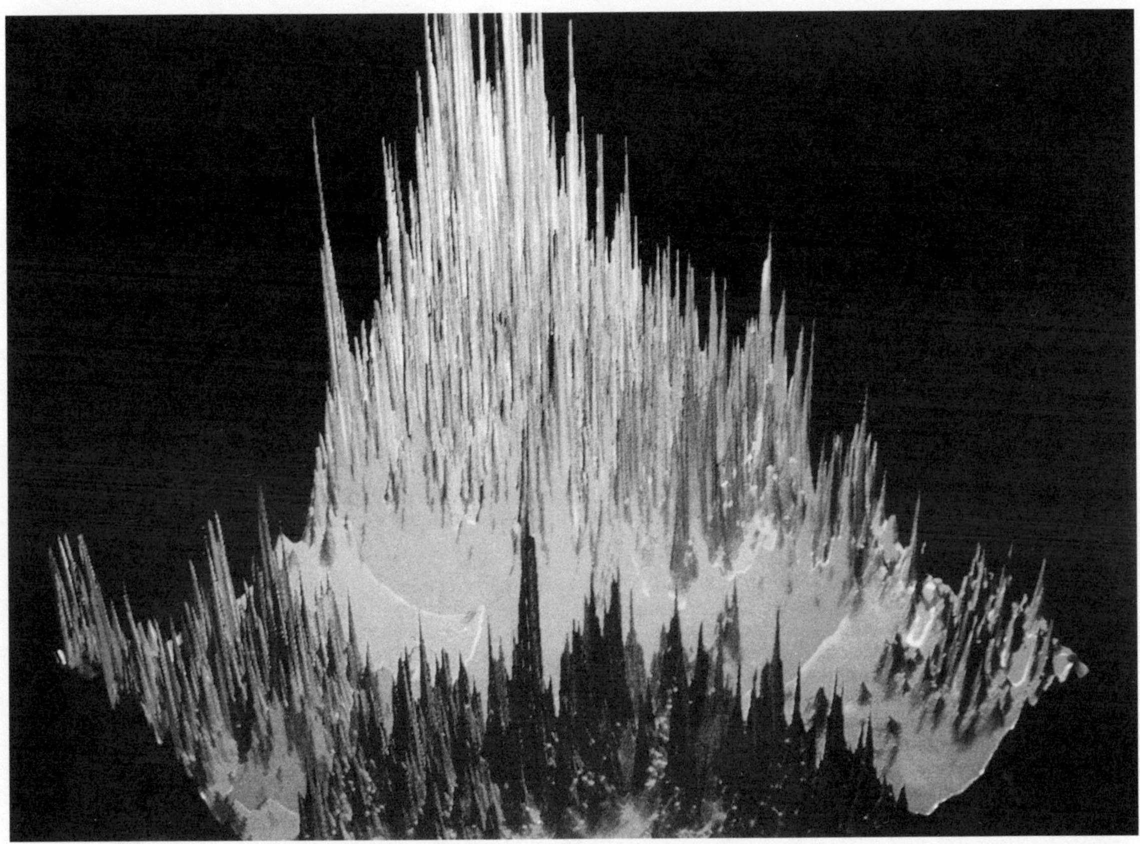

to his theories. Linde himself tells me that his pictures are attempts to show his theory is as "profoundly simple" and "beautiful" as the standard Big Bang model that it seeks to modify. The theory itself is extremely complicated when it is argued, but simple and convincing when it is shown—or rather, when something apparently derived from it is shown. The pictures, Linde hopes, will instill in readers a sense that the "rather formal and complicated" mathematics of his theory really expresses a "simple truth." To that end, the more beautiful a picture can be, the more forcibly it can convey the hidden simplicity of the mathematics.[34]

Linde also illustrates his theory with space-time graphs of partly disjunct "universes" (Figures 78 and 79). Figures 78 and 79 show three versions of a single picture: the first was sketched by his son and sometime coauthor, the second was worked up for *Scientific American*, and the third is a further refined version for the cover of the same issue. Each picture looks naturalistic—each "bubble" represents a universe—but Linde claims that the universes are mutually unknown to one another, so no such view

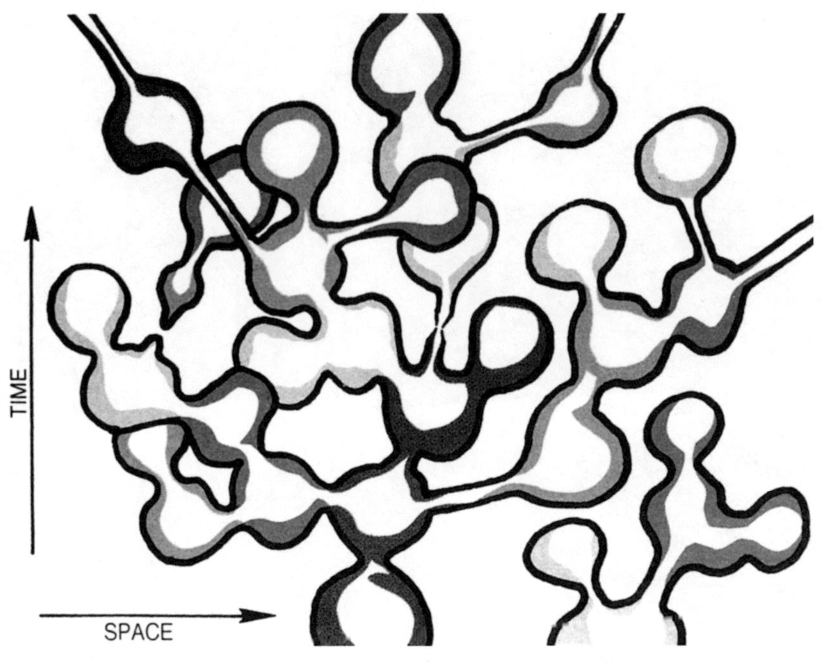

FIGURE 78

Andrei Linde's two versions of a picture showing the global structure of a chaotic, self-reproducing inflationary universe as an extended branching of inflationary bubbles. Top: From Linde, "Particle Physics and Inflationary Cosmology," *Physics Today* 40 (September 1987): 61–68, fig. 6 on p. 67. Bottom: From Linde, "The Self-Reproducing Inflationary Universe," *Scientific American* (November 1994): 55.

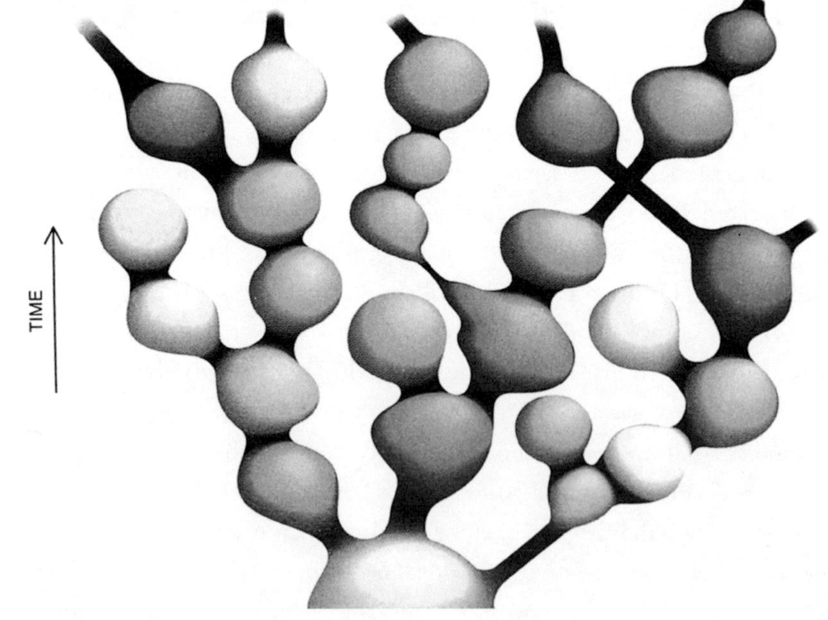

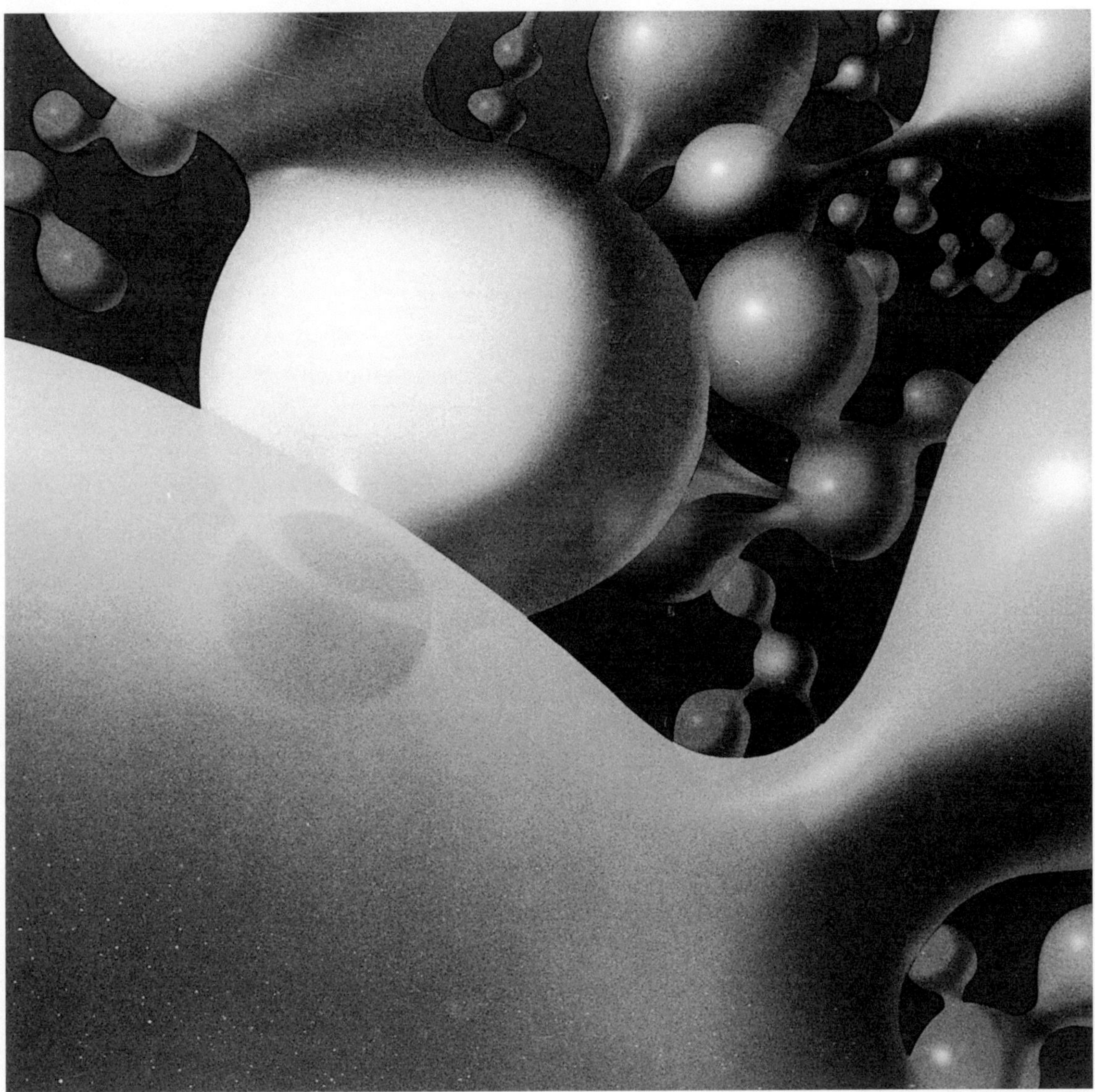

could ever be possible (not least because there would be no way to be "between" universes). The relative scales are also wrong, because in Linde's theory the threads between universes are permitted to shrink to the Planck length or expand to 10^{10} light years. The pictures also make use of symbolic pictorial devices that are inimical to realistic depiction: the universes are roughly spherical, which evokes their space-time symmetries; and—as David Kaiser has pointed out to me—the "necks" between universes and threads are reminiscent of the familiar tube shapes

FIGURE 79

Andrei Linde's third version of the pictures in Figure 78, as shown on the cover of *Scientific American*. From Linde, "The Self-Reproducing Inflationary Universe," *Scientific American* (November 1994): cover illustration.

used to denote singularities such as black holes. The spheres and necks are not exactly "a Baedecker's guide for what you'll see 'out there,'" as Kaiser puts it, but they are "a mnemonic or hint about conceptual similarities" between familiar things and the objects Linde is positing. In that way Linde's pictures have a modicum of use—but only a modicum.[35] They fail to represent quantifiable logical properties of the theories, and in doing so they become at once indispensable to the rhetorical framing of the theory and largely useless for the exposition of the theory.[36]

The less interesting questions to be asked about Linde's images have to do with his success in the physics community, which might be due in some measure to his pictorial proselytizing. It is more intriguing, I think, to ask how a theory is served by pictures that advertise themselves as realistic representations when even the theory says they can't be. What is the relation between a physical theory of absolutely unpicturable elements and the versatile and even mesmerizing means Linde uses to help communicate it? (And how did *Physical Review D*, a professional journal, come to accept an article with images that cannot be used to calculate, predict, or model?) Andreas Albrecht, who is sympathetic with Linde's views, sees the pictures as ways to "help people to think about" the fact that they have to "come to grips" with Linde's claims.[37] What they think and how they think after seeing the pictures is unclear—but that may be the best evidence that the pictures work as Linde intends them.

From the point of view of normal physics practice, the images I have been considering in this chapter are evidence of a kind of disease of image interpretation. In physics, images of all sorts are taken to be propositional: they are used as efficient carriers of information that can be put in other forms—as numbers or equations. Normally that hardly leaves room for nonpropositional content such as extra colors, illusionistic spaces, and added textures. Graphs, charts, and measured and measurable photographs are the stock in trade of physics journals, and the exceptions are photographs of scientists, pictures of experimental apparatus, or cartoons.

Figures 76–79 behave like fine-art images in the sense that their functions and purposes are more or less unclear; as such, it can be difficult to justify them in the context of a discipline that is relentlessly bent on calculation. Like some paintings and fine-art photographs, these images refute the notion that all meaning is propositional meaning. Yet they also do something significantly more radical than any postmodern art: they work at such a remove from their ostensible subjects that they actually call for apologies and retractions. (No painting has ever traveled so far from the idea of adequate representation as to warrant such an apology.) It may seem as if conceptual art has done something analogous because it abandons representation altogether. To me, the comparison reveals a weakness in conceptual art. Instead of rejecting picture making, the physics practices I am discussing cling to it, as if they were trying to wring useful meaning from the visual. In that regard, some conceptual artists are complacent about images. Works such as On Kawara's date paintings (see Figure 1) present themselves as wholly adequate to their task, even when that task is to conjure things that cannot be represented. The physicists who make the pictures I am considering in this chapter are more disaffected, more pessimistic about representation, than conceptual artists such as On Kawara.

Why are physicists, teachers, and students driven to try to use such images? I have suggested it may be because the idea of realistic pictures is historically connected with the idea of self-evident truth; it may also be because the objects that are being studied are so outlandish, so far beyond the reach of ordinary picture making, that it seems any recuperation of sense is helpful. Even images regarded by members of the physics community as patently inadequate or misguided can also come to seem potentially helpful. This is what happens to pictures when everything they are supposed to represent is already deep in the realm of the unpicturable.

50   *The Incommensurate*

When a representation falls so far short of its object, it is useful to talk about its incommensurability with that object. Linde's pictures of expanding universe bubbles are incommensurate with what they represent because they lie about perspective: those "universes" are not objects that could be seen from any place, or at any time. They are also incommensurable in a deeper sense because they cannot be quantitatively tied to their accompanying theories, no matter what kind of translation is proposed to transfer information from the pictures to the equations. Quark flow diagrams are incommensurate because they are schematic pictures of events that are actually known only through statistical trials and because they masquerade as pictures of static objects disposed in space. As in Linde's pictures, quark flow diagrams cannot be quantitatively linked to the theories they picture, the objects they represent, or even the Feynman diagrams that ultimately inspired them.

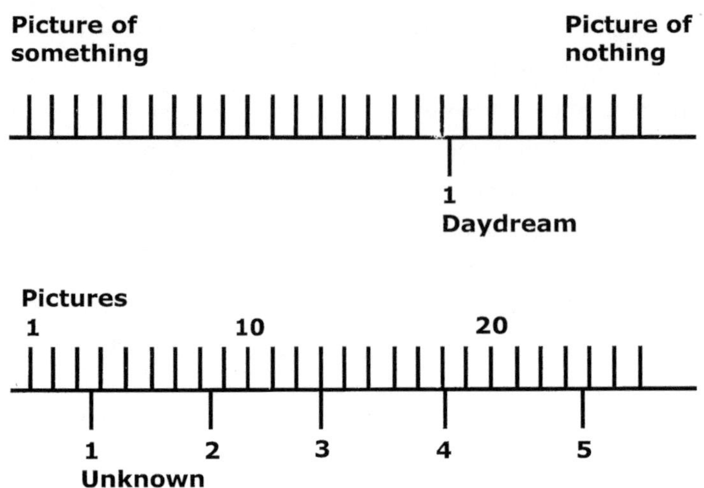

In art, incommensurability can be a theme as well as an everyday fact
of depiction. Stephanie Brooks is an artist who makes dysfunctional
scales: in one piece, a line is divided into two scales—one marked above
and the other below, as on a slide rule.[38] The one on top is calibrated in
miles, and the one underneath has a single tick mark, labeled "1 Day-
dream." The work converts miles to daydreams, along an incommensurate scale. The images I have been considering in this chapter pose the
same kind of problem: it is not at all clear how they correspond to actual
objects, or which pictures come closest and which ones miss altogether.
Figure 80, which illustrates Brooks's signature style, is an example of
what I mean. Each scale in the figure proposes an impossible measurement; pictures like Linde's only propose the *idea* of a measurement—they
"help people to think about" how they have to "come to grips" with the
theory, rather than helping them actually come to grips with it.

All the images I have considered so far in this chapter answer to my
working definition of the unpicturable: they try, and fail, to bring objects
into pictures, even though those objects cannot be imagined *as* pictures.
I have nothing in my visual imagination corresponding to a baryon with
its valence and sea quarks. Nor do I have a mental picture of an electron,
even without its allowed quanta of spin angular momentum. Linde's inflationary universe bubbles and fractal maps showing regions of physical laws
represent the mathematics that generated them rather than any picturable
objects in space. Each of these things exists as mathematical equations, and

that is how I picture them when I need to imagine how they might behave in a given case.

Throughout this book, and up to this chapter, I have been looking at images that can be described as increasingly desperate attempts to capture unrepresentable forms. In this chapter, that ethos of failure does not apply; the images do not even come close to adequate representation, because the objects they mean to depict are unpicturable. What can be brought into a picture, and what physicists want to bring into pictures, are incommensurate with one another.

51 *The Inconceivable*

The universes Linde theorizes cannot be imagined as pictures, and neither, apparently, can the insides of hadrons or the "structures" of electrons. Those are unpicturable objects—not merely unrepresentable, but also unavailable to the imagination as pictures. Beyond the unpicturable there is one last possibility: objects that are inconceivable, that are nothing but voids around which representation gathers. The example I'll offer is not some esoteric research at the margins of science: it is the widely known wave-particle duality, which has been a central question in quantum mechanics since the inception of the theory in the early 1920s.

There is a large amount of literature on the question of what, exactly, quantum mechanics describes, and I do not want to open with that topic.[39] But I would like to suggest that pictures have an especially under-theorized relation to the problem. Essentially, the question is whether quantum mechanics is a statistical description of a fundamentally statistical reality or an incomplete model of something deeper. Pictorial metaphors have been put to work on both sides of the question. Schrödinger, who discovered quantum mechanics' fundamental wave equation, was unsure about the "blurring" it entails: if a particle is best described by a wave equation, then it is spread out over some range of time and space, "blurred" in a way that "seems simply wrong" for everyday objects. Schrödinger's famous thought experiment of a wave equation that describes a cat as a linear combination of "live" and "dead" eigenstates—a cornerstone in philosophic debates about quantum mechanics—suggests that an ordinary cat might be "mixed or smeared out" between life and death.[40] Blurring and smearing are fundamentally pictorial possibilities, and the wave equation itself leads to solutions that

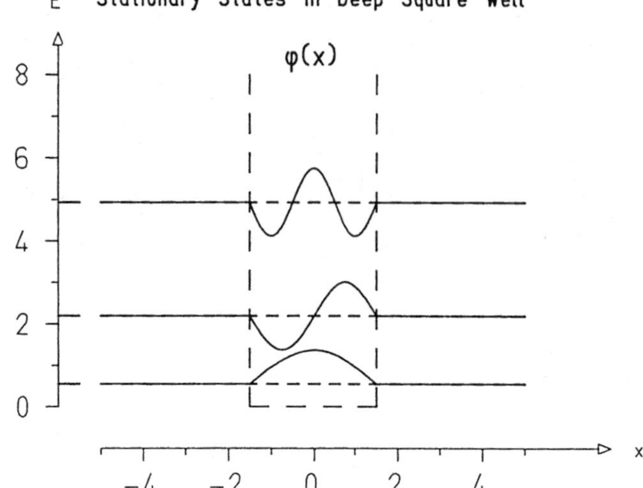

can be graphed as waves—on the model of waves of water, with their
peaks and troughs.

Pictures of solutions to the wave equation are usually flat: a sine wave
in a U-shaped trough, for instance, represents a standing "matter wave" in
a square well of potential energy (Figure 81). Figure 82 is the same kind
of image as Figure 81, but it is slightly more elaborate; it was used as an
illustration for an essay by Sheldon Goldstein defending David Bohm's
interpretation of quantum mechanics. In the Bohmian view, the wave
nature and the particle nature of elementary particles are reconciled by
imagining that Schrödinger's wave function "choreographs" or guides
the particle along waving paths. Schrödinger's equation for a system of n
particles, $\psi = \psi\,(q_1, \ldots, q_n)$, is augmented by another equation determining the particles' positions:

$$\frac{dQ_k}{dt} = v_k(\psi; Q_1, \ldots, Q_n)$$

where v_k is the velocity of the kth particle in the system, and Q represents
the particles' positions, so that the change in the position of a given particle is a function of its velocity, the positions of the other particles in the
system, and the "pilot wave" ψ.[41]

Bohm's interpretation is meant to capture both aspects of a particle's
behavior in a single theory: the wavelike nature is explained by the pilot

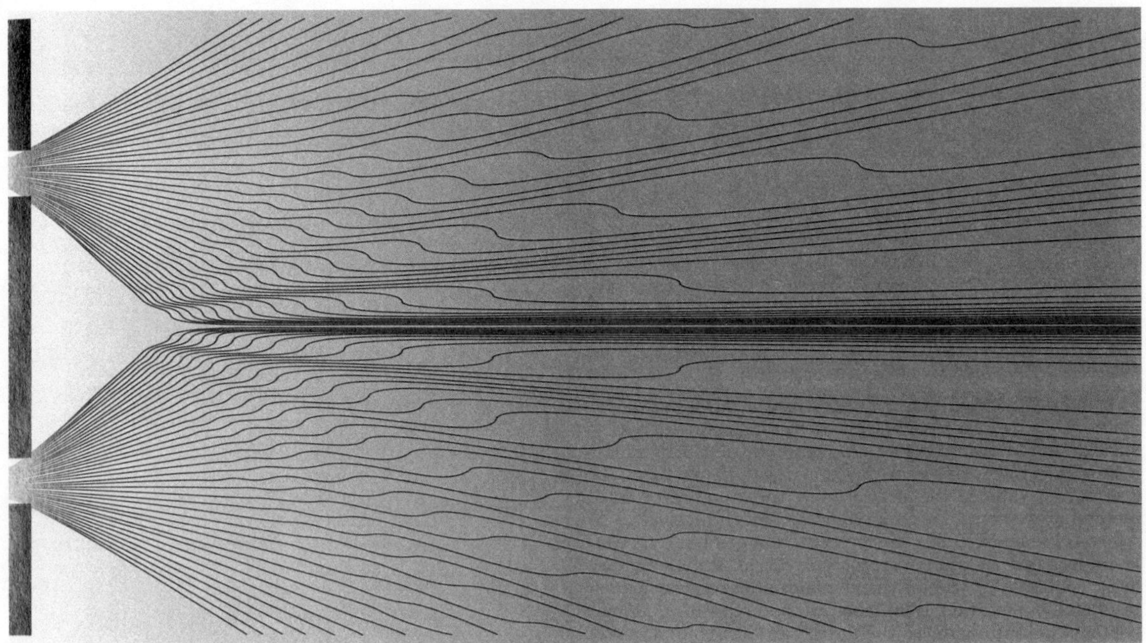

FIGURE 82

Gernot Bauer's ensemble of trajectories for the two-slit experiment. From Sheldon Goldstein, "Quantum Theory Without Observables—Part Two," *Physics Today* 51 (April 1998): 39. Copyright 1998, American Institute of Physics.

wave, and the particle-like nature is determined by separate equations predicting the locations of the particle.[42] Figure 81 is intended to help explain the so-called two-slit experiment, another centerpiece of philosophic debates in quantum mechanics. If water waves are sent toward a wall pierced by two holes, they will spread out in concentric circles as they pass through the two holes. The two sets of circular waves will cross one another, forming a lacy pattern of peaks and troughs. If there is a second wall beyond the first, the waves will strike it in a characteristic interference pattern. In the two-slit experiment, photons are sent, one by one, toward the same setup with two walls. Intuitively, one might expect that a single photon would either bounce back or go through one of the two holes; the photons that go through the holes would be expected to fan out evenly, covering the far wall with two hazy patches. Common sense would suggest that individual photons could not form an interference pattern on the far wall (after all, how can a single particle interfere with itself?). But they do: an accumulation of photons (or any other particles, atomic or subatomic) shot at the pierced wall will somehow gather in the typical interference pattern. Figure 82 is intended to show how a single particle might go through just one slit but be influenced in its trajectory by a pilot wave that goes through both slits. The picture is necessarily somewhat elaborate because Goldstein needs to show the interference pattern developing on the far wall; the picture is flat in order to exemplify how the pilot wave produces a series of undulating tracks for the particles.

Color plate 22 is a screen shot from a CD that accompanies a book called *Visual Quantum Mechanics*, written by Bernd Thaller. The author sees no problem at all in producing visual equivalents to quantum phenomena, simply because the mathematics itself enables such graphs to be made. "In the strange world of quantum mechanics," he says, "the application of visualization techniques is particularly rewarding, for it allows us to depict phenomena that cannot be seen by any other means."[43] Thaller assumes that phenomena can and should be depicted, no matter how different they are from the naturalistic techniques used to represent them. In *Visual Quantum Mechanics*, he maps colors on the complex plane in order to be able to depict the phase of the wave packet: red, green, and blue represent the angles $\phi = 0$, $2\pi/3$, and $4\pi/3$, respectively. As the phase changes, the initial wave packet cycles around the origin of the complex plane—hence the stripes in Color plate 22. The chroma (intensity) of the color varies according to the absolute value of the vector $z = x + iy$, so that each complex number is assigned a specific chroma and hue. As the wave packet disperses, the colors fade.[44] By animating his colored graphs, Thaller puts more information into his depictions than other physicists have done, making the wave function seem amenable to full depiction—but how odd to mix a representation of x-y space with a color representation of the phase space of the wave equation, almost as if complex values were *naturalistically* represented by hue and chroma. (In fact, Thaller's pictures make as much naturalistic sense as billiard-ball pictures of particles with "colored" quarks: both mingle realism with arbitrary conventions.)

Flat, monochrome, unanimated graphs of wave equations and wave packets are the norm in quantum mechanics, both for professional papers and for teaching. The images used in teaching, such as Figure 81, are the simplest: they give the impression that the wave function is purely a matter of calculating energies and probabilities, and they say little about what the function might actually represent. Another kind of illustration relies more heavily on analogies to optics and results in pictures that look convincingly three-dimensional. Figure 83 uses the same wave equation as figures 81 and 82 and Color plate 22, the previous three plates, but with

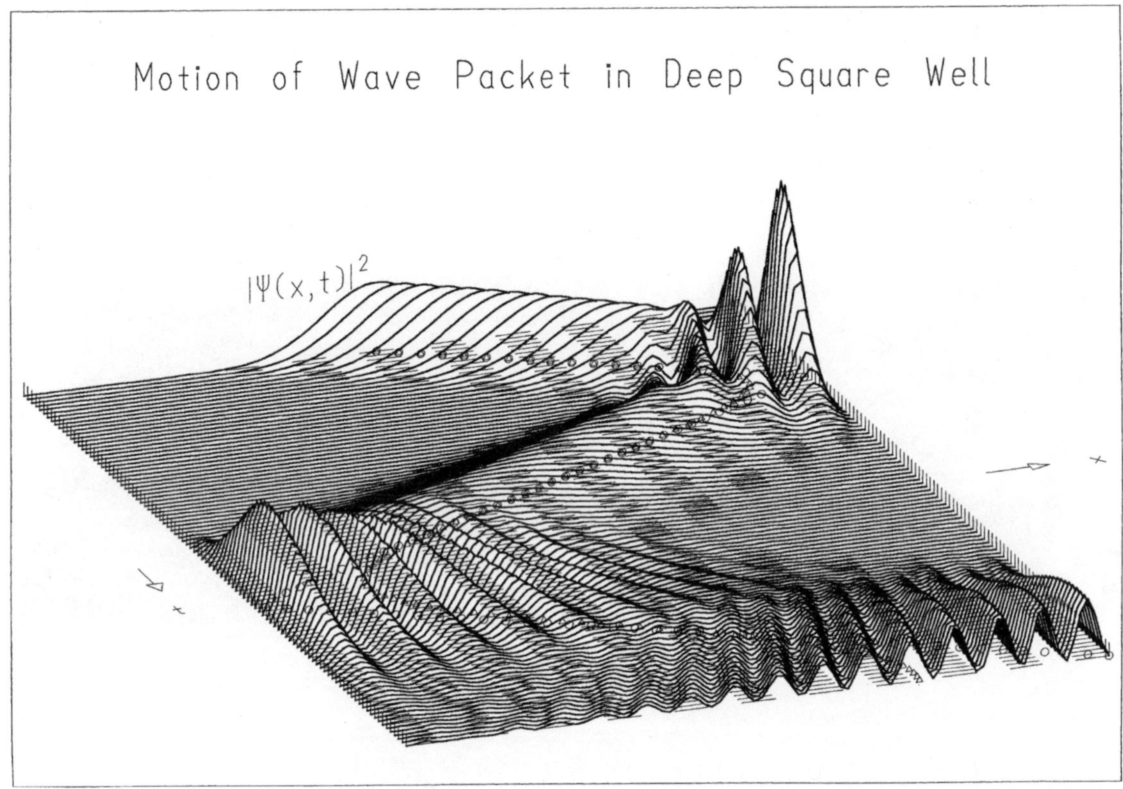

higher resolution and a wider range for position and time. Figure 83 reprises Figure 81: both represent a wave packet shuttling back and forth inside a well of potential energy; but in the case of Figure 83, the software includes routines for hidden-line removal (so the waves look solid) and for linear perspective—two essential properties of naturalistic representation. Seen this way, the wave function appears to resemble water waves, and it begins to look as if quantum mechanics is fundamentally a theory describing waves.

But in quantum mechanics nothing is what it seems, and all appearances are deceptive. Figure 83 is not a picture of the wave packet itself, but of the probability of finding a particle in a given location. The function $\Psi(x)$, as in Figure 81, gives a wave function, but $|\Psi(x,t)|^2$, as in Figure 83, is a probability function. The height of the "waves" in any given place corresponds to the probability of finding the actual particle in that spot. Figure 83 includes a zig-zag trail of little circles, indicating places where the particle would most likely be found at any given time. The

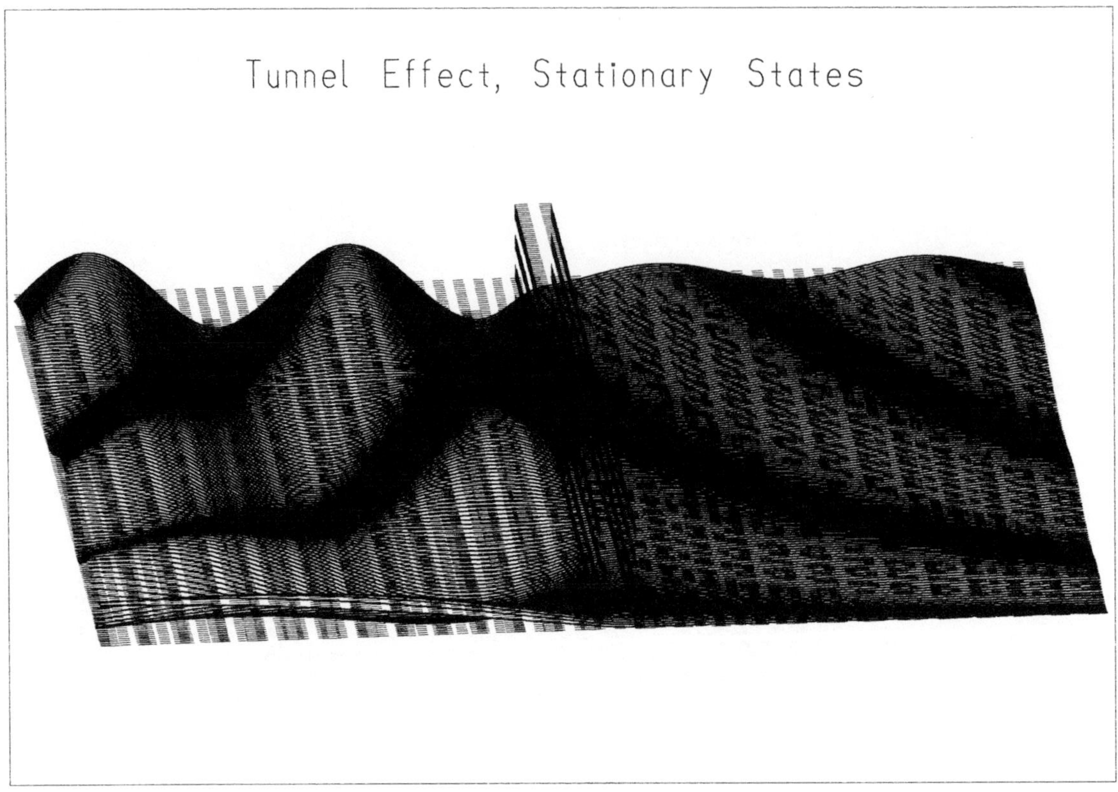

difference between $\Psi(x)$ and $|\Psi(x,t)|^2$ goes to the root of epistemological questions in quantum mechanics. Is Figure 83 a gussied-up version of a probability graph, or is it a naturalistic depiction of a fundamentally wavelike phenomenon? In the first case, the physical reality being described by the theory would be inherently unpicturable; in the second case it could readily be pictured. The paradox that quantum mechanics invariably presents to the intuition is that the physical reality proposed by the theory is both unpicturable and picturable. The actual object in question, I would say, is inconceivable, because it is only visualized by conceptualizing the unresolvable tension between what can and cannot be pictured.

This is my small contribution to a long and very intricate debate: the pictures that are used to teach and visualize the wave equation are themselves entangled in the epistemological problems that plague the theory.[45] In one person's hands they look like bare graphs with no pretense of resembling anything real. Another physicist might make them into naturalistic graphs of waves or pilot-wave tracks. The naturalistic image tradition and the statistical logic tradition intersect here, not as separate kinds of pictures but as intimately related aspects of a single formalism. Pictures of the Schrödinger wave function are the most vexing images I know in any historical tradition: they are powerful examples of the fact that pictures can bear traces of properties that are, in effect, inconceivable (Figure 84).

The objects of the last few figures are entirely outside the realm of visibility and visualization: the objects are not represented except as dilemmas, conflations of two incommensurate possibilities—the picturable and the unpicturable. Useless, incommensurate, misleading, unrepresentable, unpicturable, inconceivable: how far representation has fallen in these last, and most radical, images.

Conclusions

The six stories I have presented are a diverse group—united, I hope, across disciplines by their concern with the limits of the visual. In accord with the methodological strictures I set out in the Introduction, I am not going to draw any general conclusions. The six chapters are about six sets of image-making practices during the period from 1980 to 2000, and each practice has its own methods for working at the limits of representation. Let me list the six disciplines and their themes as a way of recalling the diversity that I want to stress:

Chapter 1, on painting, makes use of the sublime and the metaphors of the bracket and the ladder.

Chapter 2, on photography, is about formlessness, blur, darkness, the grid, and—in general—the anti-optical.

Chapter 3, on astronomy, concerns the management of magnitudes, smoothing routines, and confusion limiting.

Chapter 4, on microscopy, concentrates on the production of contrast, defocus, and the contrast transfer function.

Chapter 5, on particle physics, explores the lingering effects of realism and the calculations of particles' momentum, energy, and velocity.

Chapter 6, on quantum mechanics, is driven by the concepts of uselessness, calculability, the unpicturable, and the unrepresentable.

That is my summary, or anti-summary. I have not tried to isolate the chapters from one another, but I have worked to avoid giving the impression that people in these six fields can be seen as collaborating on a single project. It would have been possible to write this book as a history of a certain kind of image (it could have been called *Skepticism About the Image in Science and Art, 1980–2000*), but that would have suggested a unifying theme—skepticism—which many scientists and artists in this book presumably don't share. (The late twentieth century was the first period in which people making a number of kinds of images were concerned *primarily* with inadequacies of representation, but I do not know a historical narrative that would do justice to that idea without herding the individual image makers into some generic account of modernist anti-opticality.) Likewise, I could have cast the book as a philosophic text (perhaps called *A Critique of the Concept of Adequate Representation*), but that would have flattened the specific procedures in scientific texts, replacing them with words like *representation*. Hence this final form of the book, which declines parts of both history and philosophy.

I like to think of this situation using the familiar metaphor of language: particle physics is a language (or several languages: the classification does not need to be exact), and so are the contemporary discourses on photography, on painting, and on astrophysics. Each language has its vocabulary, whether it is composed of words such as *sublime* or *informe*, or equations such as the contrast transfer function or the routines encoded in Fourier transformations. The language metaphor helps emphasize the differences between the fields I have sampled in this book: like languages, they are independent and at times mutually unintelligible. (The language metaphor is explained and extended in another book, *Visual Practices Across the University*, which also deals with scientific images.[1]) Thinking of languages also helps forestall philosophy's sophophagic tendencies—it eats concepts, it devours what it takes as the intellectual foundations of disciplines—because the language metaphor reminds us that concepts like realism or limen exist principally in humanist criticism, not in astronomy or microscopy.

Latent in this book's methodology, I think, is a serious challenge. It is fairly easy to get used to the notion that concepts like the sublime cannot do justice to pictorial practices outside of certain traditions of fine art. It is far more difficult to look at a mathematical model like the contrast transfer function and see it as a way of thinking about images that is potentially just as rich and fruitful as the more familiar terminology that is used to talk about nonscientific images. I find the concepts and equations in chapters 3, 4, 5, and 6 of this book just as sensitive to the specifics of pictures as the humanist vocabulary of representation. The scientific vocabularies are clearly articulated, versatile, and responsive to actual images (as opposed to generalizations about images). Why not set aside the usual terms of philosophy and art history for a while and try to think about some fine-art images using the contrast transfer function? It is an artificial exercise, but it is also a good counterbalance to the assumption, common in the humanities, that philosophy and history provide the ultimate terms that govern our understanding of images in general.

Consider Fourier transforms, which are as ubiquitous in scientific image analysis as the notion of perspective is in Western realism. Fourier transforms are used in each of the four scientific disciplines I have surveyed in this book. A scientific image is often given along with its Fourier transform because it is understood that the image itself tells only part of the story. The transform (see Figures 35 and 36) does not look like the image. It relinquishes realism in order to provide information that is invisible but inherent in the original image. As a metaphor for what can and cannot be seen, Fourier transforms are just as suggestive as theories of the sublime, and they are considerably more exact about forms in individual images.

The contrast transfer function is another example of a scientific method that could substitute for a humanist analysis. It is used whenever an electron microscope image is just one of a series of images recording different properties of an object. Electron microscopes do not have a single plane where focus is sharpest (like ordinary cameras), nor do they have a unitary concept of blur (as I mentioned in Chapter 2). The contrast transfer function specifies relations between contrast, sharpness, resolution, and spatial frequency: it determines what information is present, and what is lost, in any given image. As a model of the photographic image, the contrast transfer function is significantly more complex than the blur circles or Airy disks of conventional photographic theory and criticism.

What prevents art historians and critics from adopting these alternate languages? Nothing more profound, I think, than our lack of good answers to Snow's and Sokal's rude questions. It is entirely possible to carry on interesting and significant conversations with a scientist if you do not know the relevant science, just as it is possible to carry on a conversation with an art historian if you do not know the relevant history. But there is a point where it becomes necessary to learn the foreign language. Imagine for instance that you are an art historian talking to a physicist friend.

The physicist wants to understand anti–art, which she has just encountered in a magazine article. The conversation that follows will depend heavily on whether or not she knows about Marcel Duchamp. Without that name as a touchstone, the conversation can certainly go forward, but if she already knows about Duchamp it will be possible to open a much more complex dialogue. The same is true of a conversation about anti-matter: it's possible to talk in general terms about matter and antimatter, but without some understanding of particle physics the concept won't be able to do much work. It isn't necessary for humanists and scientists to know each other's fields at the professional level, but without *some* common ground in the professional-level literature, the conversation has to take place in a third discipline, which I would argue is always, in the end, philosophy.

I've come back to Snow and Sokal in order to underscore the root-level problem. The challenge I am proposing—to leave the familiar ways of interpreting images behind, and to adopt, for a while, the viewpoint of some other discipline   is only possible when parts of the second discipline have been learned. I have gotten a great deal of pleasure out of researching and writing this book because it has let me ponder entirely new ways of attending to pictures. Few of the new approaches I've mentioned would have been possible without studying the actual algorithms used in the different disciplines. It is necessary, in other words, to read *past* popularized science and bowdlerized humanities, and not to assume that yours is the discipline with interpretive power over all others.

52   *The Light of Day*

Writing this book felt like wandering in a dark wood: being an outsider to most of these disciplines, I usually had no idea where I was headed, and every new topic was a surprise. Now, at the end, I feel like I'm coming out of the forest and onto a well–lit road. All around are the images we live with: television, ordinary fine art, movies, advertisements. Daily images are not part of this book, and neither is the broader visual culture that is so much a part of life. Now I begin to remember the many other concerns that drive picture making in the glare of politics and everyday life.

By comparison, I admit, the images in this book are frail. They are emotionally distant or inaccessible: many do little, and say even less. As art, they tend to be hermetically sealed inside their makers' chilled imaginations (I am thinking of Agnes Martin's solitary life after 1967, living alone on top of a mesa) or surrounded by a thickened armature of arcane science.

The images I have chosen are nearly empty and often dim or dull. In comparison to the pictures common in art and mass culture, what I have shown here is terribly weak. How could such faltering voices possibly be heard over the garish inventions of John Currin, Jeff Koons, Robert Mapplethorpe, Cindy Sherman, and Damien Hirst (to name some artists who became celebrities in the period in question) or even the Technicolor noise of astronomy's calendar pictures? They can't. And yet it is possible to get tired of the blaring lights of the art world. I know what Henry Staten means when he says that Günter Umberg produces "the most stunningly opaque and dry-looking paintings that have ever been made."[2] (I especially like the phrase "dry-looking"; what paintings before Umberg's *tried* to be powdery instead of oily like normal paintings?) The pictures I have been considering are quiet because they concentrate on their purpose. They are narrow because they are intent on capturing just one reliable phenomenon. Over the course of the twentieth century it became increasingly difficult to make pictures that were not complacent about what they could depict, and by the end of the century that meant that the most interesting images could end up looking cold and feeble.

To me, the images in this book are among the best that have been made in the last several decades. I am entranced by Hsiao-Wen Chen's image of an ugly gray strip with just four dark spots because I know it is an attempt to represent an object that is irreparably distant and hopelessly removed from anything that can be understood or experienced (see Figure 25). I love Umberg's small panels of thin metal, patiently painted and repainted with a perfect mixture of nearly black colors until they become impossibly dense and dry (see Color plates 2 and 3). I am especially fascinated by some pictures of the tunnel effect that look like waves of smooth, dark oil surging over a barrier, because I know they represent nothing that can ever be seen, adequately pictured, or understood (see Figure 84). What could be more mesmerizing? What could be more serious, more committed to the image? Alongside pictures like these, ordinary paintings and photographs can look like three-ring circuses.

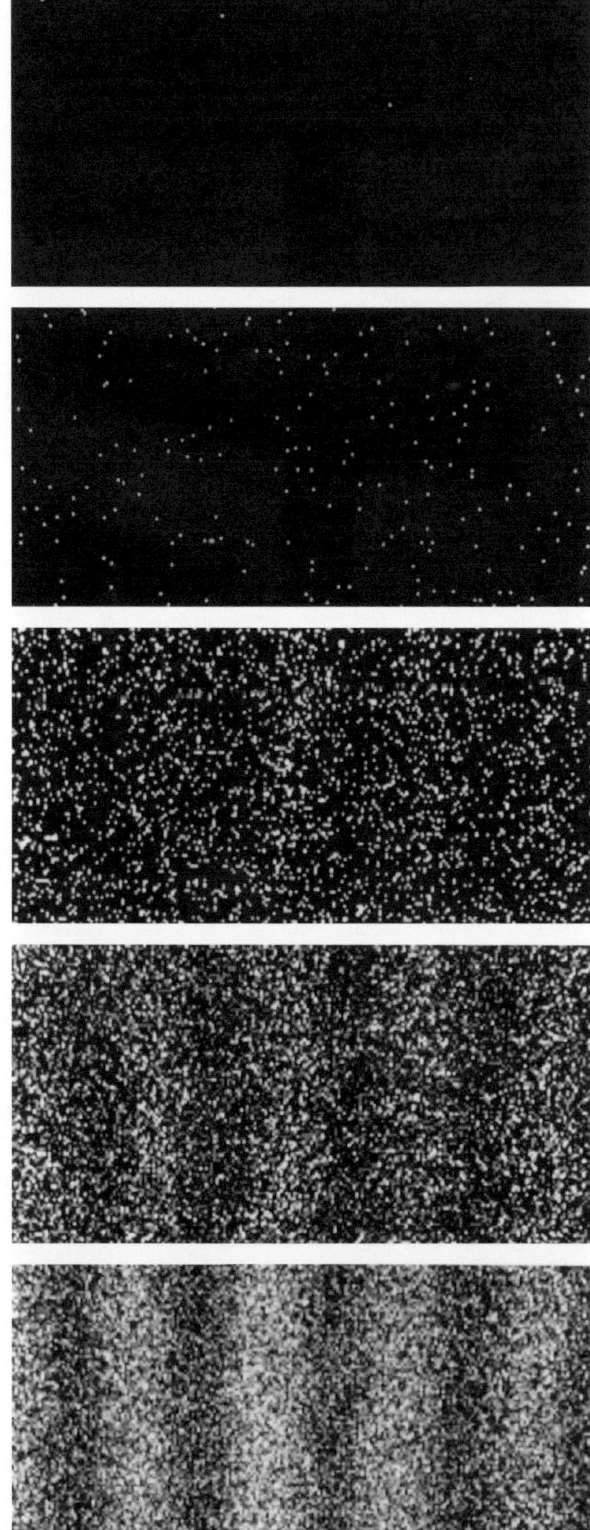

Experimental demonstration of matter waves. First image: 10 electrons recorded; second image: 100; third image: 3,000; fourth image: 20,000; fifth image: 70,000. From A. Tonomura, J. Endo, T. Matsuda, T. Kawasaki, and H. Exawa, "Demonstration of Single-Electron Buildup of an Interference Pattern," *American Journal of Physics* 57 (1989): 117. Courtesy *American Journal of Physics* and Akira Tonomura.

53   *Envoi*

And so, to close, here are five images by themselves (Figure 85): a sequence of pictures in which each dot of light is made by a single particle passing though one of two holes in a screen. (These are, once more, scientific images. Again science is the unavoidable subject when images are *in extremis*.) The particles go through one by one, and yet somehow they interfere with one another, eventually forming the pattern in the fifth frame. Everything here follows in an orderly manner from the wave equations of quantum mechanics, and each spot of light is produced in an impeccably rational fashion by just one particle. In the first frame there are 10 electrons; then there are 100; then 3,000; then 20,000; and finally 70,000.[3] At first they seem to scatter randomly, as you might expect bullets or stars to do, but then, when there are more of them, they reveal a hidden pattern.

The same phenomenon is modeled in Figure 82, where a single wave packet—the equivalent of the single particles in these pictures—somehow interferes with itself to produce a pattern of probable trajectories. In Figure 85, actual particles accumulate, replicating the interference pattern. It may be a solace that the mathematics is known with such precision, but the mathematics has little connection to the exact patterns in Figure 85: the equations only describe the ideal distribution, not the individual objects.

No one can say what this sequence records. What *are* those objects that we call electrons, that behave in a way so utterly at odds with anything that can be imagined? What sense of pictures makes us want to find something realistic here?

There is almost nothing left to look at in these pictures. The object has departed: it has gone far away, beyond representation, leaving only its hollow and inadequate traces.

Notes

Notes

Introduction

1. I am not interested here in the "science wars," the question of whether science is culturally relative, so that its results are socially constructed. The arguments I will be making, especially in Chapter 6, concerning the undependability of physics images, are not intended to raise doubts about the images' degree of independence from social constructions of truth or reality. For social construction see Ian Hacking, *The Social Construction of What?* (Cambridge, MA: Harvard University Press, 1999). Another source for the "two cultures" debate in the same period, which I am not considering here because it is too idiosyncratic (both in terms of science and contemporaneous art), is Paavo Pylkkänen, ed., *Bohm-Biederman Correspondence: Creativity and Science* (London: Routledge, 1999). I thank Melanie Bollman for drawing this to my attention. Her master's thesis on the correspondence, "Art and Science: Examining the 'Two Cultures' Debate" (2006), is on file in the John Flaxman Library at the School of the Art Institute of Chicago.

2. It is pertinent here that theorizing about the unity of the university—from Cardinal Newman to Jaroslav Pelikan, Jacques Derrida, and Bill Reading—has taken place in the humanities, with very little attention paid to science. This is explored in the introduction to *Visual Practices across the University*, ed. James Elkins (Munich: Wilhelm Fink Verlag, 2007). The gap is exemplified by the essays in another book I have edited, *Visual Literacy* (New York: Routledge, 2007); its contributors are almost exclusively theorists in the humanities. The two books are linked, and came from a single project. Their division into two books was the result of editorial decisions made with an eye to potential publics—those who read in the humanities, and those who read, or also read, outside the humanities. The editorial decisions

are discussed in the introduction to the first book and the afterword to the second.

3. I will not be addressing the question of whether "cultures" is the right word here: it is sufficiently clear as a way of naming the faculty who teach in the relevant departments. On the one side, art history, history of science, philosophy, and sometimes art practice; on the other, physics, chemistry, microbiology, biology, and astrophysics.

4. Snow, *The Two Cultures: And a Second Look* (Cambridge: Cambridge University Press, 1963), 20.

5. Ibid., 21. It is interesting that Alan Sokal notes an exact parallel—an instance—of Snow's unasked question, in Paul Virilio's "The Third Interval: A Critical Transition," in *Rethinking Technologies*, ed. Verena Conley (Minneapolis: University of Minnesota Press, 1993), 5: "it now seems appropriate to reconsider the notions of acceleration and deceleration (what physicists call positive and negative speeds)." Sokal and Jean Bricmont comment: "The reader who does not find this uproariously funny (as well as depressing) is invited to sit in on the first two weeks of Physics I." Sokal and Bricmont, "Transgressing the Boundaries: An Afterword," *Dissent* 43, no. 4 (1996): 93–99, quotation in n. 15.

6. Sokal and Bricmont, "Transgressing the Boundaries," 93.

7. The literary critic Stanley Fish, who wrote on the Sokal hoax, would say that it is. See Fish, "Professor Sokal's Bad Joke," *New York Times* (May 21, 1996), available on Sokal's Web site, www.physics.nyu.edu/faculty/sokal/fish.html.

8. Paul Boytinck, *C. P. Snow: A Reference Guide* (Boston: G. K. Hall and Co., 1980).

9. See Sokal's Web site, www.physics.nyu.edu/faculty/sokal, accessed September 15, 2004. The essential texts are: Sokal and Bricmont, *Fashionable Nonsense: Postmodern Intellectuals' Abuse of Science* (New York: Picador, 1999), which reprints the original essay, "Transgressing the Boundaries: Toward a Transformative Hermeneutics of Quantum Gravity," along with a commentary on its intentional errors; and Sokal, "A Physicist Experiments with Cultural Studies," *Lingua Franca* 6, no. 4 (1996): 62–64; *Fashionable Nonsense* originally appeared as *Impostures Intellectuelles* (Paris: Editions Odile Jacob, 1997). The book *The Sokal Hoax: The Sham That Shook the Academy*, edited by the editors of *Lingua Franca* (Lincoln: University of Nebraska Press, 2000), also contains Sokal's original essay and *Lingua Franca* admission, but is marred by the inclusion of a large number of ephemeral essays by journalists and the exclusion of responses in more scholarly journals.

10. I think Snow and Sokal would both agree, with qualifications. For Sokal's take on this see Jean Bricmont and Alan Sokal, "Reply to Turnbull, Krips, Dusek, and Fuller [reviewing *Fashionable Nonsense*]," *Metascience* 9, no.

3 (2000): 347–95. The rancor of the debate, and Sokal and Bricmont's earnestness, earned them Derrida's characterization as "pas sérieux." One of the unfortunate aspects of the Sokal debates is the distance Derrida decided to keep. Derrida's own long-standing critique of the ideology and effects of seriousness (beginning with the work on Husserl, and continuing through the exchange with Searle) would have made him an ideal interlocutor, but when he wrote a response to Sokal and Bricmont he decided to speak to those who already know his position, by simply—and nonsensically—accusing Sokal and Bricmont of being unserious. Of course, Sokal and Bricmont took his accusations entirely literally. See Sokal and Bricmont, "Les critiques de Derrida et de Dorra ratent leur cible," published as "Réponse à Jacques Derrida et Max Dorra," *Le Monde* (December 12, 1997), 23. The question of seriousness and humor is the hardest and perhaps most essential subject in the debates. Another illuminating moment occurred when Bruno Latour's sociological intervention in relativity theory (ridiculed by Sokal and Bricmont) was defended by the physicist David Mermin; Mermin's defense was then attacked by other physicists on the grounds that ambiguity and playfulness have no place in logic discourse. See *Physics Today* 52, no. 8 (1999): 82–83. The same question surfaces in Latour's response to Sokal and Bricmont, "Y a-t-il une science après la guerre froide?" *Le Monde* (January 31, 1997), 15.

11. Recent encounters are enumerated in my *Visual Practices across the University*.

12. Katz, letter in *Physics Today* 52, no. 10 (1999): 120. The letters continue in ibid., 11–15, 120–22, in response to Mermin's review of *Fashionable Nonsense* in *Physics Today* 52, no. 4 (1999): 70.

13. There are important exceptions in both debates—contributions that are modulated and well-informed regarding issues on both sides. See for example Bricmont and Sokal's "Reply to Turnbull, Krips, Dusek, and Fuller."

14. There are exceptions. Latour has communicated some of his ideas directly to scientists; see for example Latour, "From the World of Science to That of Research," *Science* 280 (April 10, 1998), 208. Some relevant books by the authors I have named are: Serres, *Hermes: Literature, Science, Philosophy*, ed. Josué Harari and David Bell (Baltimore, MD: Johns Hopkins Press, 1982); Stafford, *Visual Analogy: Consciousness as the Art of Connecting* (Cambridge, MA: MIT Press, 1999); Latour and Steve Woolgar, *Laboratory Life: The Construction of Scientific Facts* (Princeton, NJ: Princeton University Press, 1986); Galison, *How Experiments End* (Chicago: University of Chicago Press, 1987); and other texts cited in Chapter 6; Hacking, *The Social Construction of What?* and other texts cited in Chapter 4; Pais, *Niels Bohr's Times: In Physics, Philosophy, and Polity* (New York: Oxford University Press, 1991); Hübner, *Critique of Scientific Reason*, trans. Paul Dixon (Chicago: University of Chicago Press,

1983), which was also translated into Russian; Feyerabend, *Conquest of Abundance: A Tale of Abstraction Versus the Richness of Being* (Chicago: University of Chicago Press, 1999); and Pickering, *Constructing Quarks: A Sociological History of Particle Physics* (Chicago: University of Chicago Press, 1984).

15. Nagel, "The Sleep of Reason," *The New Republic* (October 12, 1998), 32–38.

16. Sokal and Bricmont, *Fashionable Nonsense*, 180.

17. Latour and Serres, *Conversations on Science, Culture, and Time*, trans. Roxanne Lapidus (Ann Arbor: University of Michigan Press, 1995).

18. He is mentioned several times in passing, for example in Jean Bricmont and Alan Sokal, "Sokalratic Debate Continues, Fueled by Latour and Copenhagen Interpretations," *Physics Today* 52 (August 1999): 15, 82: "the works of Jacques Lacan . . . and others . . . constitute a genre in their own right, characterized principally by name-dropping and the display of erudition" (quotation on p. 82).

19. C. W. F. Everitt and Anna Muza, "History, Theory, and the Ziggurat of Physics," review of Galison, *Image and Logic* (Chicago: University of Chicago Press, 1997), in *Isis* 91 (2000): 310–13, quotation on p. 313. Wolfgang Kurt Hermann Panofsky is to be distinguished from the art historian Erwin Panofsky; the former is the author of, for example, *Classical Electricity and Magnetism* (Reading, MA: Addison-Wesley, 1962).

20. Trimpi, *Muses of One Mind: The Literary Analysis of Experience and Its Continuity* (Princeton, NJ: Princeton University Press, 1983).

21. Trimpi could have extended his parallel, because the few Egyptian mathematical and geometric texts are clearly distinct from the other genres, and the same could be said of the ancient Near East in general. On the other hand, I am not proposing this as a sufficient model. Cynthia Pyle's work, especially "Historical and Philological Method in Angelo Poliziano and Method in Science: Practice and Theory," in *Poliziano nel suo tempo, Atti del VI Convegno internazionale (Chianciano-Montepulciano 18–21 luglio 1994)*, ed. Luisa Secchi Tarugi (Florence: Franco Cesati Editore, 1996), 371–86, is especially convincing on the connection between Renaissance philology and the development of science: she proposes that a set of hermeneutic rules carried over from the analysis of classical texts to the beginnings of modern natural science. It would also be possible to argue against the concept of disciplines, as Stephen Toulmin does in *Return to Reason* (Cambridge, MA: Harvard University Press, 2001), but it remains unsaid, in Toulmin's account, how the *existing* discourses could be made to speak to one another.

22. This is further described in my *Domain of Images* (Ithaca, NY: Cornell University Press, 1999), 46–51, 233.

23. *The Sokal Hoax*, 61; thanks to Teri Reynolds.

24. See, in this respect, the interesting distinction between philosophy

and theory in Fredric Jameson's review of Slavoj Žižek's work: Jameson, "First Impressions," review of Žižek, *The Parallax View*, in *London Review of Books* (September 7, 2006), 7–8.

25. Crary, *Suspensions of Perception: Attention, Spectacle, and Modern Culture* (Cambridge, MA: MIT Press, 1999).

26. Baxandall, "The Language of Art Criticism," in *The Language of Art History*, ed. Salim Kemal and Ivan Gaskell (Cambridge: Cambridge University Press, 1991), 73–74, quoted in David Summers, review of Crary, *Suspensions of Perception*, in *The Art Bulletin* 83, no. 1 (March 2001): 158.

27. E. H. Gombrich, *In Search of Cultural History* (Oxford: Clarendon Press, 1974 [1969]); my essay, "Art History Without Theory," *Critical Inquiry* 14 (1988): 354–78; and the reply to Gombrich, *Critical Inquiry* 14 (1988): 893.

28. I have tried to work out relationships between art history and science in more detail in a review of Martin Kemp, *The Science of Art: Optical Themes in Western Art from Brunelleschi to Seurat* (New Haven, CT: Yale University Press, 1990), in the *Zeitschrift für Kunstgeschichte* 54, no. 4 (1991): 597–601. For interdisciplinarity in general, see for example my "Nine Modes of Interdisciplinarity in Visual Studies," which is in reply to Mieke Bal's essay, "Visual Essentialism and the Object of Visual Culture," *Journal of Visual Culture* 2, no. 2 (2003): 232–37.

29. Personal communication, 2005.

30. For an interesting attempt to find "nonsense" in scientists' writings to complement the "nonsense" Sokal finds in post-structuralist writers, see Mara Beller, "The Sokal Hoax: At Whom Are We Laughing?" *Physics Today* 51, no. 9 (1998): 29.

31. Mulligan, "The Symptoms of Gödel-Mania," review of Sokal and Bricmont's *Impostures intellectuelles* (the original French edition of *Fashionable Nonsense*), in *The London Times Literary Supplement* (May 1, 1998), available at naturalscience.com/ns/books/book04.html.

32. His near-contemporary, Alfred Döblin, wrote such a book in *Berlin Alexanderplatz*, though the equations in that book do not need to be read.

33. I am thinking of Peter Galison's observation that physicists communicate with one another, across subspecialties, by the use of simplified languages he calls creoles. Galison does not apply his account of pidgins and creoles to his own writing, but it seems to me the extension is inevitable, because his writing is also a kind of talk that hopefully takes place between theorists and experimentalists. Galison, *Image and Logic: A Material History of Twentieth-Century Microphysics* (Chicago: University of Chicago Press, 1997), 48–51.

34. This is explored in more detail in my review of Hal Foster, *Compulsive*

Beauty (Cambridge, MA: MIT Press, 1993), in *The Art Bulletin* 76, no. 3 (1994): 546–48.

35. For the suggestion that Galileo's sense of art influenced his rejection of Keplerian elliptical orbits—made by Aby Warburg and later by Erwin Panofsky—see the excellent summary in Horst Bredekamp, "Gazing Hands and Blind Spots: Galileo as Draftsman," *Science in Context* 13, no. 3–4 (2000): 423–62, especially pp. 454–57.

36. For example, Thomas DaCosta Kaufmann's work, such as *Court, Cloister, and City: The Art and Culture of Central Europe, 1450–1800* (Chicago: University of Chicago Press, 1995); and Kaufmann, *The Mastery of Nature: Aspects of Art, Science, and Humanism in the Renaissance* (Princeton, NJ: Princeton University Press, 1993).

37. See my review of Reeves, *Painting the Heavens: Art and Sciences in the Age of Galileo* (Princeton, NJ: Princeton University Press, 1997), in the *Zeitschrift für Kunstgeshcichte* 62 (1999): 580–85.

1. Painting

1. For an interesting case study of the tacit adequacy of representation of images whose subject is, by definition, unrepresentable, see Herbert Kessler, *Spiritual Seeing: Picturing God's Invisibility in Medieval Art* (Philadelphia: University of Pennsylvania Press, 2000).

2. Which is not to say that understanding it is trivial; see the excellent study by Rachael DeLue, *George Inness and the Science of Landscape* (Chicago: University of Chicago Press, 2004).

3. For more on this problem, see *Landscape Theory*, ed. James Elkins and Rachael DeLue, vol. 6 of *The Art Seminar* (New York: Routledge, 2007).

4. The central account of Richter's method, by Benjamin Buchloh, reads it in political terms. Given Richter's preeminence, it may seem perverse not to center this chapter on him. I am doing so because Richter's position, which was largely assured by the end of the twentieth century, was partly constructed by Buchloh's persuasive advocacy. It would therefore be arguing against the grain to speak of Richter's painting as interested primarily in aspects of the inadequacy of painted (and photographed) representation, or to argue that Richter's canvases are best described as responses to the history of self-reflective painting, rather than as signs of political and biographical concerns. It is for some future text, I think, to find a way to speak persuasively about Richter as a painterly painter, one in continuous dialogue with previous painting, without implying that the political dimension of the work is somehow secondary. See Buchloh, *Gerhard Richter*, exh. cat. (Bonn: Kunst- und Ausstellungshalle der Bundesrepublik Deutschland, 1993).

5. This is pursued, more sympathetically and with a different purpose, in

my "Two Forms of Judgement: Forgiving and Demanding (The Case of Marine Painting)," *Journal of Visual Art Practice* 3, no. 1 (2004): 37–46.

6. I thank Naomi Campbell for insight into On Kawara's technique; she copied one of his paintings, repeatedly, at the Art Institute of Chicago, and wrote a paper about the existing evidence for their construction (unpublished, 2002).

7. *Farewell to an Idea* is very elliptic—coy would probably be the best word—about Longinus. See Clark, *Farewell to An Idea: Episodes from a History of Modernism* (New Haven, CT: Yale University Press, 1999): Longinus is listed in the index twice—once for his appearance on the last page of the book, where Clark says he's quoting him "for the last time," although he hadn't mentioned Longinus earlier; and on p. 314, the line "He has tortured his line into conformity with the impending disaster" is quoted without comment or reference, and the word sublime doesn't even appear. (The talk is all of pathos, pessimism, and atony.)

8. Lyotard, "The Sublime and the Avant-Garde," trans. Lisa Liebmann, *Artforum* 22 (April 1984): 36–43, quotation on p. 38. Readings of Longinus should start with G. Grube, "Notes on the ΠΕΡΙΥΨΟΥΣ," *American Journal of Philology* 78 (1957): 355–74, which makes a number of good points about Longinus's uncertainty, concluding that "his treatise should be known as 'Longinus On Great Writing,'" with "great" understood in a very general sense (ibid., 360). Other readings of Longinus may be found in Neil Hertz, *The End of the Line: Essays on Psychoanalysis and the Sublime* (New York: Columbia University Press, 1985), 1–20, and Peter De Bolla, *The Discourse of the Sublime: Readings in History, Aesthetics, and the Subject* (London: Basil Blackwell, 1989), 35–40 and passim.

9. Vico, *Estetica, estratti dall' Autobiografia e dalla Scienza Nuova*, ed. Riccardo Dusi (Torino: Società Editrice Internazionale, 1945); Boileau, *L'Art poétique, épitres, odes, poésies diverses, épigrammes* (Paris: Bibliothèque Nationale, 1877); Longinus in Boileau's translation, *On the Sublime = The Peri Hupsous*, introduction by William Johnson (Delmar, NY: Scholars' Facsimiles and Reprints, 1975). Longinus was translated into Latin as early as 1572; see Samuel Monk, *The Sublime: A Study of Critical Theories in Eighteenth-Century England*, with a new preface by the author (Ann Arbor: University of Michigan Press, 1960 [1935]). On the seventeenth-century French reception of Longinus, see Francis Goyet, "Le Pseudo-sublime de Longin," *Études litteraires* 24, no. 3 (1991–1992): 105–20. For Vico in this context see also Nella Cotrupi, "Vico, Burke, and Frye's Flirtation with the Sublime," in *Giambattista Vico and Anglo-American Science: Philosophy and Writing*, ed. Marcel Danesi (Berlin: Mouton de Gruyter, 1995), 35–49.

10. Barnett Newman, "The Plasmic Image" [1943–45], in *Abstract Expressionism: Creators and Critics*, ed. Clifford Ross (New York: Abrams, 1990), 125.

11. The prehistory of some ideas I explore in this chapter is given in Kate Flint, *The Victorians and the Visual Imagination* (Cambridge: Cambridge University Press, 2000), especially chapter 1, "The Visible and the Unseen."

12. The argument I have condensed here is in Immanuel Kant, *Critique of Judgment*, trans. W. S. Pluhar (Indianapolis, IN: Hackett, 1987), §§ 23–29. In outline: the two species of sublime are given in § 24; the mathematical sublime is defined in §§ 25–27, and the dynamical in §§ 28–29.

13. In particular I omit Kant's interest in ethics, because it does not pertain, as such, to painting or physics images. See the account in Salim Kemal, *Kant's Aesthetic Theory* (London: St. Martin's Press, 1997), 113, which argues against Paul Crowther's account on the grounds that it underestimates Kant's ethical and wider cultural concerns. I thank Raja Halwani for bringing this to my attention.

14. Kant, *Critique of Judgment*, trans. J. H. Bernard (New York: Hafner, 1951), §26, p. 94, "a supersensible substrate," rendering "ein übersinnliches Substrat."

15. Hans Blumenberg, *Shipwreck with Spectator: Paradigm of a Metaphor for Existence*, trans. Steven Rendall (Cambridge, MA: MIT Press, 1997).

16. Kant, *Critique of Judgment*, 112.

17. Kant's pre-Critical book *Observations on the Feeling of the Beautiful and Sublime* [*Beobachtungen über das Gefühl des Schönen und Erhabenen*, 1764], trans. John Goldthwait (Berkeley: University of California Press, 1960), is full of informal examples. See especially 47: "Tall oaks and lonely shadows in a sacred grove are sublime; flower beds, low hedges and trees trimmed in figures are beautiful. Night is sublime, day is beautiful. Temperaments that possess a feeling for the sublime are drawn gradually, by the quiet stillness of a summer evening as the shimmering light of the stars breaks through the brown shadows of night and the lonely moon rises into view, into high feelings of friendship, of disdain for the world, of eternity."

18. Kant, *Critique of Judgment*, 93, translation slightly modified. Kant: "Das gegenebe Unendliche aber dennoch ohne Widerspruch *auch nur denken zu können*, dazu wird ein Vermögen, das selbst übersinnlich ist, im menschlichen Gemüte erfordert."

19. Kant never saw a hurricane or a volcano, so his descriptions of their sublimity are already descriptions of texts—that is, of art. What he describes as natural phenomena are parts of the stock-in-trade of late-eighteenth-century poetry and prose. I take that not as a flaw in his argument, but as evidence that the concept of the sublime is not only applicable to artworks—it is grounded in them.

20. Marin, *To Destroy Painting*, trans. Mete Hjort (Chicago: University of Chicago Press, 1995), and Derrida, "The Truth in Pointing," in *The Truth in Painting*, trans. Geoff Bennington and Ian McLeod (Chicago: University of Chicago Press, 1987). Similar themes are pursued in my *On Pictures and the*

Words That Fail Them (Cambridge: Cambridge University Press, 1998) especially chapters 1–3.

21. Edmund Burke, *A Philosophical Enquiry*, ed. Adam Phillips (New York: Oxford University Press, 1990), part 3, section 2, pp. 84–86. In the wonderfully titled section "Proportion Not the Cause of Beauty in Vegetables," Burke doubts that proportion and harmony are properties of the beautiful. But the properties and qualities he chooses—smallness and smoothness, but especially gradual variation, and delicacy—are articulable in terms of coherence, harmony, and proportion. The properties are listed in part 3, sections 12–18 *et seq.*

22. Masheck, "Ross Bleckner's *Chamber* and the Limits of Sublimity," in *Modernities: Art-Matters in the Present* (University Park: Pennsylvania State University Press, 1993), 269–73. For Bleckner see also *Ross Bleckner*, ed. Lisa Dennison, with an essay by Thomas Crow (New York: Guggenheim Museum, 1995); and *Ross Bleckner*, exh. cat., with an essay by Peter Schjeldahl (London: Waddington Galleries, 1988).

23. Discussed in my *Why Are Our Pictures Puzzles? On the Modern Origins of Pictorial Complexity* (New York: Routledge, 1999), 89.

24. Crow, "Surface Tension: Ross Bleckner and the Conditions of Painting's Reincarnation," in *Ross Bleckner*, ed. Lisa Dennison, 101–13, quotation on p. 101. Crow skirts the issue of the sublime, finding a quotation from Burke (on "gloomy pomp") "not out of place"—though his main purpose is to link Bleckner to other traditions in modernist painting (p. 112).

25. Masheck, "Ross Bleckner's *Chamber*," 273, quoting Burke, *A Philosophical Enquiry into the Origin of Our Ideas of the Sublime and Beautiful*, ed. J. Boulton (London: Routledge and Kegan Paul, 1958), 62.

26. Crow, "Surface Tension: Ross Bleckner and the Conditions of Painting's Reincarnation," in *Ross Bleckner*, ed. Lisa Dennison, 100–113; Schjeldahl, "Spiritual Materialism: Ross Bleckner's New Paintings," in *Ross Bleckner*, 4–7.

27. Lacoue-Labarthe, "Sublime Truth," in *Of the Sublime: Presence in Question*, ed. Jeffrey Librett (Albany: State University of New York Press, 1993), 71–108, quotation on p. 84.

28. Reinhardt, "Art as Art," *Art International* (December 1962): 37.

29. Quoted in Lawrence Alloway, *Agnes Martin* (Philadelphia: Institute of Contemporary Art, 1973), 10. Both this and the Reinhardt quotation are in Marja Bloem, "An Awareness of Perfection," in *Agnes Martin: Paintings and Drawings 1974–1990* (Amsterdam: Stedelijk Museum, 1991), 32–41, quotations on p. 36.

30. The quotation is taken, without specific citation, from "Martin's private notes given to the Institute of Contemporary Art, University of Pennsylvania," in Bloem, "An Awareness of Perfection," 40–41.

31. Ibid., 39.

32. Ashton, "Agnes Martin and . . . ," in *Agnes Martin: Paintings and Drawings 1957–1975* (London: Arts Council of Great Britain, 1977), 7–14, quotation on p. 7. "Qualities of mind" is from ibid., 14.

33. Krauss, "Agnes Martin: The /Cloud/" (1993), in *Bachelors* (Cambridge, MA: MIT Press, 1999), 75–90, quotation on p. 78.

34. Krauss's quotations from Linville are in "Agnes Martin," 79; the original is Linville, "Agnes Martin: An Appreciation," *Artforum* 9 (June 1971): 72.

35. Ibid. "[T]he /cloud/ remains bracketed" is from Krauss, "Agnes Martin," 89.

36. Krauss, "Agnes Martin," 77–78, quoting from Carter Ratcliff, "Agnes Martin and the 'Artificial Infinite,'" *Art News* 72 (May 1973): 26–27.

37. Rosenblum, *Modern Painting and the Northern Romantic Tradition: Friedrich to Rothko* (New York: Harper and Row, 1975); Rosenblum, "The Abstract Sublime," *Art News* 59 (February 1961): 56–58.

38. McEvilley, "'Grey Geese Descending': The Art of Agnes Martin," *Artforum* 25 (Summer 1987): 98.

39. For the postmodern sublime see *Of the Sublime*. I do not recommend *Sticky Sublime*, ed. Bill Beckley (New York: Allworth Press, 2001), because it ranges much more widely than the concept of its title.

40. This analogy is developed in "Nine Steps Down the Ladder of Disorder," in my *On Pictures*, 213–40.

41. See *La Couleur seule, l'éxperience du monochrome*, exh. cat., ed. Maurice Besset and Thierry Raspail (Lyon: Musée St. Pierre, 1988).

42. On minimalism and the sublime see also Paul Beidler, "The Postmodern Sublime: Kant and Tony Smith's Anecdote of the Cube," *Journal of Aesthetics and Art Criticism* 53 (1995): 177–86.

43. One of the best accounts is Yve-Alain Bois's essay in *Ad Reinhardt*, exh. cat. (New York: Rizzoli, 1991), and the sources cited in note 44 below.

44. See Cheetham, *The Rhetoric of Purity: Essentialist Theory and the Advent of Abstract Painting* (Cambridge: Cambridge University Press, 1991), and Caroline Jones, *Eyesight Alone* (Cambridge, MA: MIT Press, 2006).

45. Caoimhín Mac Giolla Léith, "Spots of Time," in *Callum Innes* (Birmingham: Ikon Gallery, 1998), makes related points about instantaneity and the duration of viewing.

46. It isn't easy to find Reinhardt's "black paintings" in good condition. The one in the Art Institute of Chicago has been cleaned, but is again in danger of being degraded by scratches and fingerprints; one in the Museum of Modern Art, New York, is shown under glass, which makes its surface invisible and interposes the viewer's reflection. (I have been told that there is a new, high-tech nonreflective glass for such cases, but it was not installed as of November 2006.) The conceptual ideal, of a grid of nine squares that emerges slowly and ambiguously after long looking, is nearly impossible to realize.

47. Henry Staten, "Deconstructing Figurality: On 'Radical Painting,'"

in *Günter Umberg*, ed. Hannelore Kersting (Cologne: W. König, 1989), 104 and 96, respectively.

48. See Fried's review, "Joseph Marioni, Rose Art Museum, Brandeis University," *Artforum* 37 (1998): 149–50, quotations on p. 149. For more on Marioni, whom I won't be describing further here, see Barry Schwabsky, "Colors and Their Names," *Art in America* 87, no. 6 (June 1999): 86–91.

49. Joseph Marioni and Günter Umberg, *Outside the Cartouche: Zur Frage des Betrachters in der radikalen Malerei* (Munich: Neue Kunst Verlag, 1986). For Umberg see also *Günter Umberg: Raum für Malerei*, ed. Nicola von Velsen, with essays by Eva Schmidt and Friederich Meschede (Stuttgart: Reihe Cantz, 1994); *Günter Umberg: Esser-ci*, ed. Martina Corgnati (Milan: Goethe-Institut, 2001); *Günter Umberg*, ed. Giovanni Maria Accame (Milan: A Arte Studio Invernizzi, 1997).

50. Staten, "Deconstructing Figurality," 99.

51. Peter Nisbet, "'Getting Closer to Painting,' a Conversation with Günter Umberg [in 1997]," translated from the German and revised with the approval of the artist, in *Günter Umberg* (Deurle: Museum Dhondt-Dhaenens, 1998), 22–41, quotations on p. 33.

52. For Krane's paintings, see www.mariekranebergman.com, accessed November 2006.

53. See for example *Caspar David Friedrich und sein Kreis*, exh. cat., with essays by Irma Emmrich et al. (Dresden: Gemäldegalerie Neue Meister, 1974); *Traum und Wahrheit: Deutsche Romantik aus Museen der Deutschen Demokratischen Republik*, exh. cat., with essays by Willi Geismeier et al. (Bern: Kunstmuseum, 1985); and *La Peinture allemande à l'époque du Romantisme*, exh. cat. (Paris: Orangerie des Tuileries, 1977). The work by Katz I have in mind is in Bruce Hainley, "Five Words for Alex Katz," exh. cat. (Chicago: Arts Club, 2000).

2. Photography

1. Krauss and Bois, *Formless: A User's Guide* (Cambridge, MA: MIT Press, 1997 [1996]).

2. Ibid., 5. Bataille's "Critical Dictionary" was originally published in 1929–30; see Bataille, *Documents*, ed. Bernard Noël (Paris: Mercure de France, 1968).

3. Krauss and Bois, *Formless*, 15.

4. Reviewers of the book were mainly interested in the *October* model and its influence; for an extreme case, see *Formless: Ways in and Out of Form*, ed. Patrick Crowley and Paul Hegarty (Oxford: Peter Lang, 2005).

5. *Jiri Georg Dokuopil: The Madrid Paintings*, exh. cat. (Kyoto: Kyoto Shoin, 1985).

6. Bernard Blistène, "Conversation with Edward Ruscha," in *Edward Ruscha: Paintings/Schilderijen* (Rotterdam: Museum Boymans-van Beuningen, 1989), 136. See also Bill Berkson, "Sweet Logos," *Artforum* (January 1987): 98–101.

7. As an example of the nonhistorical use of the Kantian sublime, it would be tempting—but anachronistic—to say that the painting has some of Edmund Burke's "gloomy pomp," which he praised as a kind of sublime obscurity. See Burke, *A Philosophical Enquiry*, ed. Adam Phillips (Oxford: Oxford University Press, 1990), part 2, section 3, pp. 54–55.

8. Plagens, "Ed Ruscha, Seriously," in *The Works of Edward Ruscha* (New York: Hudson Hills Press, 1982), 32–40, especially p. 39.

9. Burke, *A Philosophical Enquiry*, part 2, section 4, p. 58, and compare p. 56: "In reality a great clearness helps but little towards affecting the passions, as it is in some sort an enemy to all enthusiasms whatsoever."

10. The original image and an animation are at www2.keck.hawaii .edu/news/wr104.html. See Peter Tuthill, John Monnier, and William Danchi, "A Dusty Pinwheel Nebula Around the Massive Star WR 104," *Nature* 398 (April 8, 1999): 487–89. The binary system, known as WR 104, is analyzed in more detail in my *Visual Practices Across the University* (Paderborn: Wilhelm Fink Verlag, 2007).

11. For the Berkeley Infrared Spatial Interferometry group, see isi.ssl. berkeley.edu/isi.html. For information on mid-infrared interferometry, see M. Bester, W. Danchi, and C. Townes, "Long Baseline Interferometer for the Mid-Infrared," *Proceedings of the SPIE* [International Society for Optical Engineering] 1237 (1990): 40–48; and for information specific to the Keck telescope, see P. Tuthill, J. Monnier, W. Danchi, and C. Haniff, "Michelson Interferometry with Keck I," in *Proceedings of the SPIE 3350, Astronomical Interferometry*, ed. R. Reasenberg and J. Breckinridge (Bellingham, WA: SPIE, 1998), 839–46.

12. See Tuthill, Monnier, and Danchi, "A Dusty Pinwheel Nebula," 487–89, fig. 1.

13. See my *The Domain of Images* (Ithaca, NY: Cornell University Press, 1999), 10–12.

14. Carl Heiles, "H I Shells, Supershells, Shell-Like Objects, and 'Worms,'" *Astrophysical Journal Supplement Series* 55, no. 4 (August 1984): 585–95, quotation on p. 585. For the "cold front": A. Vikhlinin et al., "A Moving Cold Front in the Intergalactic Medium of [the Galaxy Cluster] A3667," arxiv .org/abs/astro-ph/0008496 (August 30, 2000).

15. In addition, many can only be seen at radio wavelengths. See for example R. K. Anantharamaiah, A Pedlar, R. Ekers, and W. Goss, "Radio Studies of the Galactic Centre—II. The Arc, Threads, and Related Features at 90 cm (330 MHz)," *Monthly Notices of the Royal Astronomical Society* 249 (1991): 262–81.

16. Heiles, "H I Shells," 589.

17. More recent examples of images of the galactic center are in my *Visual Practices Across the University*.

18. B. Stappers, B. Gaensler, and S. Johnston, "A Deep Search for Pulsar Wind Nebulae Using Pulsar Gating," arxiv.org/abs/astro-ph/9904056 (April 5, 1999), fig. 5, p. 8.

19. The method is adapted for astronomical images from the SOLA method, used in helioseismology. See Pijpers, "Unbiased Image Reconstruction as an Inverse Problem," arxiv.org/abs/astro-ph/9904075 (April 6, 1999), 2 ff.

20. Pijpers, "Unbiased Image Reconstruction," 5 and fig. 7, p. 7. I thank Frank Pijpers for a conversation on his work in Aarhus, Denmark, in 1999.

21. Burke, *A Philosophical Enquiry*, part 2, section 6, p. 65.

22. I thank James Hathaway for this; the original reference is lost.

23. *Mō hitotsu no shashin: shashintekinaru mono o megutte, Akioka Miho, Berunāru Borugyō, Via Seruman, Miwa Mitsuko, Morimura Yasumasa, Sūzan Rankaitēsu* (Another Photography: Rethinking the Concept of Photography, Miho Akioka, Bernard Borgeaud, Vija Celmins, Mitsuko Miwa, Yasumasa Morimura, Susan Rankaitis), exh. cat. (Tokyo: Tokyo Metropolitan Museum of Photography, 1996), 22–26.

24. The phenomenology of Rothko's paintings in Houston is a subject of my *Pictures and Tears: A History of People Who Have Cried in Front of Paintings* (New York: Routledge, 2001), 1–39.

25. McEvilley, " 'Grey Geese Descending': The Art of Agnes Martin," *Artforum* 25 (Summer 1987): 94–99, quotation on p. 96. The ellipses omit a phrase equating the canvases to membranes—an image I don't find apposite to Martin's dry, inorganic, inflexible surfaces.

26. See the brief history and theory of graphs in my *Domain of Images*, 229–30, as well as Thomas Hankins, "Blood, Dirt, and Nomograms: A Particular History of Graphs," *Isis* 90 (1999): 50–80, esp. p. 52. I thank Roald Hoffmann for bringing this to my attention.

27. See also nssdc.gsfc.nasa.gov/imgcat/html/object_page/v02_304b88 .html.

28. See Lawrence Krauss and Michael Turner, "Geometry and Destiny," arxiv.org/abs/astro-ph/9904020 (April 1, 1999).

29. As of July 2004, FITS formatted data could be accessed on personal computers through www.spacetelescope.org/projects/fits_liberator.

30. "Counterfeit" is a central term in T. J. Clark's antisemiotic account of cubism; see Clark, *Farewell to an Idea: Episodes from a History of Modernism* (New Haven, CT: Yale University Press, 1999), 169–224.

31. Doubt about pictorial representation has been ingrained in Western thought since Byzantine and medieval debates about sacred images. The Christian sense of doubt about representation is deeply ingrained in Western image making and is the principal historical precedent for (secular) "anti-optical"

artworks throughout modernism. In the enormous literature, essential readings would include Joseph Koerner, *The Reformation of the Image* (Chicago: University of Chicago Press, 2004), and Hans Belting, *Likeness and Presence: A History of the Image Before the Era of Art*, trans. Edmund Jephcott (Chicago: University of Chicago Press, 1994); my "Visual Culture: First Draft," review of *Iconoclash!* ed. Bruno Latour and Peter Weibel, in *Art Journal* 62, no. 3 (2003): 104–7; Peter Geimer, "Das Undarstellbare: Über die Irrtümer der Bilderstürmer," *Neue Zürcher Zeitung* (March 25, 2006); my *The Strange Place of Religion in Contemporary Art* (New York: Routledge, 2005), and James Elkins, ed., *Re-Enchantment*, vol. 7 of *The Art Seminar* (New York: Routledge, forthcoming).

32. For Breuer see for example *Marco Breuer: SMTWTFS*, exh. cat. (New York: Roth Horowitz, 2002).

33. They are blind drawings, literally, made in darkness and therefore blind in the sense that Derrida has described classical drawing in *Memoirs of the Blind*. Derrida, *Memoirs of the Blind: The Self-Portrait and Other Ruins*, trans. Pascal-Anne Brault and Michael Naas (Chicago: University of Chicago Press, 1993).

34. The famous acheiropoietai (images made without human hands), such as the Veronica, the Shroud, and the Mandylion, are anti-optical exceptions to the predominantly optical tradition. Acheiropoietai resemble what they represent, but the condition of their veracity is that they were said to be neither drawn or painted: they were made by direct contact with Jesus, whom they represent. By the standards of the acheiropoietai, any other pictures of God—ones involving skill, manual work, techniques, materials, and methods—must fail. See for example Robin Jensen, *Face to Face: Portraits of the Divine in Early Christianity* (Minneapolis, MN: Fortress Press, 2005).

35. Most of these are Chicago photographers. For Stivers and Geesaman, see edelmangallery.com. Thanks to Jim Hugunin for these references. Readman is reproduced in *Canadian Art* (Winter 2000): 35.

36. "Photography's Fugitive Figure, An Interview with Lori Newdick," and "The Revolutionary Two-Step, An Interview with Barbara Astman," both in *Border Crossings* 23, no. 1 (2004): 58–63 and 44–50, respectively.

37. For Pamela Bannos see *Imaging Space*, exh. cat. (Evanston, IL: Mary and Leigh Block Gallery, 2001).

38. *Photovision* 30 (2001), special issue, "Big Crunch," 10, 27, 51, 80, respectively. Coolen was born in 1955; Tromeur in 1946; Rispa in 1964; Misrach in 1949.

39. Gober's book, *Robert Gober, 1978–2000*, is unpaginated and untitled; it is identified by title in the accompanying calatogue, *Robert Gober: The United States Pavilion, 49th Venice Biennale*, ed. James Rondeau (Chicago: Art Institute, 2001).

40. For example, *Hiroshi Sugimoto: Architecture of Time*, exh. cat., ed. Ekhard Schneider (Cologne: Distributed Art Publishers, 2002).

41. For Jacobson's blurred "History Series," see www.artnet.com/artwork/423856266/_Bill_Jacobson_Untitled_From_the_History_Series.html. Thanks to Catherine Edelman for drawing my attention to Jacobson and Weber, and to the commercial photographer Richard Howard for confirming my suspicion that *Martha Stewart Living* started the shallow-focus fad.

42. See play-create.com/xmas for an example.

43. Borgeaud, "Notes de carnets (1994–96)," in *Another Photography: Rethinking the Concept of Photography* (Tokyo: Metropolitan Museum of Photography, 1996), 23, translation mine.

44. The images were available at www.phys.unsw.edu.au/southpolediaries/webcam.html, accessed May 2003. The site and South Pole camera have changed, and they may now be accessed through www.archive.org.

45. See www.pesharpe.com, accessed November 2006.

46. Juan Aliaga, "Interview [with Tuymans]," in *Luc Tuymans*, ed. Ulrich Loock, Juan Aliaga, and Nancy Spector (London: Phaidon, 1996), 12.

47. Aliaga, "Interview," 13–14; Tuymans is speaking of Friedrich, de Chirico, and Léon Spillaert.

48. Philippe Lacoue-Labarthe, "Sublime Truth," in *Of the Sublime: Presence in Question*, ed. Jeffrey Librett (Albany: State University of New York Press, 1993), 71–108, quotation on p. 84.

49. The state of the conceptualization of photography is sampled in *Photography Theory*, ed. James Elkins, vol. 2 of *The Art Seminar* (New York: Routledge, 2007).

50. This is taken up in my "What Do We Want Photography to Be? [reply to Michael Fried's "Barthes's *Punctum*]," *Critical Inquiry* 31, no. 4 (2005): 938–56.

51. Piero Rosati, S. A. Stanford, Peter Eisenhardt, Richard Elston, Hyron Spinrad, Daniel Sten, and Arjun Day, "An X-ray Selected Galaxy Cluster at $z = 1.26$," preprint, arxiv.org/abs/astro-ph/9903381 (March 24, 1999).

52. Rosati informed me that Chandra, an X-ray satellite, made a fifty-three-hour exposure of the same region, resulting in a thirty-fold increase in resolution. (Personal communication, December 9, 1999.)

53. See www.batse.msfc.nasa.gov/ for the Burst and Transient Source Experiment, attached to NASA's Compton Gamma-Ray Observatory.

54. See www.umich.edu/~rotse/ for the Robotic Optical Transient Search Experiment; also see E. E. Fenmore et al., "RGB 990123: Evidence That the Gamma Rays Come from a Central Engine," *Astrophysical Journal* 518 (June 20, 1999): L73–L76.

55. For another gamma ray burst (GRB), with typical optical photograph showing a pointlike source, see J. P. Halpern et al., "The Rapidly Fading Optical Afterglow of GRB 980519," arxiv.org/abs/astro-ph/9903418 (March 26, 1999).

56. J. S. Bloom et al., "The Host Galaxy of GRB 990123," arxiv.org/abs/

astro-ph/9902812 (February 13, 1999), submitted to *The Astrophysical Journal Letters*; M. S. Briggs et al., "Observations of GRB 990123 by the Compton Gamma Ray Observatory," *Astrophysical Journal* 524 (October 10, 1999): 82–91; Andrew Fruchter et al., "Hubble Space Telescope and Palomar Imaging of GRB 990123: Implications for the Nature of Gamma-Ray Bursts and Their Hosts," *Astrophysical Journal* 519 (July 1, 1999): L13–L16; and Andrew Fruchter et al., "HST and Palomar Imaging of GRB 990123: Implications for the Nature of Gamma-Ray Bursts and Their Hosts," arxiv.org/abs/astro-ph/9902236 (February 17, 1999).

57. The estimation of GRB magnitudes also depends on the cosmological model that is adopted; see Daniel Enström, Sverker Fredriksson, and Johan Hansson, "Luminosities of High-Redshift Objects in an Accelerating Universe," arxiv.org/abs/astro-ph/9904262 (April 20, 1999).

3. Astronomy

1. Edgerton and Lynch's work is discussed in my *Domain of Images* (Ithaca, NY: Cornell University Press, 1999), 10–12. See also recent work by Elizabeth Kessler, especially her "Astronomy's Landscapes. Romanticism and the Hubble Space Telescope Images" (unpublished diss., University of Chicago), chapter 3, "Balancing Science and Art."

2. Bevilacqua La Masa Foundation, Galleria di Piazza San Marco, Venice, June 16–October 15, 2006.

3. *Vija Celmins*, ed. Judith Tannenbaum, with essays by Douglas Blau and Dave Hickey (Philadelphia: Institute of Contemporary Art, 1992); *Vija Celmins*, ed. James Lingwood (Madrid: Museo Nacional Centro de Arte Reina Sofia, 1996); also see *Mō hitotsu no shashin: shashintekinaru mono o megutte, Akioka Miho, Berunāru Borugyō, Via Seruman, Miwa Mitsuko, Morimura Yasumasa, Sūzan Rankaitēsu* (Another Photography: Rethinking the Concept of Photography, Miho Akioka, Bernard Borgeaud, Vija Celmins, Mitsuko Miwa, Yasumasa Morimura, Susan Rankaitis), exh. cat. (Tokyo: Toyko Metropolitan Museum of Photography, 1996), 28–34. It is not irrelevant that Celmins's photos in *Another Photography* immediately follow a presentation of photos by Bernard Borgeaud.

4. An example of a fine-art photographer who uses astronomical images is Pamela Bannos; see *Pamela Bannos and Farhad Zadeh: Imaging and Imagining Space, A Collaboration Between Art and Science* (Evanston, IL: Mary and Leigh Block Museum of Art, 2001).

5. Oliver Morton, review of Michael Light's *Full Moon* (London: Cape, 1999), in *The London Times Literary Supplement* (August 20, 1999), 16–17, quotation on p. 16.

6. Clark, *Farewell to an Idea: Episodes from a History of Modernism* (New Haven, CT: Yale University Press, 1999), 215.

7. Images like the ones Sagan showed are archived at www.nineplanets.org/pxjup.html and voyager.jpl.nasa.gov/image/jupiter.html.

8. The various lights of the night sky are described in my *How to Use Your Eyes* (New York: Routledge, 2000), 212–17.

9. On this topic see the careful accounts in Louis Roy, "Kant's Reflections on the Sublime and the Infinite," *Kant-Studien* 88 (1997): 44–59, and Malcolm Budd, "Delight in the Natural World: Kant on the Aesthetic Appreciation of Nature, Part III, The Sublime in Nature," *British Journal of Aesthetics* 38, no. 3 (1998): 233–50.

10. Edmund Burke, *A Philosophical Enquiry*, ed. Adam Phillips (Oxford: Oxford University Press, 1990), part 2, section 13, p. 71: "Besides, the stars lye in such apparent confusion, as makes it impossible on ordinary occasions to reckon them. This gives them the advantage of a sort of infinity."

11. Rennan Barkana, Roger Blandford, and David Hogg, "A Possible Gravitational Lens in the Hubble Deep Field South," arxiv.org/abs/astro-ph/ 981273 (January 7, 1999). The lens is also discussed briefly in Barkana and Hogg, "Gravitational Lensing of High Redshift Sources," arxiv.org/abs/ astro-ph/0001325 (January 19, 2000).

12. For the geometry of gravitational lenses, see Joachim Wambsganss, "Gravitational Lensing: A Universal Astrophysical Tool," arxiv.org/abs/astro-ph/0012423 (December 20, 2000). An interesting meditation on related optical phenomena is Lawrence Mertz, *Excursions in Astronomical Optics* (New York: Springer, 1996), 128–37.

13. The central point is a singularity, with infinite density, so that formally there is no fifth point in the model. I thank Rennan Barkana for clarifying this.

14. For other examples see especially P. Wozniak, C. Alard, A. Udalski, M. Szymanski, M. Kubiak, G. Pietrzynski, and K. Zebrun, "The Optical Gravitational Lensing Experiment Monitoring of QSO 2337+0305," arxiv .org/abs/astro-ph/9904329 (April 26, 1999).

15. On the other hand, the mathematics can easily outstrip what is observable, if only ideal conditions (other than irregular real-life galaxies) are considered. See for example K. Virbhandra and George Ellis, "Schwarzschild Black Hole Lensing," arxiv.org/abs/astro-ph/9904193 (April 15, 1999). The authors predict that in the region of a black hole there will be multiple partly unresolved images of the entire universe.

16. For a cogent counterargument advocating the use of the sublime, see Martha Fleming, "Painting by Numbers: Image, Data and the Mathematical Sublime in Late 20th Century Astrophysics" (unpublished paper delivered to the History of Scientific Observation Seminar, Ideas and Practices of Rationality, Department II, Max Planck Institute for the History of Science, Berlin, Germany, May 31, 2006); martha@marthafleming.net.

17. Burke, *A Philosophical Enquiry*, part 2, section 7, p. 66.

18. Enlightenment responses to microscopy are discussed in my *Pictures of the Body: Pain and Metamorphosis* (Stanford, CA: Stanford University Press, 1999), 205–44.

19. Philip and Phylis Morrison and the Office of Charles and Ray Eames, *Powers of Ten: A Book About the Relative Sizes of Things in the Universe and the Effect of Adding Another Zero* (New York: Scientific American Library, 1994). The nearly forgotten original, not sufficiently credited in the new incarnations, is Kees Boeke, *Redelijke ordening van de mensengemeenschap* (n.p., 1945), translated as *Cosmic View: The Universe in Forty Jumps,* trans. Arthur Compton (New York: J. Day, 1957). It is relevant to the themes of other chapters in this book that *Powers of Ten* was also reprinted in an art photography journal, *Apterture* 157 (1999), in a special issue, "Steps in Space."

20. See Louis Roy, "Kant's Reflections on the Sublime and the Infinite," *Kant-Studien* 88 (1997): 44–57, especially p. 53.

21. Malcolm Budd, "Delight in the Natural World: Kant on the Aesthetic Appreciation of Nature, Part III, The Sublime in Nature," *British Journal of Aesthetics* 38, no. 3 (1998): 233–50, especially pp. 237–38. I am not taking into account Kant's interest in the field of view, and how near or far from an object a person needs to be: those considerations have to do with actual objects (Kant's example is the pyramids and their rows of stones) rather than images. Budd argues, I think correctly, that Kant's analysis of magnitude is muddled by his introduction of the infinite (ibid., 240).

22. Kant's theory lends itself to sublimities of time as well as space: see David Cook, "The Last Days of Liberalism," in *Postmodernism: A Reader,* ed. Thomas Docherty (New York: Columbia University Press, 1993), 120–27, especially p. 126.

23. Gerald Sussman and Jack Wisdom, "Chaotic Evolution of the Solar System," *Science* 257 (July 3, 1992): 56–62, quotation on p. 61.

24. Dirac, "The Mathematical Foundations of Quantum Theory," in *Mathematical Foundations of Quantum Theory,* ed. A. R. Marlow (New York: Academic Press, 1978), 2.

25. For Davis see also Steve Nadis, "Science for Art's Sake," *Nature* 407 (October 12, 2000): 668–70; "Microvenus," *Art Journal* 55, no. 1 (1996): 70–74; W. Wayt Gibbs, "Art as a Form of Life," *Scientific American* (April 2001). The supercode is published as Davis, "Romance, Supercodes, and the Milky Way DNA," in *Next Sex, Ars Electronica 2000* (Vienna: Springer, 2000), 217–35. Davis is one of a very loosely affiliated group of artists, including Eduardo Kac (www.ekac.org), Adam Zaretsky (see Nadis, "Science for Art's Sake"), Ionat and Oron Catts (www.tca.uwa.edu.au/), and Ben Delaney ("Art Is Where You Grow It," *IEEE Multimedia* [January–March 2001]: 4–7). At the time this book went to press, Davis's project was under way, thanks to a private collector and to Davis's collaborator, Dana Boyd, who wrote automatic DNA supercoding software.

26. The image is from the COBE satellite; see aether.lbl.gov/www/projects/cobe/. For further information see *The Center, Bulge, and Disk of the Milky Way,* ed. Leo Blitz (Dordrecht: Kluwer, 1992).

27. Personal correspondence, April 1999.

28. Davis, "Milky Way DNA," unpublished draft, revised January 29, 1999.

29. Davis comments on the whole process (personal communication, November 2001) that once it is possible to overwrite supercoded DNA "infogenes" with sequences that code for known proteins, "we could then embark on a kind of 'genome project of the world.' Theoretically, all recorded knowledge could then be incorporated into routine DNA synthesis carried out for science and commerce and this extra information could be added at little or no cost. DNA supercode . . . [could then] be used to 'underwrite' non-expressed, or passive data into sequences that are 'overwritten' with sequences that have desired translation products": a cosmic information program if ever there was one.

30. The HDF-N Web page is at www.stsci.edu/ftp/science/hdf/hdf .html. The Deep Fields held the record for "deepest" views for almost five years; since then, the Advanced Camera for Surveys (ACS) has made more sensitive images. An eighty-four-hour view of a small area near M31 broke the HDF achievement in 2003.

31. In addition to the other sources I cite in this chapter, see Stefano Casertano et al., "WFPC2 Observations of the Hubble Deep Field South," *Astrophysical Journal* 120 (December 2000): 2747–2824; Robert Williams et al., "The Hubble Deep Field South: Formulation of the Observing Campaign," *Astrophysical Journal* 120 (December 2000): 2735–46.

32. An animation that produces this same sequence for the HDF-N is available at oposite.stsci.edu/pubinfo/pr/96/01/HDF.html. Unfortunately, it uses a blurry image for the penultimate enlargement.

33. Roberto Abraham, Nial Tanvir, Basilio Santiago, Richard Ellis, Karl Glazebrook, and Sidney van den Bergh, "Galaxy Morphology to $I=25$ Mag in the *Hubble Deep Field*," *Monthly Notes of the Royal Astronomical Society* 279 (1996): L47; preprint at arxiv.org/abs/astro-ph/9602044 (February 9, 1996). Among recent classification attempts is Roberto Abraham and Michael Merrifield, "Explorations in Hubble Space: A Quantitative Tuning Fork," *Astronomical Journal* 120 (December 2000): 2835–42.

34. Sidney van den Bergh, Roberto Abraham, Richard Ellis, Nial Tanvir, Basilio Santiago, and Karl Glazebrook, "A Morphological Catalogue of Galaxies in the *Hubble Deep Field*," arxiv.org/abs/astro-ph/9604161 (April 26, 1996), pp. 5, 6.

35. An example is NGC3991; see for example crux.astr.ua.edu/gifimages/ ngc3991.gif. Cited in van den Bergh et al., "A Morphological Catalogue," 6.

36. Richard Ellis, "The Hubble Deep Field: Introduction and Motivation," arxiv.org/abs/astro-ph/9708075 (August 8, 1997).

37. Among many others, see Dan Maoz, "The Dectectability of High-Redshift Ellipticals in the Hubble Deep Field," arxiv.org/abs/astro-ph/9704173 (October 3, 1997).

38. Gerhardt Meurer, "The Case for Substantial Dust Extinction at $z \approx 3$," arxiv.org/abs/astro-ph/9708163 (August 18, 1997).

39. Stephen Schectman, "The Small-Scale Anisotropy of the Cosmic Light," *Astrophysical Journal* 188 (1974): 233–42.

40. J. A. Tyson, "Deep CCD Survey: Galaxy Luminosity and Color Evolution," *Astronomical Journal* 96 (1988): 1–23, quotation on p. 1.

41. Michael Vogeley, "Fluctuations in the Extragalactic Background Light: Analysis of the Hubble Deep Field," arxiv.org/abs/astro-ph/9711209 (November 18, 1997).

42. Ibid., 3; the autocorrelation function is on p. 6.

43. Ibid., 3–4.

44. van den Bergh et al., "A Morphological Catalogue," 6.

45. These images were made using Photoshop, but the same results can be achieved with the original NASA FITS files and professional software such as ImageJ or Ouroboros.

46. Park and Kim, "Diffuse Dark and Bright Objects in the Hubble Deep Field," arxiv.org/abs/astro-ph/9712039 (December 3, 1997).

47. Ibid., 1–2.

48. Ibid.

49. Ibid., 8.

50. Siang Peng Oh, "Observational Signatures of the First Luminous Objects," arxiv.org/astro-ph/9904255 (April 20, 1999), p. 1. For the next generation of instruments, see for example Torsten Böker and Ronald Allen, "Imaging and Nulling with the *Space Mission*," *Astrophysical Journal Supplement Series* 125 (November 1999): 123–42.

51. Carl Gibson, "Dark Matter at Viscous-Gravitational Schwarz Scales: Theory and Observations," in *Proceedings of the International Workshop on Dark Matter in Astro- and Particle Physics*, ed. H. Klapdor-Kleingrothaus (River Edge, NJ: World Scientific, 1997), 409–16.

52. Their results have not been doubted, as far as I am aware (I also thank the authors for this information), though it is now superseded by more recent deep field images and by plans for the Next Generation Space Telescope. More recently, one of the authors has studied spatial distribution in HDF images: Changbom Park, Richard Gott III, and Y. J. Choi, "Topology of the Galaxy Distribution in the Hubble Deep Fields," arxiv.org/astro-ph/0008353 (August 23, 2000).

53. Piero Madau, "Starlight in the Universe," arxiv.org/astro-ph/9902228 (February 16, 1999), p. 1, and see p. 2 on the "possible existence of a large population of faint galaxies still undetected at high redshifts." Also see Madau, "After the Dark Ages: The Evolution of Luminous Sources at $z < 5$," arxiv.org/abs/astro-ph/9812087 (December 4, 1998).

54. For a distant object in the HDF-S, see M. Stiavelli et al., "VLT and HST Observations of a Candidate High Redshift Elliptical Galaxy in the

Hubble Deep Field South," arxiv.org/abs/astro-ph/9812102 (December 4, 1998).

55. Zsolt Paragi et al., "VLBI Imaging of Extremely High Redshift Quasars at 5 GHz," preprint at arxiv.org/abs/astro-ph/9901396 (January 28, 1999), 3, 6. The essay later appeared in *Astronomy and Astrophysics* 348 (1999): 910. Another example is Wil van Breugel, Carlos De Breuck, S. Stanford, Daniel Stern, Huub Röttgering, and George Miley, "A Radio Galaxy at $z = 5.19$," arxiv.org/abs/astro-ph/9904272 (April 21, 1999).

56. Hsiao-Wen Chen, Kenneth Lanzetta, and Sebastian Pascarelle, "A Spectroscopically Identified Galaxy of Probable Redshift $z = 6.68$," arxiv .org/abs/astro-ph/9904161 (April 14, 1999). The observation was later doubted; see Hsiao-Wen Chen, Kenneth Lanzetta, Sebastian Pascarelle, and Noriaki Yahata, "The Unusual Spectral Energy Distribution of a Galaxy Previously Reported to Be at Redshift 6.68," arxiv.org/abs/astro-ph/0011558 (November 30, 2000).

57. Chen et al., "A Spectroscopically Identified Galaxy," 5.

58. The particular valence of this melancholy is elaborated (without reference to these materials) in Max Pensky, "Tactics of Remembrance: Proust, Surrealism, and the Origin of the *Passagenwerk*," in *Walter Benjamin and the Demands of History*, ed. Michael Steinberg (Ithaca, NY: Cornell University Press, 1996), 164–89.

59. R. B. Alley and I. M. Willians, "Changes in the West Antarctic Ice Sheet," *Science* 254 (November 15, 1991): 959–63, fig. 3, p. 961.

60. This image has many parallels: in electron microscopy the newest imaging systems produce vast "landscapes" of atoms. When they are folded, they can look irresistibly like fabric. A lovely example is in Ken-ichi Yamamoto et al., "Formation Process of the Amorphous MgNi by Mechanical Alloying," *Journal of Electron Microscopy* 47, no. 5 (1998): 461–70, fig. 7, p. 467.

4. Microscopy

1. One of the best introductions is R. Bouyer, "Resolution," in *The Encyclopedia of Microscopy and Microtechnique*, ed. Peter Gay (New York: Van Nostrand Reinhold, 1973), 504b–515a.

2. See William Carpenter, *The Microscope and Its Revelations*, 1st ed. (London: J. & A. Churchill, 1856; Philadelphia: Blakiston, 1856), and compare the slightly less well-resolved photo in Wilhelm Behrens, *Leitfaden der botanischen Mikroskopie* (Braunschweig: H. Bruhn, 1890), 60. Diatoms are discussed as resolution objects in Hans Hermann Julius Hager, *Das Mikroskop und seine Anwendung* (Berlin: Springer, 1866) and many subsequent editions; Sigmund Schertel, *Das Mikroskop* (Stuttgart: Union Deutsche Verlagsgesellschaft, 1904); and for a contemporary introduction see David Walker, "Counting the Dots: Giving Microscopes a 'Workout' Using Diatom Test Slides," www

.microscopy-uk.org.uk/mag/artoct99/dwdiatom.html. I thank Martin Mach for drawing my attention to these sources.

3. Roy Morris Allen, *Photomicrography*, 2nd ed. (Princeton, NJ: D. Van Nostrand, 1958), 275. See also Hendrik Christoffel Hulst, *Light Scattering by Small Particles* (New York: Wiley, 1957).

4. Dippel, *Das Mikroskop und seine Anwendung*, 2 vols. (Braunschweig: F. Vieweg und Sohn, 1892, 1898). I am indebted to Martin Mach for much of this section on diatoms in nineteenth- and twentieth-century research. See also Dippel, *Diatomeen der Rhein-Mainebene* (Braunschwieg: F. Vieweg und Sohn, 1904).

5. The electron microscope images are available in Kurt Krammer, *Kieselalgen: Biologie, Baupläne der Zellwand, Untersuchungsmethoden* (Stuttgart: Kosmos, 1986).

6. See the extended discussion comparing SEM and optical images of a diatom: Theodore Rochow et al., "Light and Electronmicroscopical Studies of *Pleurosigma angulatum* for Resolution of Detail and Quality of Image," *The Microscope and Crystal Front* 15, no. 5 (1966): 177–201; Kurt Krammer, "Zur Deutung der Diatomeen-Feinstrukturen im Lichtmikroskop," *Mikrokosmos* 68, no. 3 (1979): 66–72, and the rebuttal by Rochow, "Erwiderung zu Kurt Krammers Aufsatz 'Zur Deutung der diatomeen-Feinstrukturen im Lichtmikroskop,'" *Mikrokosmos* 70 (1981): 164–69. Much of the debate over *Pleurosigma* turned on the choice of focal planes because the diatom has a complicated 3-D structure. See also Bouyer, "Resolution," 514b.

7. Jeremy Burgess, Michael Marten, and Rosemary Taylor, *Microcosmos: Under the Microscope, a Hidden World Revealed* (New York: Cambridge University Press, 1987), 187–88.

8. See *Handbook of Microscopy: Applications in Materials Science, Solid-State Physics and Chemistry*, ed. Severin Amelinckx et al. (Weinheim: VCH, 1997). There are many sources that review the choices I outline here: for example, Edward B. Brain and Arnold Richard Ten Cate, *Techniques in Photomicrography* (Princeton, NJ: D. Van Nostrand, 1963); Roger Loveland, *Photomicrography: A Comprehensive Treatise*, 2 vols. (New York: John Wiley and Sons, 1970); Roy Morris Allen, *Photomicrography* (Princeton, NJ: Van Nostrand, 1958); and D. J. Thomson and Savile Bredbury, *An Introduction to Photomicrography* (Oxford: Oxford University Press, 1987).

9. Carpenter, *The Microscope and Its Revelations*, 60, lists the objective as a Zeiss with *N.A.* 1.6, the flint glass as *N.A.* 1.72, and the magnification as 3,000.

10. Bouyer, "Resolution," 510b.

11. Roy Morris Allen's *Practical Refractometry by Means of the Microscope: With Listings of Index Liquids, and Other Aids for Mineralogists* (New York: R. P. Cargille, 1954) has instructions for determining *N.A.* to three significant figures.

12. Abbé's papers are regularly cited, but I think seldom read. Physicists have a tradition of citing the original essays but learning diffraction theory from contemporary textbooks. The most recent reprint is Ernst Abbé, *Abhandlungen über die Theorie des Mikroskops* (Hildesheim: Georg Olms Verlag, 1989), reprinting Abbé, *Gesammelte Abhandlungen*, vol. 1, *Abhandlungen über die Theorie des Mikroskops* (Jena: Gustav Fischer, 1904).

13. It is often assumed that Köhler illumination is the only method of producing coherent illumination in the object-space; but see Bouyer, "Resolution," 513a.

14. For introductions to phase microscopy see Alva Herschel Bennett, *Phase Microscopy: Principles and Applications* (New York: Wiley, 1951); and see F. Zernicke, "The Concept of Degree of Coherence and Its Application to Optical Problems," *Physica* 5, no. 8 (August 1938): 785–95; and Hermann Beye, *Theorie und Praxis des Phasenkontrastverfahrens* (Leipzig: Akademische Verlagsgesellschaft Geest und Portig, 1965).

15. Savile Bradbury and Peter Evennett, *Contrast Techniques in Light Microscopy* (Herndon, VA: BIOS Scientific Publishers, 1996), 68.

16. Van Fraassen, *The Scientific Image* (Oxford: Clarendon Press, 1980). I have added the diffraction theory of microscope image formation to Van Fraassen's argument because, as Ian Hacking argues, it is part of the same larger framework of epistemological objections. See Hacking, *Representing and Intervening: Introductory Topics in the Philosophy of Natural Science* (Cambridge: Cambridge University Press, 1983), 189. I thank David Kaiser for drawing my attention to this book.

17. Hacking, *Representing and Intervening*.

18. J. Padawer, "The Nomarski Interference-Contrast Microscope: An Experimental Basis for Image Interpretation," *Journal of the Royal Microscopical Society* 88 (1967): 305–49. Nomarski's system also alters earlier interference techniques in such a way as to produce an effect of relief rather than interference figures that could be used to measure the thickness of objects; that history is not my concern here. Microscopists can still find the earlier systems: Jamin-Lebedeef systems come on the market (on eBay, LabX, and Dovebid) intermittently, and there are also interference objectives available (for reflected light). I have experimented with, and I own, several intereference systems; they do not raise issues analogous to the ones I am exploring here because they all produce interference patterns that are *counted* (i.e., they do not produce images that ask to be seen as three-dimensional).

19. Robert Hoffman, "The Modulation Contrast Microscope: Principles and Performance," *Journal of Microscopy* 110, no. 3 (1977): 205–22, especially p. 217; see also Robert Hoffman and Leo Gross, "The Modulation Contrast Microscope," *Nature* 254 (April 17, 1975): 586–88, and Hoffman and Gross, "Modulation Contrast Microscope," *Applied Optics* 14 (1975): 1169.

20. For example Thomas Lkar and Stefan Hell, "Subdiffraction Resolution

in Far-Field Fluorescence Microscopy," *Optics Letters* 24, no. 14 (July 15, 1999): 954–56; A. M. van Oijen et al., "3-Dimensional Super-Resolution by Spectrally Selective Imaging," *Chemical Physics Letters* 292 (July 31, 1998): 183–87; and the introductory report by Rob van den Berg, "Molecular Imaging Beats Limits of Light," *Science* 281, no. 5377 (July 31, 1998): 629–30.

21. Michael Schmidt, Matthias Nagomi, and Stefan Hell, "Subresolution Axial Distance Measurements in Far-Field Fluorescence Microscopy with Precision of 1 Nanometer," *Review of Scientific Instruments* 71, no. 7 (July 2000): 2742–45; Thomas Klar et al., "Fluorescence Microscopy with Diffraction Resolution Barrier Broken by Stimulated Emission," *PNAS* [Proceedings of the National Academy of Sciences, Washington, DC] 97, no. 15 (July 18, 2000): 8206–10; and J. Michaelis et al., "Optical Microscopy Using a Single-Molecule Light Source," *Nature* 405 (May 18, 2000): 325–28.

22. Shinya Inoué, *Video Microscopy* (New York: Plenum, 1986).

23. Siham Sabri et al., "Interest of Image Processing in Cell Biology and Immunology," *Journal of Immunological Methods* 208 (1997): 1–27. The *Journal of Electron Microscopy* has been interested in resolution limits since its inception in 1997; see for example II. Jiang et al., "Crystal Structure Determination by Image Deconvolution in Combination with Image Simulation," *Journal of Electron Microscopy* 46, no. 5 (1997): 375–80; A. I. Kirkland, W. O. Saxton, and G. Chand, "Multiple Beam Tilt Microscopy for Super Resolved Imaging," *Journal of Electron Microscopy* 1 (1997): 11–22.

24. Available for Macintosh computers at rsb.info.nih.gov/nih-image/index.html. IBM PC versions are available commercially.

25. Christoph Böttcher, Kai Ludwig, et al., "Structure of Influenza Haemagglutinin at Neutral and at Fusogenic pH by Electron Cryo-Microscopy," *FEBS* [Federation of European Biochemical Societies] *Letters* 463 (1999): 255–59. For the relevant image analysis, see Marin van Heel and Marina Stöffler-Meilicke, "Characteristic View of *E. coli* and *B. stearothermophilus* 30S Ribosomal Subunits in the Electron Microscope," *EMBO Journal* [European Molecular Biology Organization] 4 (1985): 2389–95.

26. P. A. Thuman-Commike and W. Chiu, "Reconstruction Principles of Icosahedral Virus Structure Determination Using Electron Cryomicroscopy," *Micron* 31 (2000): 687–711, especially pp. 690–91.

27. In the case of electrons, $\lambda = \frac{h}{g}$, where h is the Planck constant (6.62×10^{-34} Joule-seconds) and g is the energy of the beam: the higher the energy, the better the resolution.

28. Maximilian Haider et al., "Toward 0.1 nm Resolution with the First Spherically Corrected Transmission Electron Microscope," *Journal of Electron Microscopy* 47, no. 5 (1998): 395–405, especially pp. 396, 399.

29. Phillips, "High Resolution Electron Microscope Observations of Precipitation in Al-3.0% Cu Alloy," *Acta Metallurgica* 23 (June 1975): 751–67.

See also the earlier work, Phillips and J. A. Hugo, "The Resolution of Lattice Planes and Lattice Defects in Semiconductors Including Observation on Boundaries," *Acta Metallurgica* 18 (January 1970): 123–35; both articles are mentioned in D. E. Newbury and D. B. Williams, "The Electron Microscope: The Materials Characterization Tool of the Millennium," *Acta Materialia* 48 (2000): 323–46, especially pp. 328–29.

30. Tsuyoshi Matsuda et al., "Morphology and Structure of Biogenic Magnetite Particles," *Nature* 302 (May 31, 1983): 411–12.

31. In 2000 the record was 0.109 nm, held by a group in the Graduate School of Frontier Sciences at the University of Tokyo. H. Ichinose, "Crystal Interface and High-Resolution Electron Microscopy—The Best Partner," *Science and Technology of Advanced Materials* 1 (2000): 11–20, especially p. 12, fig. 1. For resolution limits in organic chemistry, see Richard Leapman and Nancy Rizzo, "Towards Single Atom Analysis of Biological Structures," *Ultramicroscopy* 78 (1999): 251–68.

32. A good comparative survey is H.-J. Butt, R. Guckenberger, and Jürgen Rabe, "Quantitative Scanning Tunneling Microscopy and Scanning Force Microscopy of Organic Materials," *Ultramicroscopy* 46 (1992): 375–93. For information on high temperatures see T. Kamino et al., "High Resolution Electron Microscopy In Situ Observation of Dynamic Behavior of Grain Boundaries and Interfaces at Very High Temperatures [e.g. 1773K]," *Microscopy and Micoanalysis* 3 (1997): 393.

33. Holger Stark, Friedrich Zemlin, and Christoph Böttcher, "Electron Radiation Damage to Protein Crystals of Bacteriorhodopsin at Different Temperatures," *Ultramicroscopy* 63 (1996): 75–79, quotation on p. 75. An example of a specialized solution is Ki Hean Kim et al., "High-Speed, Two-Photon Scanning Microscope," *Applied Optics* 38, no. 28 (October 1, 1999): 6004–9, showing euglena unharmed in a sequence of images.

34. This is from the excellent introduction to TEM optics written by Gina Sosinsky and others. See Sosinsky, T. S. Baker, G. M. Hand, and M. H. Elliman, "The Electron Microscopy Outreach Program: A Web Based Resource for Research and Education," *Journal of Structural Biology* 125 (1999): 246–52; also available at em-outreach.sdsc.edu. I thank Gina Sosinsky for these citations.

35. Why not just build stronger lenses? It is possible to construct lenses made with superconducting coils, cooled by liquid helium; they generate fields of 50,000 gauss and have correspondingly high magnifications, but the resolution does not increase proportionately as the lens gets more powerful. After a certain point it does not even help to stain objects, as in light microscopy, because the visibility of the stained tissues is limited by the size of the stain molecules themselves, which are typically on the order of 0.5–1.0 nm.

36. Through-focus series can be simulated using commercially available software. The National Center for Electron Microscopy offers a free package

designed to accompany the commercial software Digital Micrograph. See ncem.lbl.gov/frames/computing.htm.

37. For a similar series and analysis see C. B. Boothroyd, "Quantification of High-Resolution Electron Microscope Images of Amorphous Carbon," *Ultramicroscopy* 83 (2000): 159–68, especially p. 162, fig. 1.

38. For the measurement of C_S and defocus from Fourier transforms, see O. L. Krivanek, "A Method for Determining the Coefficient of Spherical Aberration from a Single Electron Micrograph," *Optik* 45 (1976): 97–101.

39. N. Tanaka et al., "An 'On-Line' Correction Method of Defocus and Astigmatism in HAADF-STEM," *Ultramicroscopy* 78 (2000): 103–10; also see Isabel Angert, Endre Majorovits, and Rasmus Schröder, "Zero-Loss Image Formation and Modified Contrast Transfer Theory in EFTEM [Energy Filtering Transmission Electron Microscopy]," *Ultramicroscopy* 81 (2000): 203–22. A related technique been proposed for measuring the CTF using a single micrograph made with intentionally high astigmatism; see G. Möbius and M. Rühle, "A New Procedure for the Determination of the Chromatic Contrast Transfer Envelope of Electron Microscopes," *Optik* 93 (May 1993): 108–18; and Möbius et al., "Quantitative Diffractometry at 0.1 nm Resolution for Testing Lenses and Recording Media of a High-Voltage Atomic Resolution Microscope," *Journal of Electron Microscopy* 46, no. 5 (1997): 381–95.

40. For further information on the CTF see F. H. Li, "Image Processing Based on the Combination of High-Resolution Electron Microscopy and Image Diffraction," *Microscopy Research and Technique* 40 (1998): 86–100.

41. For more on Ronchigrams see J. A. Lin and J. M. Cowley, "Calibration of the Operating Parameters for an HB5 STEM [Scanning Transmission Electron Microscope] Instrument," *Ultramicroscopy* 19 (1986): 31–42; and E. M. James and N. D. Browning, "Practical Aspects of Atomic Resolution Imaging and Analysis in STEM," *Ultramicroscopy* 78 (1999): 125–39.

42. O. L. Krivanek, N. Dellby, and A. R. Lupini, "Towards Sub-Å Electron Beams," *Ultramicroscopy* 78 (1999): 1–11, especially p. 6.

43. See H. W. Zandbergen and D. van Dyck, "Exit Wave Reconstructions of Surfaces and Interfaces Using Through Focus Series of HREM Images," *Solid State Ionics* 131 (2000): 35–49, quotation on p. 38. More exactly, ρ_I is "the maximal diffracted beam angle that is still transmitted with appreciable signal-to-noise ratio" (ibid., p. 39). See also J.C.H. Spence, "The Future of Atomic Resolution Electron Microscopy for Materials Science," *Materials Science and Engineering* 26 (1999): 1–49, especially pp. 6–10.

44. Jingyue Liu, "Contrast of Highly Dispersed Metal Nanoparticles in High-Resolution Secondary Electron and Backscattered Electron Images of Supported Metal Catalysts," *Microscopy and Microanalysis* 6 (2000): 388–99.

45. See Newbury and Williams, "The Electron Microscope"; for CFMs, see Aleksandr Noy, Dmitri Vezenov, and Charles Lieber, "Chemical Force

Microscopy," *Annual Review of Materials Science* 27 (1997): 381–421. I thank Jie Liu for the latter reference. Also helpful is Stefan Kirstein, "Scanning Near-Field Optical Microscopy," *Current Opinion in Colloid and Interface Science* 4 (1999): 256–64.

46. These and others are cited in H. Kumar Wickramasinghe, "Progress in Scanning Probe Microscopy," *Acta Materialia* 48 (2000): 347–58, n. 60.

47. For capacitance microscopes see J. Schmidt et al., "Microwave-Mixing Scanning Capacitance Microscopy of *pn* Junctions," *Journal of Applied Physics* 86, no. 12 (December 15, 1999): 7094–99.

48. For an introduction see Mitch Jacoby, "New Tools for Tiny Jobs," *Chemical and Engineering News* 78, no. 42 (October 16, 2000): 33–35.

49. A good introductory text is *Scanning Tunneling Microscopy I: General Principles and Applications to Clean and Adsorbate-Covered Surfaces*, ed. Hans-Joachim Güntherodt and Roland Wiesendanger (New York: Springer, 1994). I thank Jie Liu for this reference.

50. J. Tersoff and D. R. Hamann, "Theory and Application for the Scanning Tunneling Microscope," *Physical Review Letters* 50, no. 25 (June 20, 1983): 1998–2001.

51. See the incisive essay by Eric Scerri, "Have Orbitals Really Been Observed?" *Journal of Chemical Education* 77, no. 20 (2000): 1–3. I thank Davis Baird for drawing this to my attention. It is possible to locate individual chemical bonds within single molecules: Barry Stipe, "Tuning In to a Single Molecule: Vibrational Spectroscopy with Atomic Resolution," *Current Opinion in Solid and Materials Science* 4 (1999): 421–28; B. C. Stipe, M. A. Rezaei, and W. Ho, "Single-Molecule Vibrational Spectroscopy and Microscopy," *Science* 280 (June 12, 1998): 1732–35.

52. "Quantitative Scanning Tunneling Microscopy," 389; quotation on pp. 381–82.

53. NSOMs can also be used to measure spectroscopic properties, polarization, and local dielectric contrasts. See P. M. Adam et al., "Polarization Contrast with an Apertureless Near-Field Optical Microscope," *Ultramicroscopy* 71 (1998): 327–31 with further references.

54. H. Kumar Wickramasinghe, "Apertureless Near-Field Optical Microscope," U.S. Patent 4,947,034, April 1989, col. 4: "The resultant detected signal at the difference frequency (f_1-f_2) . . . is very localized at the end of the tip since it is only at that location where the light will be phase and/or amplitude modulated by both dither motions."

55. An early paper is Eduard Chilla et al., "Probing of Surface Acoustic Wave Fields by a Novel Scanning Tunneling Microscopy Technique: Effects of Topography," *Applied Physics Letters* 61, no. 26 (December 28, 1992): 3107–9.

56. T. Hesjedal, E. Chilla, and H-J. Fröhlich, "Direct Visualization of the Oscillation of Au (111) Surface Atoms," *Applied Physics Letters* 69, no. 3 (July

15, 1996): 354–57; also see the same authors' "Scanning Acoustic Tunneling Microscopy and Spectroscopy: A Probing Tool for Acoustic Surface Oscillations," *Journal of Vacuum Science and Technology B* 15, no. 4 (1997): 1569–72. As of this writing, the group has a forthcoming paper that is also relevant: Peter Voigt, Eduard Chilla, S. Krauß, and Reinhold Koch, "Surface Acoustic Wave Investigation by Ultrahigh Vacuum Scanning Tunneling Microscopy," *Journal of Vacuum Science and Technology A: Vacuum, Surfaces, and Films* 19 (2001): 1817–21.

57. *Électrons et microscopes, vers les nanosciences,* ed. Peter Hawkes (Paris: SNRS Editions, 1995), 115.

58. For the application of topographic maps to genetic mapping see my *The Domain of Images* (Ithaca, NY: Cornell University Press, 1999), 37–38. For quantum Hall experiments see Gleb Finkelstein et al., "Topographic Mapping of a Quantum Hall Liquid Using a Few-Electron Bubble," *Science* 289 (2000): 90–94; and Finkelstein et al., "Imaging of Low-Compressibility Strips in the Quantum Hall Liquid," *Physical Review B* 61, no. 24 (June 15, 2000): 323–36. I thank Gleb Finkelstein for introducing me to this material.

59. Harrison Fuller and Murray Hale, "Determination of Magnetization Distribution in Thin Films Using Electron Microscopy," *Journal of Applied Physics* 31 (1960): 238–48, and the same authors' "Domains in Thin Magnetic Films Observed by Electron Microscopy," *Journal of Applied Physics* 31 (1960): 1699–1705.

60. R. H. Wade, "Lorentz Microscopy or Electron Phase Microscopy of Magnetic Objects," in *Advances in Optical and Electron Microscopy,* vol. 5, ed. R. Barber and V. E. Cosslett (London: Academic Press, 1973), 239–96, especially p. 243.

61. H. Boersch et al., "Antiparallele Weißasche Bereiche als Biprisma für Elektroneninterferenzen, II," *Zeitschrift für Physik* 167 (1962): 72–82; and Dieter Wohlleben, "Diffraction Effects in Lorentz Microscopy," *Journal of Applied Physics* 38 (1967): 3341–52 (with an illustration, fig. 8, p. 3347, uncannily close to Marco Breuer's photograph reproduced here as Figure 14).

62. For Lorentz microscopy in superconductors see Marco Beleggia et al., "Recent Results in the Interpretation of Interference and Lorentz Images of Vortices in Superconductors," *Materials Characterization* 42 (1999): 209–20; Akira Tonomura, "Observation of Plastic Flows of Vortices in Superconductors at the Depinning Threshold by Lorentz Microscopy," *Micron* 30 (1999): 479–84; Tonomura et al., "Real-Time Observation of Individual Vortices in High-T_c Superconductors by Lorentz Microscopy," *Physica C* 332 (2000): 45–50; Tonomura, "Real-Time Observation of the Dynamics of Vortices by Electron Microscopy," *Materials Characterization* 42 (1999): 201–8; and Tonomura, "Direct Observation of Vortex Motion in High-T_c Superconductors by Lorentz Microscopy," *Physica B* 280 (2000): 227–28.

63. Tetsuo Shimizu and Hiroshi Tokumoto, "Single Silicon Atom Detection on a Tungsten Tip," *Japanese Journal of Applied Physics* 38 (1999): 3860–62: "We think the change of the digital FIM images [is] caused by the local change of electric field distribution on the tip apex. The single atom evaporation by the pulse voltage application induced the change of the electric field distribution. The change influenced the spot structure of the FIM images and dark and invisible atoms can be seen by the concentration of the electric field. Further bright spots became dark by no concentration [i.e., a decreased concentration] of the electric field" (p. 3862).

64. J.-O. Bovin, R. Wallenberg, and David Smith, "Imaging of Atomic Clouds Outside the Surfaces of Gold Crystals by Electron Microscopy," *Nature* 317, no. 6032 (1985): 47–49, quotation on p. 49.

65. J.-O. Bovin and J.-O. Malm, "The Real-Time Growth of Atom Planes on Ru, Rh, and Sn Microcrystals Observed at Atomic Resolution," *Journal of Crystal Growth* 89 (1988): 165–70, quotation on p. 169; and see also Bovin, "Real-Time Atomic-Resolution Imaging of Polymorphic Changes in Ruthenium Clusters," *Angewandte Chemie, International Edition in English* 27, no. 4 (April 1988): 555–58.

66. Bovin informs me that he has tried image analysis software, but the time-averaged nature of the images limits its usefulness (personal communication, November 2000). There are a number of apparently independent attempts to write software for image tracking: see for example Benjamin Painter and Robert Behringer, "Symmetry Breaking and Convection in a 2-Dimensional Horizontally Shaken Granular System," submitted to *Physical Review Letters*, November 2000. I thank Emily Longhi for drawing this to my attention.

5 Particle Physics

1. Galison, *Image and Logic: A Material Culture of Microphysics* (Chicago: University of Chicago Press, 1997).

2. Ibid., 19, 807, and see the further definitions (having to do with the "passivity of the image experimenter") on p. 555.

3. Ibid., 19, 755, 907.

4. Galison gives further definitions of each tradition and introduces two additional terms, *homologous* and *homomorphic*, to characterize them. I won't be using those terms or his full definitions here because they describe different, and partly divergent, phenomena. Galison's use of *homologous* to denote images that preserve the logical relations of their objects seems problematic: it applies very broadly, intersecting his definition of *homomorphic*. See my essay, "Logic and Images in Art History," *Perspectives on Science* 7, no. 2 (1999): 151–80.

5. Apparently all the original plates have disappeared; thanks to Hans Bethe, Peter Galison, and Victor Regener for helping me try to find them.

6. J. Naugle and P. Freier, "High-Energy Phenomena in Nuclear Emulsions," *Physical Review* 104, no. 3 (November 1, 1956): 804–11.

7. For sample figures, see *Reflections on the Fifteen Foot Bubble Chamber at Fermilab*, ed. Mark Bodnarczuk (Batavia, IL: Fermi National Accelerator Laboratory, 1989), 81–108, which lists all the experiments performed at the fifteen-foot bubble chamber.

8. Galison and Assmus, "Artificial Clouds, Real Particles," in *The Uses of Experiment*, ed. David Gooding, Trevor Pinch, and Simon Schaffer (New York: Cambridge University Press, 1989), 225–74. I thank David Kaiser for this reference.

9. Galison, *Image and Logic*, 65–142, quotation on p. 109.

10. Sources are reviewed in my *On Pictures and the Words That Fail Them* (Cambridge: Cambridge University Press, 1998), 3–46.

11. For Baroque physiognomy, see my *Pictures of the Body: Pain and Metamorphosis* (Stanford, CA: Stanford University Press, 1999), 73–82.

12. Hubert Damisch, *Traité du trait: Tractatus tractus* (Paris: Réunion des Musées Nationaux, 1995), 143.

13. Elkins, *On Pictures and the Words That Fail Them*, 42–46.

14. Lyotard, "Presenting the Unpresentable: The Sublime," trans. Lisa Liebmann, *Artforum* 20 (April 1982): 64–69, quotation on p. 64.

15. Ibid., esp. p. 65.

16. Jean-Luc Nancy, "Peface to the French Edition," in *Of the Sublime: Presence in Question*, trans. and with an afterword by Jeffrey Librett (Albany: State University of New York Press, 1993), 1–3.

17. *Les Immatériaux*, vol. 1, *Album: Inventaire*, exh. cat., ed. Jean-François Lyotard (Paris: Centre Georges Pompidou, 1985). See also Lyotard, "Les Immateriaux," *Parachute* 36 (1984): 43–44; "Presenting the Unpresentable"; and further materials and reviews at sun3.lib.uci.edu/~scctr/Wellek/lyotard/.

18. A related project is Gyorgy Kepes, *The New Landscape in Art and Science* (Chicago: Paul Theobald and Co., 1956), which includes cloud-chamber photographs (for example, pp. 200–203).

19. Goronwy Tudor Jones, "A Remarkably Rich Bubble Chamber Picture (The Competition!)," in *Bubbles 40: Proceedings of the Conference on the Bubble Chamber and Its Contributions to Particle Physics (Marking the Fortieth Anniversary of the Bubble Chamber)*, ed. G. Harigel et al., *Nuclear Physics B (Proceedings Supplements)* 36 (July 1994): 475–78. I thank Jones for additional information and annotations.

20. An introductory account is Yu. Aleksandrov, G. Voronov, V. Gorbunkov, N. Delone, and Y. Nechayev, *Bubble Chambers*, trans. Scripta Technica, Inc. (Bloomington: Indiana University Press, 1967), 289–333.

21. The units are: p in MeV/c, B in gauss, and r in centimeters.

22. Here $x = \rho_l$, where ρ is the mass density of the material in the chamber. The variable x therefore has units g cm^{-1}.

23. These equations are adapted from Robert Cahn and Gerson Goldhaber, *The Experimental Foundations of Particle Physics* (Cambridge: Cambridge University Press, 1989), 14–15.

24. For a discussion of the use of bubbles to determine velocity see Walter Barkas, "Velocity Determination in Bubble Chambers," *Review of Scientific Instruments* 38 (1967): 425–27.

25. Jones hypothesizes that the initial reaction involves a K^*: $\nu_\mu N \to \mu^- \Sigma^+ K^*$ (890).

26. There are a number of animations that play with the standard views; for example, l3.web.cern.ch/l3/scan_program/91GeV/events/jpsimm.mpg.

27. For an analysis, see W. S. C. Williams, *Nuclear and Particle Physics* (Oxford: Clarendon Press, 1995), 313–16.

28. The labeling follows Robert Gouiran, *Particles and Accelerators*, trans. W. Crozier (New York: McGraw-Hill, 1967), 83, fig. 16. Gouiran's illustration is reversed right to left, relative to the CERN original.

29. William Wimsatt, "Taming the Dimensions—Visualizations in Science," in *PSA 1990*, vol. 2, ed. M. Forbes, L. Wessels, and A. Fine (East Lansing, MI: Philosophy of Science Association, 1991), 111–35.

30. In my *The Domain of Images* (Ithaca, NY: Cornell University Press, 1999), 40–44.

31. This image has its own afterlife in the physics literature, where it is used in an unquantitative manner, as a token of the actual experiment. See for example Donald Perkins, *Particle Astrophysics* (Oxford: Oxford University Press, 2003), 14, fig. 1.6.

32. Photos are available at l3.web.cern.ch/l3/.

33. The experiment in question was one of a number designed to study Z^0 resonance. See for example B. Adeva et al., "Search for Lepton Flavor Violation in Z^0 Decays," *Physics Letters B* 271 (1991): 453–60; The LEP Collaborations: ALEPH, DELPHI, L3, and OPAL, "Electroweak Parameters of the Z^0 Resonance and the Standard Model," *Physics Letters B* 276 (1992): 247–53; and L3 Collaboration [O. Adriani et al.], "Inclusive Production in Z^0 Decays," *Physics Letters B* 288 (1992): 412–20.

34. J. Mulvey, "Legacies of the Bubble Chamber," in *Bubbles 40*, 427–44, quotation on p. 436.

6. Quantum Mechanics

1. I am only presenting these issues schematically here. They are explored at greater length in my *On Pictures and the Words That Fail Them* (Cambridge: Cambridge University Press, 1998), 250–61, and in my *Pictures of the Body: Pain and Metamorphosis* (Stanford, CA: Stanford University Press, 1999), 278–84. A revised version of these concepts, which takes into account the exposition in this book, is in my "Einige Gedanken über der Unbestimmtheit der Darstellung" [On the Unrepresentable in Pictures], in *Das

unendliche Kunstwerk: Von der Bestimmtheit des Unbestimmten in der ästhetischen Erfahrung, ed. Gerhard Gramm and Eva Schürmann (Berlin: Philo, 2006), 119–40.

2. The quickest introduction I know to practical calculations with the Schrödinger equation is N. J. B. Green, *Quantum Mechanics I: Foundations* (Oxford: Oxford University Press, 1997), which is geared toward chemistry students.

3. For the sake of full disclosure, I'll say that my own understanding of quantum mechanics peters out after J. Sakurai's first-year graduate-level text, *Modern Quantum Mechanics,* 4th ed. (Reading, MA: Addison-Wesley, 1994); I have no grasp of field theory.

4. Two examples of humanists' interest in unnecessarily complex graphics are: Edward Tufte, *The Visual Display of Quantitative Information* (Cheshire, CT: Graphics Press, 1983), which argues for compression of data but also showcases complex graphics; and the Hyperatlas at the Chicago School of Media Theory (csmt.uchicago.edu/projectshyperatlas.htm), which contains a multidimensional visual database of media that is flamboyantly complex.

5. Griffiths, *Introduction to Elementary Particles* (New York: John Wiley and Sons, 1987), 40, 41.

6. Ibid., 73, 77. The last is a notion Griffiths attributes to other writers.

7. I have found Jeffrey Bub, *Interpreting the Quantum World* (Cambridge: Cambridge University Press, 1999 [1997]), and I owe a great deal to the lucid account in R. I. G. Hughes, *The Structure and Interpretation of Quantum Mechanics* (Cambridge, MA: Harvard University Press, 1989).

8. Paul Teller, *An Interpretive Introduction to Quantum Field Theory* (Princeton, NJ: Princeton University Press, 1995), 139.

9. David Kaiser, *Drawing Theories Apart: The Dispersion of Feynman Diagrams in Postwar Physics* (Chicago: University of Chicago Press, 2005); Kaiser, "Stick-Figure Realism: Conventions, Reification, and the Persistence of Feynman Diagrams, 1948–1964," *Representations* 70 (2000): 49–86. Kaiser's work is cited, with further discussion, in my *The Domain of Images* (Ithaca, NY: Cornell University Press, 1999), 37.

10. Stephen Martin, "A Supersymmetry Primer," arxiv.org/abs/hep-ph/9709356 (April 7, 1999); the essay is available in abbreviated form in *Perspectives on Supersymmetry,* ed. G. Kane (Singapore: World Scientific, 1998).

11. As Kaiser has shown, Feynman diagrams have even found their way into areas, such as S-matrix theory, where their calculational elements are no longer operative. I only give this example to show how the diagrams find uses in ever-expanding theoretical contexts.

12. Teller, *An Interpretive Introduction,* 142.

13. Reproduced, for example, in Robert Eisberg and Robert Resnick, *Quantum Physics of Atoms, Molecules, Solids, Nuclei, and Particles* (New York: John Wiley and Sons, 1985), 258, fig. 7-12.

14. Gouiran, *Particles and Accelerators*, trans. W. Crozier (New York: McGraw-Hill, 1967), 40.

15. Personal communication, April 1999.

16. This interpretation is reviewed in my *Why Are Our Pictures Puzzles? On the Modern Origins of Pictorial Complexity* (New York: Routledge, 1999), 218–19.

17. Brian G. Duff, *Fundamental Particles: An Introduction to Quarks and Leptons* (London: Taylor and Francis, 1980), 86.

18. The relevant experiments are summarized in Bertram Schwarzschild, "Experiments at Jefferson Lab and MIT Probe for Strange Quarks in the Proton," *Physics Today* 52, no. 6 (June 1999): 21–23.

19. Bernard Frois and Marek Kaliner, "Where Does the Nucleon Spin Come From?" *Physics World* 7 (July 1994): 44–49.

20. F. M. Steffens, W. Melnitchouk, and A. W. Thomas, "Charm in the Nucleon," arxiv.org/abs/hep-ph/9903441 (March 1999), p. 9.

21. These pictures continue to get more elaborate: see Klaus Rith and Andreas Schäfer, "The Mystery of Nucleon Spin," *Scientific American* (July 1999): 42–48.

22. Edward Witten, "Duality, Spacetime, and Quantum Mechanics," *Physics Today* (May 1997): 28–33; the image and the quotation are on p. 30.

23. Duff, *Fundamental Particles*, 85; the same page has Duff's picture, fig. 6.12, of "magnetic dipoles inside the cavities of a superconductor."

24. See Nora Brambilla and Antonio Vairo, "Quark Confinement and the Hadron Spectrum," arxiv.org/abs/hep-ph/9904330 (April 14, 1999); see also Gunnar Bali, "The Mechanism of Quark Confinement," arxiv.org/abs/hep-ph/9809351 (September 12, 1998).

25. Brambilla and Vairo, "Quark Confinement and the Hadron Spectrum," figs. 11, 14, 15, and 16; quotation on p. 65, slightly modified.

26. Personal communication, April 26, 1999.

27. *Quark Confinement and the Hadron Spectrum II*, ed. Nora Brambilla and G. M. Prosperi (Singapore: World Scientific, 1997). I thank Nora Brambilla for bringing this to my attention.

28. An example from biology is Paul Ehrlich's drawings of antibodies, discussed in my *Domain of Images*, fig. 3.7. Nora Brambilla suggested I look at a parallel between a theory of molecular vortices, published by James Maxwell in 1861, and an essay on color magnetic vortices, published in 1980. The formal similarities are in fact very clear, and the physics that underlies them is comparable. See James Maxwell, "On Physical Lines of Force," Part I, "The Theory of Molecular Vortices Applied to Magnetic Phenomena," *The London, Edinburgh and Dublin Philosophical Magazine and Journal of Science*, 4th series, 21, no. 139 (March 1861), compared to J. Ambjorn and P. Olesen, "A Color Magnetic Vortex Condensate in QCD," *Nuclear Physics B* 170 (1980): 265–82.

29. Personal communication, April 1999.

30. In addition to the sources cited in note 31 below, see Andrei Linde, "Quantum Cosmology and the Structure of [the] Inflationary Universe," arxiv.org/abs/gr-qc/9508019 (August 7, 1995); and Nemanja Kaloper, Andrei Linde, and Raphael Bousso, "Pre-Big-Bang Requires the Universe to Be Exponentially Large from the Very Beginning," arxiv.org/abs/hep-th/9801073 (January 29, 1998).

31. Andrei Linde, Dmitri Linde, and Arthur Mezhlumian, "From the Big Bang Theory to the Theory of the Stationary Universe," *Physical Review D* 49 (February 15, 1994): 1783–1826, illus. on p. 1806, fig. 6. The "Pollock universe" is illustrated in ibid., p. 1805, fig. 5. See also Andrei Linde, "The Self-Reproducing Inflationary Universe," *Scientific American* (November 1994): 48–55.

32. Linde, "The Self-Reproducing Inflationary Universe," caption to fig. on p. 49.

33. See Linde's Web site at www.stanford.edu/~alinde/.

34. Personal communication, April 1999.

35. Personal communication, summer 2000.

36. A very different-looking version of the same picture—it resembles a series of elliptical maps, divided into regions and separated by arrowed lines—is in John Barrow, *Impossibility: The Limits of Science and the Science of Limits* (Oxford: Oxford University Press, 1998), 172, fig. 6.10.

37. Personal communication, April 1999.

38. For Brooks's work see www.artnet.com/artist/3131/Stephanie_Brooks.html.

39. See, among many others, Sunny Auyang, *How Is Quantum Field Theory Possible?* (New York: Oxford University Press, 1995), and Bub, *Interpreting the Quantum World.*

40. See *Quantum Theory and Measurement,* ed. John Wheeler and Woyciech Zurek (Princeton, NJ: Princeton University Press, 1983).

41. Sheldon Goldstein, "Quantum Theory Without Observables—Part Two," *Physics Today* 51 (April 1998): 38–42. See also Goldstein, "Quantum Theory Without Observables—Part One," *Physics Today* 51 (March 1998): 42–46.

42. The Bohmian interpretation has mathematical difficulties, which are still being debated. See D. Appleby, "Generic Bohmian Trajectories of an Isolated Particle," arxiv.org/abs/quant-ph/9905003 (May 3, 1999).

43. Thaller, *Visual Quantum Mechanics: Selected Topics with Computer-Generated Animations of Quantum-Mechanical Phenomena* (New York: Springer, 2000), p. v.

44. Ibid., 8. Several of the animations in the book are available on Thaller's Web site, www.kfunigraz.ac.at/imawww/vqm/. The images in color plate 22 are from the CD version of the animation at www.kfunigraz.ac.at/imawww/vqm/pages/samples/04_18a.html.

45. In addition to the literature I have cited, I found the following very helpful: Jeeva Anandan, "Are There Dynamical Laws?" *Foundations of Physics* 29, no. 11 (1999): 1647–72; Anandan, "Meaning of the Wave Function," *Physical Review A* 47, no. 6 (June 1993): 4616–26. And this is where I bow out of the debate, which appears to be deeper than even what the specialists can fathom.

Conclusions

1. *Visual Practices Across the University*, ed. James Elkins (Munich: Wilhelm Fink Verlag, 2007).

2. Henry Staten, "Deconstructing Figurality: On 'Radical Painting,' " in *Günter Umberg*, ed. Hannelore Kersting (Cologne: W. König, 1989), 96.

3. Akira Tonomura, personal communication, July 2004.

Index